SALVAGE

GAS DYNAMICS

Proceedings of the Lebedev Physics Institute
Academy of Sciences of Russia

Proceedings of the Lebedev Physics Institute
Academy of Sciences of Russia
Volume 202

GAS DYNAMICS

Edited by Yu.I. Koptev

Nova Science Publishers, Inc.
New York

Nova Science Publishers, Inc.
6080 Jericho Turnpike, Suite 207
Commack, New York 11725

Library of Congress Cataloging-in-Publication Data
available upon request

ISBN-1-56072-072-7

Book Production Manager Janet Glanzman White
Graphics   Elenor Kallberg and Debbie Lippus

Printed in the United States of America

CONTENTS

EXPERIMENTAL STUDY OF SELF-SUSTAINED WAVE PROCESS IN SHOCK HEATED XENON PLASMA

G. K. Tumakaev, Z. A. Stepanova

Abstract: Experimental data on the instability of inert gas flows behind a shock wavefront are systematized. Phenomena accompanying four types of shock heated plasma instability are outlined.

Oscillations are discovered in a radiation intensity variation of xenon plasma at $M \sim 8.3$ as well as in shock heated plasma when propagating along the low-pressure chamber of the shock tube at M = 10-15. Variation of frequency and amplitude of radiation intensity oscillation with Mach number, initial pressure of gas under study, and molecular impurities (H_2, N_2, O_2, CO_2) concentration is investigated.

Conditions are ascertained for developing third type instability in recombining xenon plasma ($M \geqslant 17.4$).

1. Systematization of Anomalous Phenomena in Shock Heated Plasma of Monoatomic Gases

At present extensive experimental data is available which demonstrates the anomalous behavior of monoatomic [4-10] and molecular [8, 11, 12] gases in flow behind the shock wavefront which does not conform with conventional conceptions [2, 3].

Anomalies differing from each other by reason of their rise and accompanying effects are observed in a wide range of incident shock wave Mach numbers: in inert gases at M = 8.3-23 in xenon [4, 5, 9, 10] and at M = 10.5-32 in argon [6, 8, 10]; in molecular gases at M = 4-16 in freons [11, 12] and at M = 14-26 in carbon dioxide [8].

As a rule anomalies arise in a threshold manner and are accompanied by some remarkable concomitant phenomena. For instance, radiation fluctuations, loss of gas dynamic stability of the flow behind the shock wavefront and stability of the shock front itself are found in inert gases. Destruction of the detached shock wave, formation of inner shocks, flow turbulization, etc. are observed in ballistic experiments [11, 12] in heavy molecular gases.

1

Analysis of the experimental data [4-10] demonstrating the anomalous rise in the flow behind the shock wavefront allows us to outline at least four groups of different phenomena for inert gases that show, respectively, four different types of shock heated plasma instability. The validity of this systematization is further substantiated by experimental results.

Systematization is carried out according to the following signs:

1. rise conditions, i.e., range of shock wave intensity;

2. location of the structural flow changes behind the shock wavefront;

3. form and intensity of the structural changes.

The first group includes phenomenon related to the anomalous fluctuation of spectral lines and continuum radiation of shock heated gas [4, 10] that appears in the narrow range of incident shock wave Mach number; at M = 8.3 ±0.5 in xenon, and at M = 10.5-11 in argon.

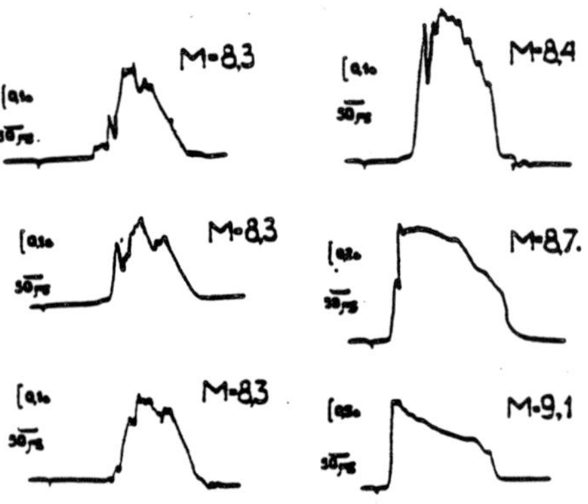

Fig. 1. Fluctuation of shock heated plasma radiation [10].

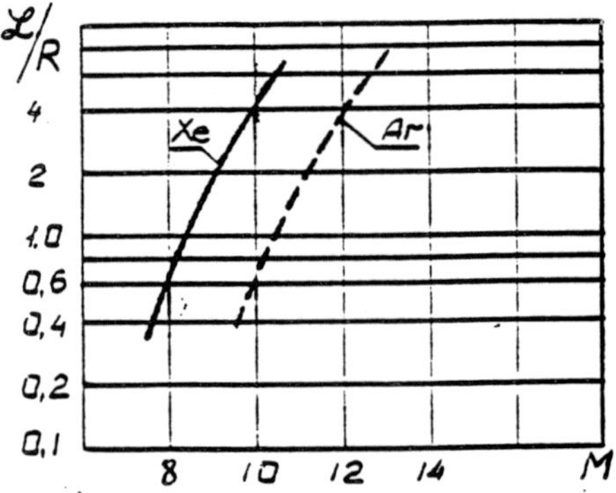

Fig. 2. L/R ratio variation with Mach number [10].

The depth of modulation of radiation intensity fluctuation in xenon at M = 8.3 (see Figure 1) is large and amounts to 30-40%. Mach number increases as the modulation decreases. At M = 8.7 radiation splash is primarily observed in the relaxation zone and becomes indiscernible against the thermal plasma radiation background when the shock wave intensity is further increased (M = 9.1).

The rise of this phenomenon correlates with the condition of equality of the radiative destruction rate ($R = n^*/\tau$) for 4p-states of argon and 6p-states of xenon atoms to the rate of de-excitation of these states by inelastic collisions with electrons ($L = \langle \sigma v \rangle \cdot n_e \cdot n^*$) in the equilibrium flow region behind the shock wavefront [10]. Here n^*, n_e are concentrations of excited atoms and electrons, τ is the lifetime of the excited state, $\langle \sigma v \rangle$ is the quenching rate constant due to electron impact. The data on $L/R = f(M)$ ratio variation presented in Figure 2 substantiate the validity of the above supposition: $L/R \sim 1$ at $M \sim 8.3$ in xenon and at $M \sim 10.5$–11 in argon.

Radiation fluctuation appears in the relaxation zone.

The second group includes phenomena evolving in the recombination zone behind the shock wavefront, but in experiments [4, 6, 7] they develop at most intensities within a rather narrow transitional flow region where, on the one hand, ionization relaxation is complete and, on the other hand, plasma parameters begin to be governed mainly by recombination processes caused by the radiative cooling of gas flow.

In this case anomalous flow behavior appears in nonmonotonic variation of plasma radiation intensity, electron, normal, and excited atom concentrations, as well as in the shock wavefront distortion. The above mentioned effects are recorded in xenon at Mach numbers somewhat exceeding $M \sim 9.2$ [4]. Shock tube interferometric research in argon and krypton [6, 7] reveal a nonmonotonic plasma optical thickness variation at $M \sim 15$. These anomalous phenomena appear for the most part as irregular flow variations.

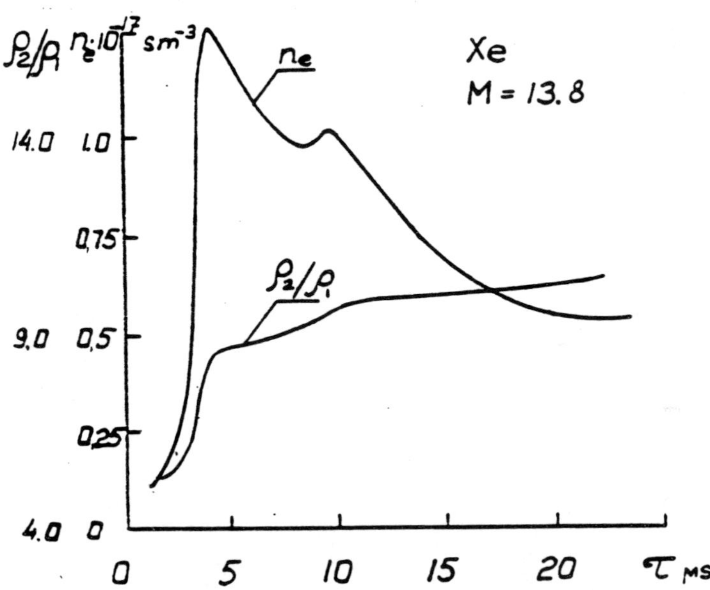

Fig. 3. Nonmonotonic variation of electron concentration and gas density in the flow behind the shock wavefront, xenon, M = 13.8 [4].

Some experimental data on electron concentration and gas compression in xenon flow behind the shock wave at M = 13.8 are presented in Figure 3. Here electron concentration in the recombination zone varies nonmonotonically with a distance from the shock wavefront. An appreciable splash in electron concentration variation appears after its initial rapid decrease. This correlates with the rapid density increase recorded in the correspondent flow region. Development of anomalies of this group depends on molecular impurity concentration in the gas under study [6, 7].

The third group covers phenomena evolving in xenon at $M \geqslant 17.4$ and in argon at $M \leqslant 21$. In this case the shock wavefront and shock heated gas flow are also perturbed. However, according to Refs. [5, 9], these perturbations exhibit greater intensity and are sometimes accompanied by a powerful radiation flash [9].

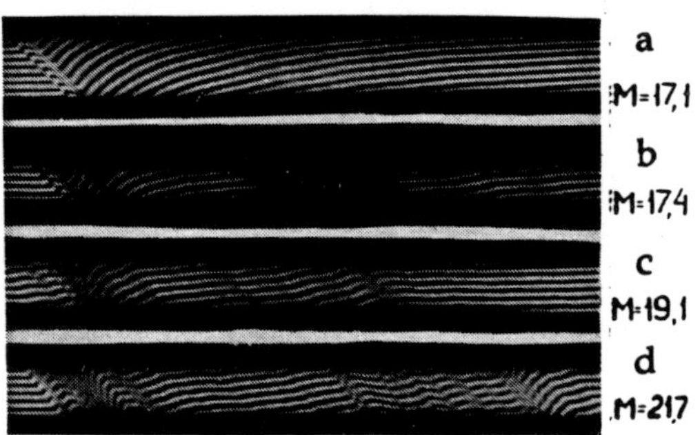

Fig. 4. Gas flow interferograms. Xenon [5].

Anomalies of this group appear in a distinct threshold manner when the incident shock wave Mach number increases. Figure 4 shows the interferograms of xenon flows at $M = 17.1-21.7$ [5]. In all interferograms the shock wavefront is noted with an abrupt shift of the interferometric fringes. It is followed by a relaxation zone with a length of ~3 mm that is characterized by small optical thickness variations in the greatest part of this flow region. Further, the extensive zone of shock heated plasma recombinating due to radiative cooling is found. Usually this zone is characterized by monotonic fringe shifting because of gradual electron concentration decrease and gas compression increase. This is seen in Figure 4 at $M = 17.1$. For more intensive shock waves ($M = 17.4$) the additional and more effective plasma recombination process develops in this area resulting in the essential increase of gas flow optical thickness. This is evidenced by an intensive inteferometric fringes shift in the vicinity of the relaxation zone. The incident shock wave Mach number increase does not effect the location of this structural change of flow. Its further increase ($M = 19.1$) is usually accompanied with deformation and, probably, destruction of the shock wavefront.

Anomalies rise in xenon at $M \geqslant 17.4$ and in argon at $M \geqslant 21$ which correlates well with the condition $dn_0/dM \leqslant 0$ for equilibrium shock heated plasma. In other words, these anomalies evolve at those Mach numbers when a decrease of the normal concentration of atoms (n_0) due to their ionization is no longer compensated for by gas compression $\rho = (n_0 + n^+)/n_1$. Here n^+ is ion concentration in the flow; n_1 is normal atom concentration ahead of the shock wave. Functions $\bar{n}_0 = n_0/n_1 = f(M)$, $\bar{n}^+ = n^+/n_1 = f(M)$ and $\bar{\rho} =$

$f(M)$ are plotted in Figure 5 for argon and xenon. In this figure the dots above the curves denote only availability of the experimental data providing information on the flow parameters behind the shock wavefront [4, 8, 9, 13]. Open circles correspond to "stable" flows, experiments showing the signs of anomalous phenomena are noted by hatching.

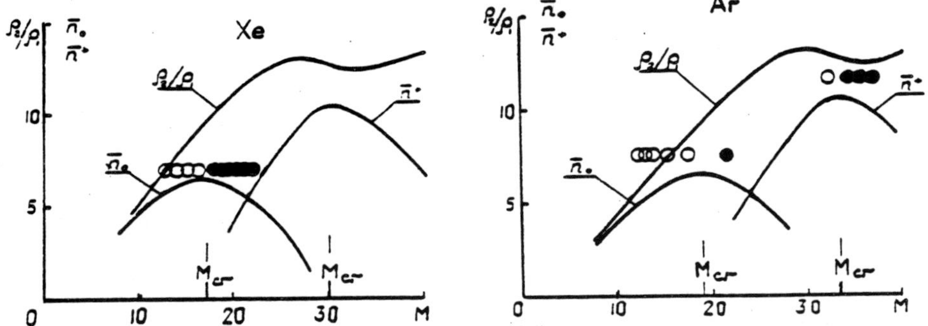

Fig. 5. Variation of normal atom and ion concentration in equilibrium flow region in argon and xenon with the shock wave intensity [5].

In accordance with data presented the condition $dn_0/dM = 0$ is satisfied at $M = M_{cr} = 17.4$ in xenon and at $M = M_{cr} = 19.5$ in argon. It should also be noted that at $M \geqslant M_{cr}$ the energy accumulated in the internal degrees of freedom exceeds the transitional energy of the heavy particles [14].

The fourth group contains the phenomenon of the shock wavefront distortion revealed at $M = 32$ for argon flows.

In accordance with Ref. [4] the instability of the shock wavefront under these conditions is related to the anomalous behavior of a *P-V* diagram. However, the results of our analysis show another possible reason for this instability development, namely, the decrease of single-charged ion concentration in the flow behind the shock wave. Here, indeed, a distinct correlation of the Mach number between instability rise and condition $dn^+/dM \leqslant 0$ may be established. According to data in Figure 5 equality $dn^+/dM = 0$ is satisfied at $M = M_{cr} = 32$ in argon and $M = M_{cr} = 31$ in xenon.

Systematization of the experimental data presented above, allows us to show, on the one hand, that the groups of phenomena under study evolve independently from each other and differ essentially with regard to the conditions of their rise (incident shock wave Mach number and location of the structural changes); and on the other hand, to reveal some general features, namely threshold character of development of the first, second, third, and fourth group anomalies. In all cases anomalous phenomena rise is related to shock heated plasma instability in the flow behind the shock wavefront.

2. Experimental Study of First, Second, and Third Type Instabilities of Shock Heated Xenon Plasma

In accordance with the above-stated systematization of anomalous phenomena, experimental study of the optical properties of shock heated xenon plasma was carried out for a wide range of the incident shock wave Mach number 7.8 < M < 19 including the regimes of the three types of instability development. The results are as follows:

1) variation of shock heated xenon plasma radiation intensity in the relaxation zone behind the shock waves at M = 8.3 ±0.5 is found to be exactly periodic;

2) oscillation of shock heated xenon plasma radiation intensity during plasma propagation along the low-pressure camera of the shock tube is found at the incident shock wave Mach number range M = 10-16.5   (the second type of instability);

3) it is shown that when condition $dn_0/dM \leqslant 0$ is satisfied the third type of instability develops only in the flows behind the shock waves attenuating with time ($du/dt < 0$) where u is shock wave velocity.

Experiments are carried out on a single diaphragm shock tube. A low-pressure chamber is made of cylindrical steel tubes with channel diameter of 100 mm. The total length of the low-pressure chamber is ~8 m. The internal channel surface is ground. The coaxiality of the shock tube sections is no less than 0.1 mm.

The high-pressure chamber is supplied with an electric heater providing gas heat up to 250-300°C with a length of 1.5 m.

The level of preliminary pumping of the system is 10^{-5} torr. Experiments have been carried out in pure xenon (concentration of foreign gas impurities in the tank is 0.002%) and in xenon with a small addition of molecular gases (nitrogen, oxygen, hydrogen, carbon dioxide, and air). The addition of molecular gas impurities was controlled within the range $5 \cdot 10^{-5} - 2 \cdot 10^{-2}$ torr. The gas mixture was prepared just before the experiment by means of successive low-pressure chamber filling, first by the impurity component and then by the gas under study. Gas mixing was achieved by means of multiple mixture pumping in a closed ring channel formed by inlet trunks and by the low-pressure chamber itself. Pumping was forced by an oil-free two-rotor compressor.

Implementation of measuring of the cross-section of the low-pressure chamber provides hermetic mounted optically transparent insets in the diagnostic tracts and a system of ionizing probes used to determine the shock wave velocity.

Radiative properties of shock heated plasma were investigated in a wide wavelength range of 400 nm — 10 μm. Observations were carried out in two ways: by scanning the luminous object (plasma) moving relative to the fixed measurement cross-section and by continuous registration of plasma luminescence variation during plasma propagation along the low-pressure chamber. Observation was performed in the direction normal to the vector of shock wave velocity in the shock tube in the first case and parallel to this vector in the second case. Optical schemes for both methods are presented in Figure 6.

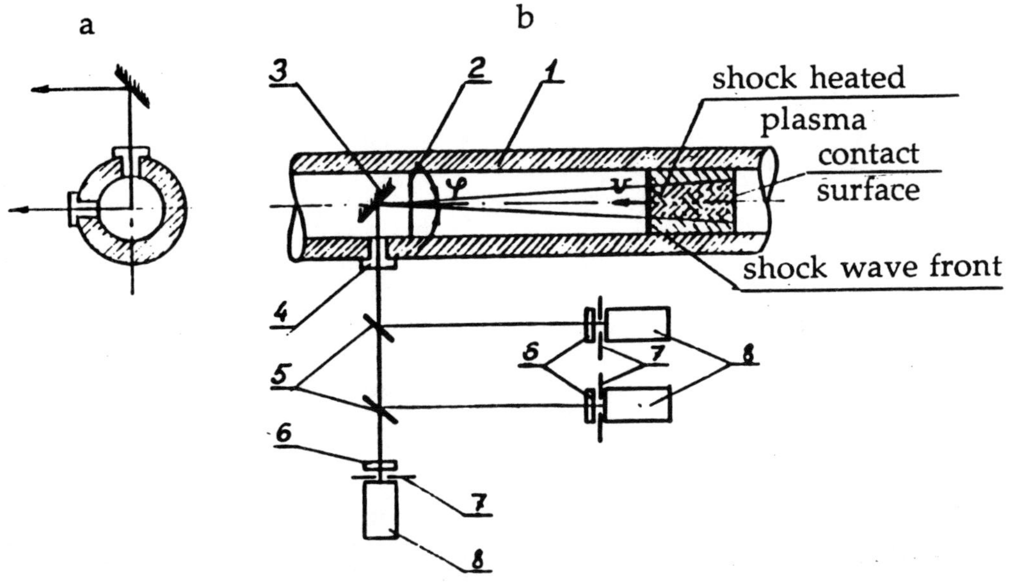

Fig. 6. Optical scheme of the facility:

a) scanned image method; b) method of continuous plasma luminescence recording during plasma propagation along the low-pressure chamber.

In the first regime (Figure 6a) the variation of plasma optical properties in three reciprocally orthogonal directions at three fixed cross-sections of the low-pressure chamber can be demonstrated. Here, shock heater plasma radiation was shown by the photomultipliers. Spatial resolution of the optical tracts is ~1 mm.

An optical scheme for investigating the time dependence of shock heated plasma radiation intensity variation during its propagation along the low-pressure chamber is shown in Figure 6b. It includes a deflecting mirror 3 mounted at low-pressure chamber axis 1; a hermetized optically transparent inset 4 made of melted quartz or lavsan film; a set of semi-transparent mirrors 5 providing lightbeam division into several channels for the simultaneous registration of different wavelength radiation, photoelectric light receivers 8, and changeable broad- and narrowband optical filters 6. Light beams on each measuring channel were collimated by means of two diaphragms with an orifice diameter of 1 mm. Field sight diaphragm 2 was mounted ahead of the deflecting mirror in the low-pressure chamber, aperture diaphragm 7 was installed on the frame of the photorecording receiver. The distance between diaphragms usually does not exceed 1 m. It was chosen so that the field of vision of the optical system on the total channel length would be determined by the field sight diaphragm at the cross-section of the cylindrical channel of the shock tube low-pressure chamber. The radiation of plasma volume bounded with the field of vision solid angle φ and the length of uniform shock heated flow was demonstrated in the experiments. This area is shown in the figure by double hatching.

a) First type instability of shock heated plasmas.

The study of the origin and development of the first type of instability in xenon is carried out in a narrow range of incident shock wave Mach number values M = 8.0 ±0.5. The initial pressure of the gas under study varied from 10 to 30 torr and the percentage of molecular gas impurity (H_2, N_2, CO_2, air) was varied from $10^{-1}\%$ to $10^{-3}\%$.

Radiative plasma properties were demonstrated by the scanned image method.

The results of early experiments [10] (see Figure 1) showed a rise of shock heated xenon plasma radiation fluctuation at M ~ 8.3 but contained no information about the possibility of its development. The present investigation, carried out with greater demands on experiment preparation and performance shows that the variation of plasma radiation intensity in the relaxation zone is not sporadic but demonstrates a periodic character that shows the self-sustaining wave nature of the phenomenon under study [15, 16].

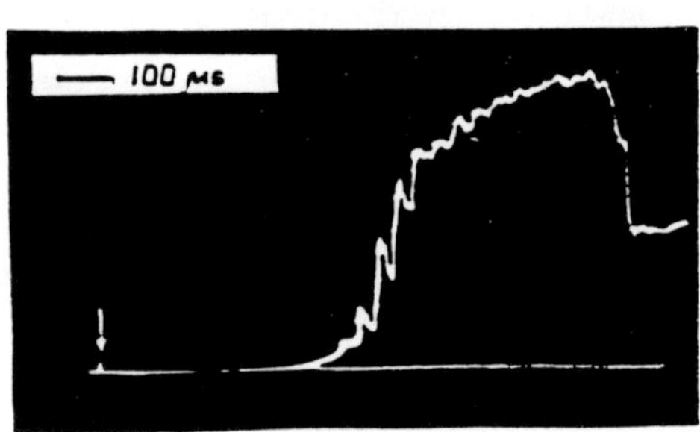

Fig. 7. Radiation oscillation in relaxation flow region.

Figure 7 shows an oscillogram of shock heated xenon plasma radiation variation behind the shock wavefront. It demonstrates all the characteristics of the flow behind a weak ionizing shock wave, i.e., the large extent of the relaxation zone (shock wavefront is noted by the arrow) with low radiation level of those parts where atom ionization is governed mainly by atom-atom collisions; insignificant plasma cooling in the equilibrium area; and finally, a sharp decrease of plasma radiation intensity in the contact mixing region. Contrary to conventional thought [2, 3] is the distinctly seen signal oscillation in the flow region where atom ionization occurs due to electron-atom collisions. The dynamics of the self-sustaining wave process development in the flow behind the shock wavefront are clearly seen in the figure as well. The degree of gas ionization in the avalanche part of the relaxation zone increasing the amplitude of radiation oscillations at first increases but then clearly shows a trend to attenuation. Note that the frequency of radiation intensity oscillations remains almost constant and equal to ~30 kHz while the depth of radiation modulation in the relaxation zone varies in a wide range (up to 40-50%).

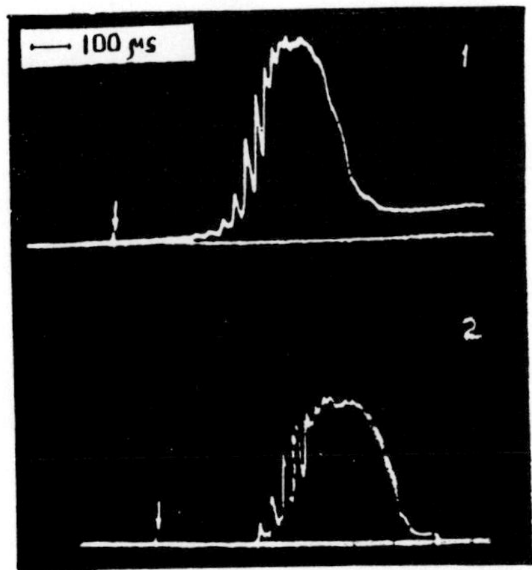
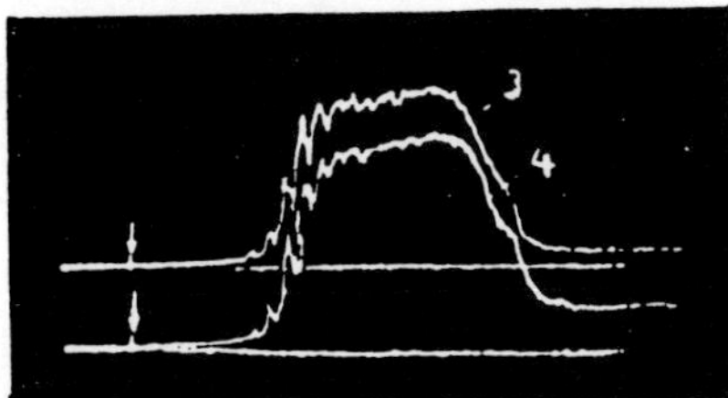

Fig. 8. Radiation oscillation.

1, 2 - in two cross-sections located a distance of 280 mm; 3, 4 - observed in two reciprocal orthogonal directions in fixed cross-section.

Figure 8 shows four oscillograms. The first two of them obtained in the same experiment provide information on shock heated plasma radiation intensity variation at two cross-sections of the low-pressure chamber located at a distance of 280 mm from each other. Oscillograms 3 and 4 were also obtained in the same experiment but showed radiation in two reciprocally orthogonal directions at fixed cross-section.

The distinct correlation of oscillating signals on frequency and amplitude clearly seen in both cases shows the invariability of dynamics and the regular development of the process in gas flow behind the shock wavefront. This experimental data allows us to suppose the bulk nature of the development of shock heated plasma instability and rise of creasy periodic structure in the relaxation flow region. They exclude the possibility that the observed instability rise is caused by a trivial violation of gas flow uniformity behind the incident shock due to, for instance, roughness of the internal surface of the low-pressure chamber channel.

The intensity of self-sustained wave process development is rather sensitive to variation of molecular impurity concentration or more exactly their dissociation product concentration in the flow behind the shock wave-front.

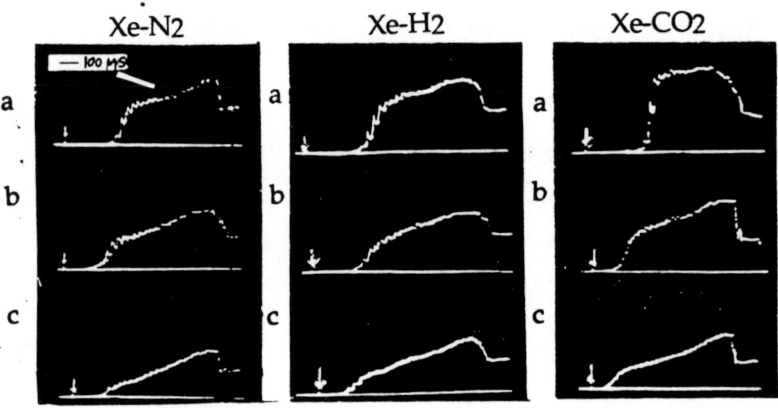

Fig. 9. Shock heated plasma emissivity variation with molecular impurity percentage:
1 - Xe + H_2; 2 - Xe + N_2; 3 - Xe + CO_2

Figure 9 presents experimental data illustrating the variation of xenon plasma luminescence in the flow behind the shock wavefront with the percentage of hydrogen (a), nitrogen (b), and carbon dioxide (c) impurities for the following mixtures:

Mixture	Percentage of impurity ξ, %		
Xe-H_2	$3.8 \cdot 10^{-2}$	$9.0 \cdot 10^{-2}$	$1.8 \cdot 10^{-1}$
Xe-N_2	$5.1 \cdot 10^{-2}$	$1.3 \cdot 10^{-1}$	$2.7 \cdot 10^{-1}$
Xe-CO_2	$5.6 \cdot 10^{-3}$	$3.4 \cdot 10^{-2}$	$1.1 \cdot 10^{-1}$

Analysis of these experimental data reveal that a molecular impurity concentration increase to $\xi \sim 0.1\%$ is accompanied by complete suppression of radiation fluctuation and shrinkage of the relaxation zone behind the shock wavefront.

Note that the results of test experiments carried out at air injection into the system within the range $5 \cdot 10^{-4} < \xi < 10^{-2}\%$ substantiate that the rise of the first type of instability in xenon is related solely to the presence of impurities in the gas under study and that maximum development of this instability takes place at $\xi \sim 4\text{-}6 \cdot 10^{-3}\%$. Decrease of air impurity concentration to $\xi \leqslant 10^{-3}\%$ results in complete suppression of radiation oscillation.

b) Second type of shock heated plasma instability [19].

This phenomenon was investigated in a wide range of the incident shock wave Mach number values M = 10-16.5. Initial xenon pressure varied from 1.5 to 15 torr, concentration of molecular gas (N_2, H_2) impurities varied from $5 \cdot 10^{-3}$% to 10^{-1}%. Here the method of continuous registration of plasma luminescence variation during its propagation along the low-pressure chamber (see Figure 6b) was mainly used for plasma diagnostics. Observations were performed in the visible and infrared wavelength range (λ = 400-10,000 nm).

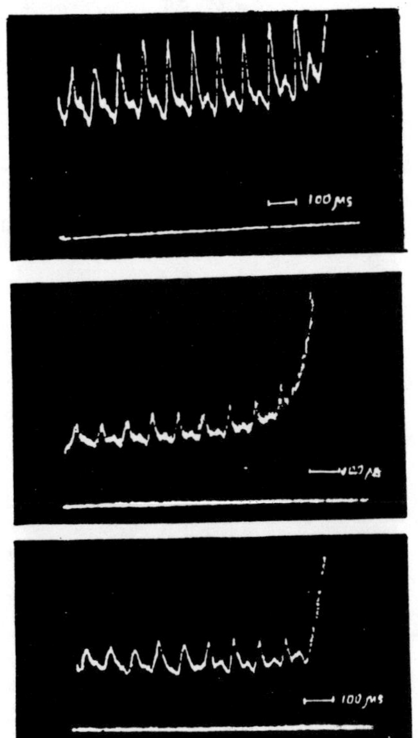

Fig. 10. Oscillation of shock heated xenon plasma radiation in different wavelength ranges (M = 11.9, p = 6.6 torr):
a - 400 < λ < 700 nm; b - 800 < λ < 1200 nm; c - 3 < λ < 10 μm.

Radiative properties of shock heated xenon demonstrating an interesting physical phenomenon were discovered. Basically, shock heated plasma radiation intensity in continuum and spectral lines in the flow behind the shock wavefront are not constant in time but demonstrate a clearly expressed periodic pulsing variation. This phenomenon, on the one hand, for the first time shows the possibility of self-sustained wave process development in shock heated plasma of inert gases at M = 10-16.5 and, on the other hand, reveals the reason for the rise and development of the second type of instability.

Frequency and modulation depth of the amplitude of plasma radiation intensity periodic variation are found to be dependent on the incident shock wave Mach number, the initial pressure of the gas under study, and impurity concentration. Oscillation frequency is varied within some tens of kh, depth of signal modulation up to 40% of recombination, and free-free continua level.

Figure 10 shows oscillograms of the signals of photoelectric receivers recording variation of shock heated plasma radiation intensity (M ~ 11.9, p = 6.6 torr) during plasma propagation in a low-pressure channel at a length of 2 m ahead of the measurement cross-sections. Oscillograms a, b, and c contain information on plasma radiation intensity variation in visible (400 < λ < 700 nm), near IR (800 < λ 1200 nm), and far IR (4 < λ < 10 μm)

spectral ranges, respectively. The moment of the incident shock wave interaction with the frame of field sight diaphragm and deflecting mirror
installed inside the low-pressure chamber is identified in the oscillograms by
a sharp increase of signal due to plasma radiation intensity variation behind
the reflected shock wave.

Contrary to conventional ideas on the radiative properties of monoatomic gas plasma in the shock tubes, radiation intensity varies in time just
periodically showing a development of self-sustained wave process in the
shock heated plasma [16].

Qualitative analysis of these experimental data shows that the plasma radiation oscillation character is almost the same in the wide spectral
range. Here, indeed, there is a clear correlation on frequency and amplitude
modulation depth variation and even on the rise of satellite radiation
splashes.

The depth of signal modulation is

$$\Delta A = \frac{J_{max} - J_{min}}{J_{max}}$$

where J_{min} and J_{max} are minimum and maximum values of radiation intensity. Under the conditions considered $\Delta A \sim 30\%$, oscillation frequency $F \sim$
12 kH.

Observing variation of moving plasma optical properties at a more
extensive base as compared, for example, to experimental results presented
in Figure 10, we can trace the evolution stages, i.e., rise and attenuation of
self-sustained wave process in the shock tube.

Figure 11a and b present experimental data that illustrate plasma
radiation intensity variation during shock tube flow formation starting from
rupture of the diaphragm dividing low- and high-pressure chambers (observation base is 7 m). Figure 11a shows the phase of the oscillation process
development in shock heated plasma. Figure 11b demonstrates a typical case
of rise and attenuation of oscillations. It should be noted that in each of
these experiments oscillation amplitude, i.e., self-sustained wave process
intensity, varies within a rather wide range during shock wave propagation
along the low-pressure chamber while oscillation frequency remains invariable (at least at large enough intervals).

Being invariable in each experiment the oscillation frequency shows
a trend to increase with the shock wave intensity and initial pressure of a
gas under study.

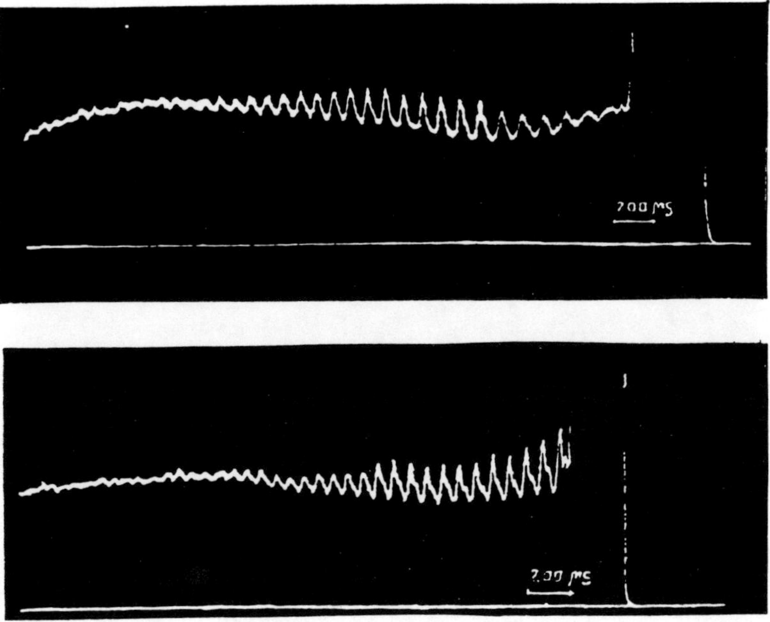

Fig. 11. Development of shock heated xenon plasma radiation oscillation in time:

a - M = 10.8, p = 6.6 torr; b - M = 12.0, p = 6.6 torr.

Figure 12 presents the data on oscillation frequency F variation with the incident shock wave Mach number at initial pressures, p = 3.3, 6.6, and 13.3 torr. In this experiment shock heated plasma radiation oscillations were seen at a base 2 m ahead of the measurement cross-section. Function F = f(M) at p = const demonstrates complicated, clearly expressed threshold behavior. At M ~ 10 oscillation frequency is limited from below by the value F ~ 9.8-10 kHz that is almost independent of initial pressure. Mach number increases as the frequency increases. However, this increase is insignificant until the shock wave intensity remains less some critical value M = M_{cr}. Later the behavior of function F = f(M) changes, at first for large pressure values and then for lower pressure values as well. Correspondent curve intervals show the sharp increase of oscillation frequency gradient with Mach number. As a result, at p = 6.6 torr, for instance, a rather small Mach number increase from M = M_{cr} = 11.9 to M $\simeq$ 12.9 results in an almost

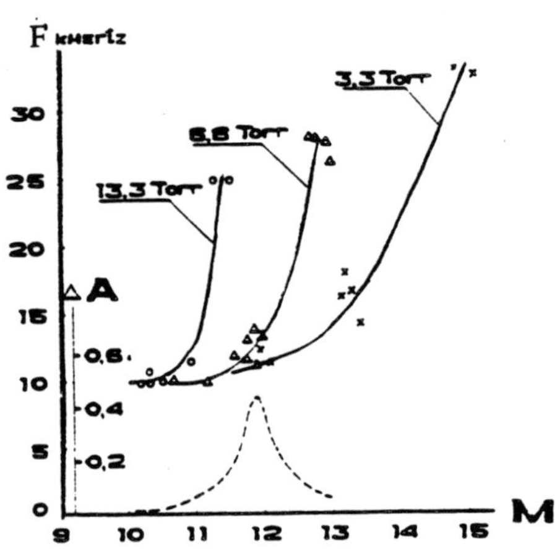

Fig. 12. Variation of xenon plasma luminescence oscillation frequency with the incident shock wave Mach number at p = var.

Dashed curve, $\Delta A = f(M)$ at p = 6.6 torr.

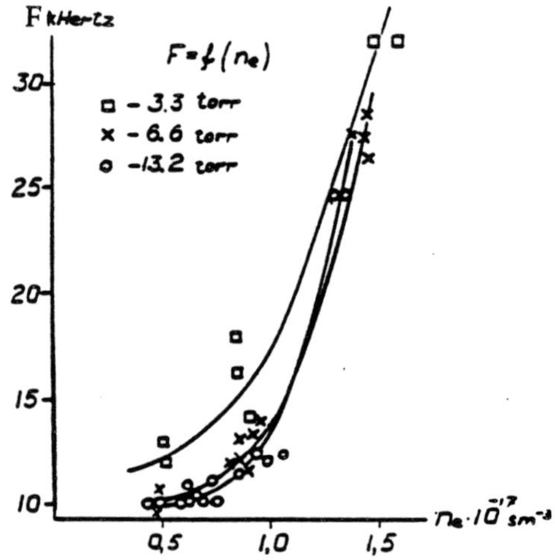

Fig. 13. Variation of plasma radiation oscillation frequency with equilibrium value of electron concentration.

threefold increase of plasma radiation oscillation frequency, at p = 3.3 torr, frequency varies from $F \sim$ 15 kHz at M = M_{cr} = 13.2 to $F \sim$ 33 kHz at M = 15.

Modulation depth of shock heated plasma radiation (near the measurement cross-section) reaches its maximum value $\simeq$30–40% at M = M_{cr}. Dependence $\Delta A = f(M)$ at p = 6.6 torr is plotted in Figure 11 by a dashed line.

The nature of the self-sustained process discovered is unambiguously related to shock heated plasma kinetic instability [15]. Particularly, this conclusion is substantiated by the functional dependence of radiation oscillation frequency variation both on electron number concentration in the flow behind the shock wavefront and on molecular impurity concentration.

The former is proven by the fact that at M > M_{cr} and p = var each fixed oscillation frequency value of functions $F_i = f(M, p_i)$ takes place (within the experimental accuracy) at the same equilibrium value of electron concentration in the flow behind the shock wavefront. For example, a self-sustained oscillation regime with radiation oscillation frequency F = 25 kHz evolves independently on initial gas pressure when electron concentration in the equilibrium flow region becomes equal to n_e = 1.4$\cdot$10^{17} cm^{-3}. Graphs F

$= f(n_e)$ plotted in Figure 13 in fact demonstrate quite good agreement of the results obtained under different experimental conditions.

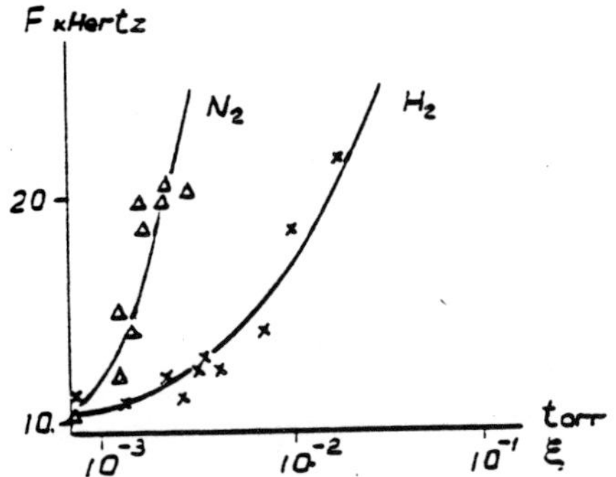

Fig. 14. Variation of shock heated xenon plasma oscillation frequency with hydrogen and nitrogen impurities percentage.

$M = 11.9, p = 6.6$ torr.

The second thesis is directly substantiated by experimental results. Variation of radiation oscillation frequency with impurity (N_2 and H_2) concentration in xenon is shown in Figure 14. These data are obtained at constant xenon pressure ahead of the shock wave, $p = 6.6$ torr and constant incident shock wave Mach number, $M = M_{cr} = 11.9$. Both hydrogen and nitrogen impurity concentrations increasing oscillation frequency increase like the function $F = f(n_e)$. However, the impurity effects on the self-sustained wave process development are different in these two cases. This effect depends essentially on the type of impurity. Oscillation frequency value $F = 20$ kHz is achieved at $\xi \sim 2 \cdot 10^{-2}\%$ in the first case and $\xi \sim 3 \cdot 10^{-3}\%$ in the second.

Without discussion of the kinetic nature of the instability rise in detail we only point out that periodic variation of shock heated plasma radiation intensity with time is related to permanent renovation of gas composition in each cycle of the oscillation process in the flow behind the shock wavefront. Within the framework of this idea it is possible to determine an extent of "active" zone of the flow responsible for oscillation generation. On the basis of data on the oscillation period (T), shock wave velocity (u), and gas compression ($\bar{\rho}$), its length Δl is determined by the simple formula $\Delta l = u \cdot T / \bar{\rho}$. In our experiments Δl varies in the range 1-4 cm depending on the incident shock wave Mach number. The length of the "active" zone (60-80 cm) is small compared to the length of shock heated gas flow region behind the shock wavefront.

Note that this result does not contradict the data on the length of strong perturbation flow region revealed by interferometric investigations [6, 7].

c) Third type of instability of shock heated plasma

Investigation of shock heated xenon plasma emissivity is carried out in the following ranges of the incident shock wave, Mach number, and initial pressure of gas under study: $15.5 < M < 19$, $1.5 \leqslant p \leqslant 5$ torr. The results substantiate the above-formulated suppositions about the formation of stable structural changes in the flow behind the shock wavefront at $M \geqslant$

17.4 and about the threshold character of the third type of instability development.

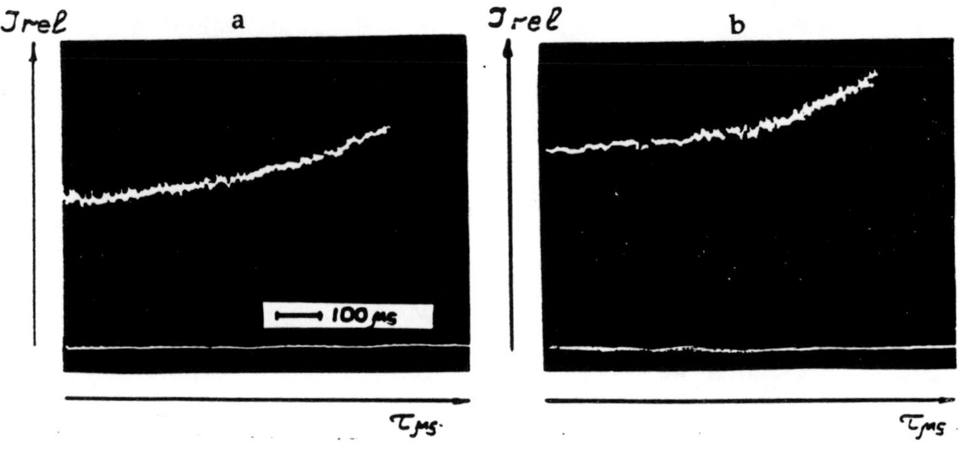

Fig. 15. Time variation of shock heated plasma radiation intensity.
a - $M = 17.0$, $p = 5$ torr; b - $M = 17.5$, $p = 5$ torr.

Figure 15 presents the results of two experiments illustrating plasma radiation intensity variation during plasma propagation along the low-pressure channel at sub-critical ($M = 17.0 < M_{cr}$) and super-critical ($M = 17.5 > M_{cr}$) Mach number values, respectively.

The character of plasma luminescence variation in both cases is identical and does not demonstrate signs of oscillating process development.

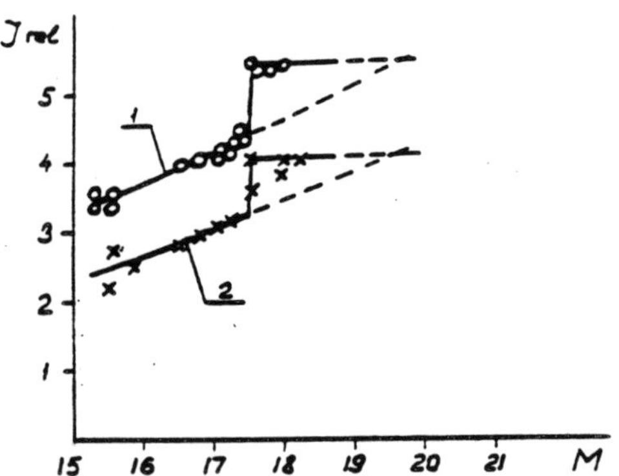

Fig. 16. Variation of shock heated plasma radiation intensity with Mach number:

$1 - \lambda = 443 \pm 5$ nm; $2 - 3 < \lambda < 10\mu$m.

Variation of relative intensity of shock heated plasma radiation in visible ($\lambda = 443 \pm 5$ nm) and IR ($3 < \lambda < 10$ μm) continua with Mach number at $p_0 = 5$ torr is shown in Figure 16. To draw these graphs plasma radiation intensity was determined from oscillograms analogous to oscillograms presented in Figure 15, and at each experiment the value was recorded that corresponded to the moment of the shock wavefront passing the measurement cross-section. Reflecting trivial temperature dependence shock heated plasma radiation intensity both in visible and IR wavelength regions increases with the incident shock wave Mach number at $M < M_{cr}$. Plasma luminescence sharply increases within 15–20% at $M = M_{cr} = 17.4$ that shows the structural changes in the flow. Radiation intensity of shock heated plasma remains constant within experimental accuracy at $M = M_{cr}$.

The last result is rather surprising. Since plasma luminescence contains information about the structural changes in the flow at $M = M_{cr}$ its constancy at $M > M_{cr}$ shows relative attenuation of this process with Mach number increase and probably for its complete suppression at $M \sim 20$, the value that is determined by the intersection point of the curves when the function $J_{rel} = f(M)$ is extrapolated.

On the one hand, this conclusion contradicts our former results (see Ref. [5]), where for the same initial gas pressure values intensive plasma instability development at $M = 17.4$–19.1 and distinct structural changes in appearance at $M = 21.7$ were seen (see Figure 4). On the other hand, it is affirmed by the results of test experiments on flow visualization by means of continuous recording of the interferometric fringes shift [17]. These results prove that under our experimental conditions the development of the structural changes is really less apparent as compared to Ref. [5]. Complete suppression of plasma instability is found at $M \sim 18$, $p \sim 1.5$ torr. Test experiment interferograms are presented in Figure 17.

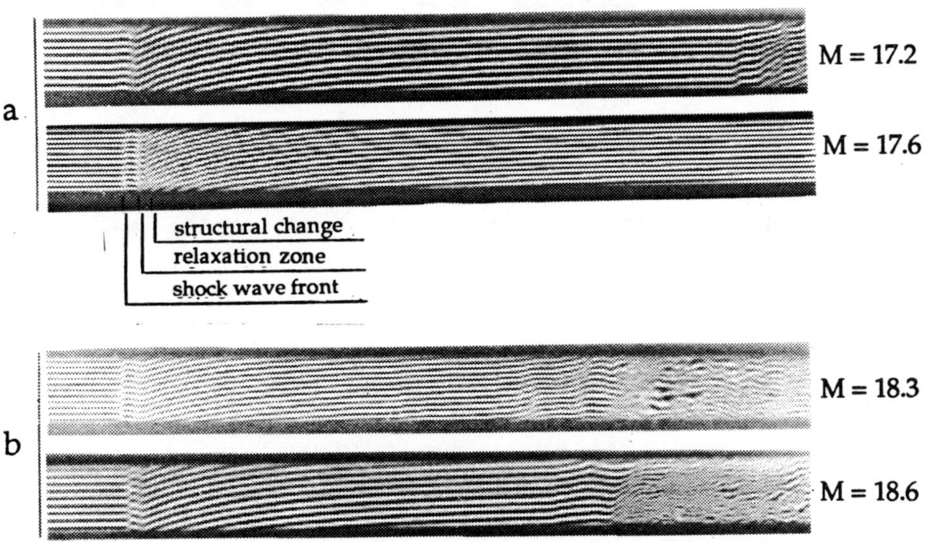

Fig. 17. Flow interferograms: a – $p = 4$ torr; b = $p = 1.5$ torr.

Analysis of the above experimental data shows that development of structural changes in the flow at $M > M_{cr}$ depends on not only absolute value of the shock wavefront velocity but also the behavior of the derivative du/dt and its sign.

It is known [18] that the shock wave velocity varies during its propagation along the low-pressure chamber. In the formation zone it rises to some maximum determined by the pressure drop between low- and high-pressure chambers and then decreases due to energy dissipation.

Formation zone lengths, $\varkappa$ of the weak shock waves usually do not exceed 40-50 shock tube diameters. However, the pressure drop, i.e., Mach number increases or gas pressure in the low-pressure chamber at M = const decreases as the formation zone extends.

According to indirect estimates on the basis of experimental data on variation of the shock wave intensity with pressure drop it is possible to assume that the length of the shock wave formation zone in xenon at $p = 5$ torr is no less than 60-65 calibers at M ~ 16 and 85-90 diameters at M ~ 23.

Comparing these values with the real length of the low-pressure chamber $L = 70$ diameters in the present work and $L = 85$-90 diameters in Ref. [5], we can conclude that condition $dn_0/dM \leqslant 0$ being satisfied, the third type of instability of shock heated plasma develops only if the shock wave velocity in the low-pressure chamber decreases $du/dt < 0$. The greater difference between L and $\varkappa$ instability is more intensive.

Instability suppression observed at M = 18, $p = 1.5$ torr is probably explained by the fact that in this case $\varkappa > L$ and therefore, $du/dt > 0$. The last conclusion must be substantiated by the results of direct measurement of the intensive shock wave formation zone length in the shock tubes.

3. Conclusions

1. Systematization of anomalous phenomenon in the shock waves in inert gases is performed.

2. Four groups of anomalous phenomena accompanying the development of four types of instabilities of the shock heated plasma of inert gases are outlined.

3. Periodic variation of shock heated plasma radiation intensity in a relaxation flow region at M ~ 8.3 is found.

4. The first type of instability development is found to be related to foreign gas impurities in xenon in an amount of $10^{-1}\% > \xi > 4 \cdot 10^{-3}\%$.

5. At Mach number range of 10 < M < 16.5 (the second type of instability) oscillation of shock heated xenon plasma radiation during its propagation along the low-pressure chamber is discovered.

6. The oscillation development dynamics in plasma motion along the low-pressure chamber channel of the shock tube is studied.

7. The dependence of radiation oscillation frequency variation on the incident shock wave intensity, the initial pressure of gas under study, and the molecular gas impurity percentage is ascertained.

References

1. G. K. Tumakaev, In: *High-Temperature Gas Dynamics, Shock Tubes and Shock Waves*, Ed. by R. I. Soloukhin, Minsk, 1983, pp. 154-160.

2. Ya. B. Zeldovich and Yu. P. Rizer, *Physics of Shock Waves and High-Temperature Hydrodynamic Phenomena*, Moscow: Nauka, 1966.

3. H. Petschek and S. Byron, *Ann. of Phys.*, vol. 1, no. 3, 1957, pp. 270-315.

4. G. K. Tumakaev and V. R. Lazovskaya, In: *Eighth Conf. on Phenom. on Ionized Gases*, Vianna, 1967, p. 464.

5. G. K. Tumakaev, V. G. Maslennikov, and E. V. Serova, *Pis'ma Zh. Tekhn. Fiz.*, vol. 6, no. 6, 1980, pp. 354-358 (in Russian).

6. I. I. Glass and W. S. Liu, *J. Fluid Mech.*, vol. 84, pt. 1, 1978, pp. 55-57.

7. I. I. Glass, W. S. Liu, and F. C. Tang, *Canad. J. Phys.*, vol. 55, no. 14, 1977, pp. 1269-1279.

8. P. W. Griffiths, R. J. Sandeman, and H. G. Hornung, *J. Phys. D, Appl. Phys.*, vol. 9, no. 12, 1976, pp. 1681-1691.

9. A. P. Ryazin, *Pis'ma Zh. Tekhn. Fiz.*, vol. 6, no. 9, 1980, pp. 516-520.

10. G. K. Tumakaev and Z. A. Stepanova, *Zh. Tekhn. Fiz.*, vol. 52, no. 11, 1982, pp. 2305-2306 (in Russian).

11. W. F. Hilton, *High-Speed Aerodynamics*, London, New York, Toronto, 1952.

12. G. K. Tumakaev, *Zh. Tekhn. Fiz.*, vol. 52, no. 4, 1982, pp. 690-694 (in Russian).

13. P. E. Oettinger, D. Bershader, *AIAA J.*, vol. 5, no. 9, 1967, pp. 1625-1632.

14. A. S. Baruishnikov, A. P. Bedin, G. I. Mishin, and G. E. Skwortsov, In: *Physical Gas Dynamic Ballistic Research*, Ed. by G. I. Mishin, Moscow: Nauka, 1980, pp. 34-44.

15. W. Ebeling, *Strukturbildung Bei Irreversiblen Prozesson*, Leipzig, 1976 (in German).

16. *Oscillations and Traveling Waves in Chemical Systems*, Ed. by R. J. Fiedd and M. Burger, New York, 1986.

17. G. K. Tumakaev and V. R. Lazovskaya, In: *Aerophysical Research of Supersonic Flows*, Ed. by Yu. A. Dunaev, Moscow: Nauka, 1967, pp. 74-104.

18. E. V. Stupochenko, S. A. Losev, and A. I. Osipov, *Relaxation Processes in Shock Waves*, Moscow: Nauka, 1965.
19. G. K. Tumakaev, Z. A. Stepanova, *Pis'ma Zh. Tekhn. Fiz.*, vol. 15, no. 2, 1989, pp. 15–22.

PRECURSOR PHENOMENA IN INERT GASES
IN SHOCK TUBES

G. K. Tumakaev, T. V. Zhikhareva, D. G. Zavarin

Abstract: **Experimental data on electron concentration distribution in the precursors of inert gases are presented for a wide range of initial conditions, namely, the incident shock wave Mach number ($M = 13$-23) and gas pressure ahead of its front ($p = 2$-6 torr). A microwave interferometric device with two-wire transmission line providing high resolution was used for plasma diagnostics.**

On the basis of experimental data and theoretical analysis ionization kinetics of monoatomic gases is ascertained. Precursor formation processes in different gases are found to be quite different due to specific atomic energy level distributions. Unusual results on the kinetics of gas ionization near the shock wavefront are presented.

1. Ionization

Today we see a certain progress in the study of elementary processes kinetics in monoatomic gases ahead of shock waves [1-3]. The dominant role of the associative ionization mechanism in charged particle production has been established [4-6]. An electron gas heating mechanism has been discussed. It has been shown that gas parameters ahead of the shock wave can be notably influenced by radiation reflection at the shock tube walls [7-10]. Finally, available data analysis reveals significant discrepancies in ionization and excitation kinetics of monoatomic gases depending on the atomic states structure. For example, in mercury vapor investigating it was found in Ref. [4] that because of different excitation energies and different atomic constant values populating the resonance 6^3P_1 state near the shock wavefront was provided mainly by inelastic electron-atom collisions whereas 6^1P_1 state excitation occurred mainly due to radiative processes.

The specific character of the atomic energy states structure influences not only the excitation mechanism in mercury vapor and inert gases, but the mechanism for molecular ion production ahead of the shock wave as well. This is why charged particle production is related to associative ionization at two excited atoms collision [4]

$$A^* + A^* \rightarrow A_2^+ + e \tag{1}$$

in the first case, and to the Molnar-Hornbeck reaction [5, 6]

$$A^{**} + A \rightarrow A_2^+ + e \tag{2}$$

in the second case. Here A^{**} denotes a highly excited state.

There is a difference between the energy state structure of argon and xenon atoms. For instance, shift values between the limits correspondent to the ion states $^2P_{3/2}$ and $^2P_{1/2}$ are different for these atoms [11]. Moreover, there is a difference in the relative disposition of the terms approaching one limit [12]. These peculiarities may influence the rates of populating highly excited states participating in the associative ionization process. However, it is difficult to discuss this question, since available experimental data on argon [13-16] is rather unusual and there is no data on xenon.

Consider in more detail the theoretical description of ionization kinetics for monoatomic gases using an instant ionization approach for highly excited atoms that participate in a Molnar-Hornbeck reaction. Ignoring both radiative and collisional processes resulting in destruction of highly excited states as well as dissociative recombination process, the set of kinetic equations for concentration of atoms at highly excited Rydberg states n_R^* and concentration of electrons n_e is written as follows

$$-v\frac{dn_R^*}{dx} = B_R - K_R n_0 n_R^* \tag{3}$$

$$-v\frac{dn_e}{dx} = \sum_R - K_R n_0 n_R^* \tag{4}$$

where v is shock wave velocity, B_R is the rate of Rydberg states populating due to absorption of equilibrium flow region radiation, K_R is the associative ionization rate constant, n_0 is the concentration of atoms at a grown state.

Since rate constant K_R is large we can assume the electron production rate to be equal to the rate of highly excited states populating. In this case electron concentration is defined by formula

$$n_e = \sum_R \frac{1}{v} \int_\infty^x B_R(x')dx' \tag{5}$$

where

$$B_R = \frac{n_R^B(T_{eq})}{3\tau_R\sqrt{\rho k_0^R x}}\left[1 - \left(1 + (x/R)^{-2}\right)^{-3/4}\right]. \tag{6}$$

Here $n_R^B(T_{eq})$ is the Boltzmann value for R-state concentration at equilibrium gas temperature, T_{eq} behind the shock wavefront; τ_R is the lifetime of the R-state; K_0^R is the absorption coefficient at the center of a line correspondent to transition from ground state to R-state; x is a distance from the shock wavefront; r is tube radius.

At $x/r \gtrsim 2$ the following asymptotic formula is valid for electron concentration

$$n_e = 0.965 \cdot 10^{19}\frac{p_1 r^2}{M}\sqrt{\frac{m_a}{\rho\gamma kT_a^3}}\sum_R \frac{g_R e^{-\frac{E_R}{kT_{eq}}}}{6g_0\tau_R\sqrt{k_0^R}x^{2/3}} \tag{7}$$

where p_1 is initial gas pressure (torr); T_a is gas temperature ahead of shock wave, K; M is Mach number; $\gamma = 5/3$ is the specific heat ratio; m_a is atomic mass; g_0, g_R are degeneracies of the ground and R-states, respectively; E_R is excitation energy for the R-state, K is the Boltzmann constant.

It is seen from Formula (7) first, that within the framework of the above model asymptotic value for a tangent of $\lg(n_e/p_1)$ slope as a function of $\lg x$ is equal to -1.5, and in the second place, if the distance from the shock wavefront x and initial temperature T_a is fixed, parameter $\dfrac{n_e M}{p_1\exp(-E_R/k T_{eq})}$ is invariant as regards to p_1 and M for the tube of a given radius r.

The present work deals with experimental investigation of precursors in monoatomic gases, namely, argon and xenon. Plasma diagnostics in the shock tube are carried out with the use of a 3 cm microwave interferometer with a two-wire transmission line providing high spatial resolution (up to 0.1 λ). Investigations are performed for a wide range of the incident shock wave Mach number and initial gas pressure. The data on electron concentration profiles at a distance up to 40 cm ahead of the shock waves are obtained at an initial pressure of $p_1 = 2$-6 torr and Mach numbers M = 13-17 for argon and M = 14-23 for xenon.

On the basis of experimental data analysis it is found that an efficiency for electron production in argon precursors is several times higher compared to xenon. This anomaly is caused by a higher Rydberg state population rate for argon atoms due to significant Van der Waals absorbing line broadening, not the formerly assumed resonance broadening [6, 9].

It is also shown that two excited atom collisions

$$A^* + A^* \begin{cases} A^+ + A + e \\ A_2^+ + e \end{cases} \tag{8}$$

provide an appreciable contribution to xenon atom ionization side-by-side with the Molnar-Hornbeck process [25, 26].

2. Experimental Method

Intensive shock waves in monoatomic gases were initiated in the shock tube [17] operating in the single-diaphragm regime. The length of the high-pressure chamber is 2.7 m. The measurement cross-section is located at a distance of 8 m from the diaphragm. The diameter of the cylindrical channel of the low pressure chamber is 100 mm. Hydrogen was used as a driver gas. In order to obtain shock waves with maximum Mach numbers, M = 17 in argon and M = 23 in xenon, driver gas pressure was raised to 300 atmosphere and its temperature was increased to 450°C with the use of an external electric heater.

Before the experiment the low-pressure chamber was pumped out to a pressure of 10^{-2} torr, washed with the gas under study, and filled with argon or xenon to a pressure of 2-6 torr. A microwave interferometer with a wave length range of 3 cm supplied with a two-wire transmission line [18, 19] was applied to determine precursor electron concentration ($n_e \sim 1 \cdot 10^{10}$ – $7 \cdot 10^{11}$ cm^{-3}). Note that the use of a two-wire transmission line provides high spatial resolution up to 0.1 λ that is inaccessible for interferometers with horn antennae.

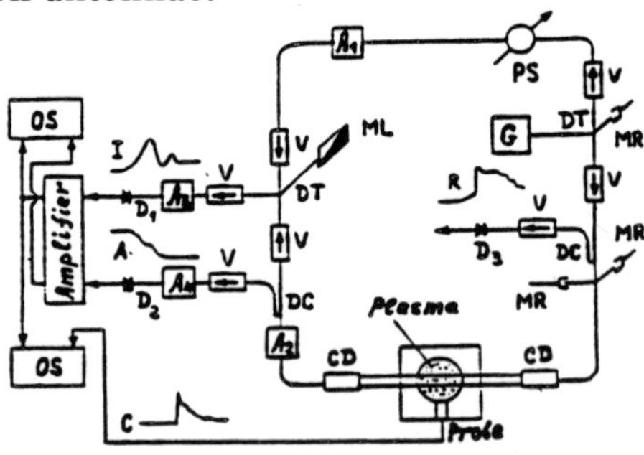

Fig. 1. Schematic of microwave interferometer.

The scheme of the microwave interferometer is presented in Figure 1. The interferometer is assembled out of the standard microwave technique elements: rectangular waveguides, directional couplers DC, double T-bends DT, attenuators A_1 - A_y, movable reflectors UR, valves V, phase-shifters PS, diodes D_1-D_3, absorber matched load UL.

The two-wire transmission line was matched with the waveguide by means of special conforming devices CD. The use of a directional coupler allows one to record simultaneously interference, absorption, and reflection of microwave plasma radiation (I, A, R-signals). The working regime of diodes recording microwave radiation was controlled and supported within the quadratic interval of the J-V-characteristic. Other elements of the device are microwave radiation generator and recording equipment, i.e., oscillographs.

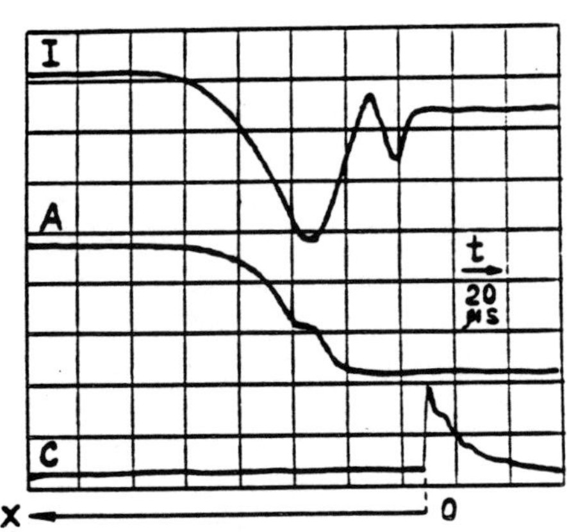

Fig. 2. Signal oscillograms.

The use of microwave plasma diagnostics with a two-wire waveguide transmission line predetermined the construction and choice of material for the shock tube working section.

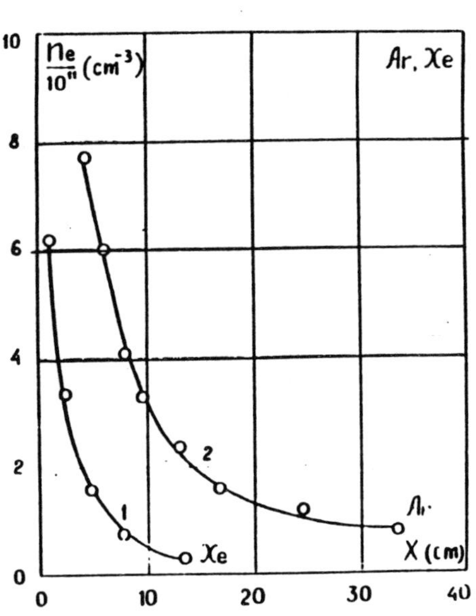

Fig. 3. Electron concentration distribution ahead of the shock wavefront in xenon (1) and argon (2) at M = 14.3, p_1 = 4.3 torr.

To suppress stationary waves in low-pressure chamber volume the working section 20 cm long was made of fluoroplastic. Since the two-wire waveguide line is destroyed at each experiment by concurrent gas flow and must be replaced by a new one, a split working section was found to be most convenient. It consists of two identical branch pipes and is supplied with double vacuum hardening and differential pumping.

The set of ionizing probes mounted along the low-pressure chamber of the shock tube was used to start the recording equipment, to determine the shock wavefront velocity and the moment when it passes through the working cross-section. As the shock front approaches, an increase of electron concentration in the shock tube working cross-section results in phase shift of the interfering beams and appreciable absorption of microwave radiation.

Figure 3 illustrates typical oscillograms of the signals of interference I absorption A of microwave radiation in the precursor. A drastic increase of plasma conduction signal (curve C) recorded by the ionizing probe occurs at the moment when the shock wavefront passes by the two-wire transmission line ($x = 0$). Phase shift is distinctly recorded in the oscillograms at significant distance from the shock wavefront ($x \sim 40$ cm). As the shock wave approaches, the rate of phase shift essentially increases and the amplitude of the interference signal monotonously decreases due to absorption of microwave radiation in plasma.

Note that at the pressure range considered the abrupt microwave signal cut-off at $n_e = n_{cr}$ is not observed because of the high efficiency of the elastic electric-atom collisions. However, microwave radiation is always completely absorbed behind the shock front where $n_e > n_{cr}$ ($n_{cr} = 1.2 \cdot 10^{12}$ cm^{-3} for $\lambda = 3$ cm).

Oscillogram processing provides the data on electron concentration variation in the precursors. Electron concentration and collision frequencies were calculated according to the formulae

$$\frac{\nu}{\omega} = \frac{2\frac{\Delta\eta}{d}\left(\frac{\omega}{c} - \frac{\Delta\varphi}{d}\right)}{\left(\frac{\Delta\eta}{d}\right)^2 - \left(\frac{\Delta\varphi}{d}\right)^2 + 2\frac{\Delta\varphi}{d}\cdot\frac{\omega}{c}} \tag{9}$$

$$\frac{n_e}{n_{cr}} = \left[\left(\frac{\Delta n}{d}\right)^2 - \left(\frac{\Delta\varphi}{d}\right)^2 + 2\frac{\Delta\varphi}{d}\cdot\frac{\omega}{c}\right]\left(1 + \frac{\nu^2}{\omega^2}\right)\frac{c^2}{\omega^2} \tag{10}$$

where $\Delta\varphi$, $\Delta\eta$ are phase shift and attenuation obtained from the experiment; ω is field angular frequency; c is light speed; and d is plasma length scale equal to the diameter of the low-pressure camera channel of the shock tube.

Expressions (9) and (10) are obtained from the following relations [20]:

$$\beta^2 = \frac{1}{2}\frac{\omega^2}{c^2}\left\{1 - \frac{\omega_0^2}{\omega^2 + \nu^2} + \left[\left(1 - \frac{\omega_0^2}{\omega^2\nu^2}\right)^2 + \frac{\nu^2}{\omega^2}\left(\frac{\omega_0^2}{\omega^2 + \nu^2}\right)^2\right]^{1/2}\right\} \tag{11}$$

$$\alpha^2 = \frac{1}{2}\frac{\omega^2}{c^2}\left\{\left[\left(1 - \frac{\omega_0^2}{\omega^2 + \nu^2}\right)^2 + \frac{\nu^2}{\omega^2}\left(\frac{\omega_0^2}{\omega^2 + \nu^2}\right)^2\right]^{1/2} - \left(1 = \frac{\omega_0^2}{\omega^2 + \nu^2}\right)\right\}$$

$$(12)$$

$$\Delta\varphi = -\left(\beta - \frac{\omega}{c}\right)d, \quad \Delta\eta = \alpha d, \quad \omega_0 = \left(\frac{4\pi n_e e^2}{m_e}\right)^{1/2} \tag{13}$$

where α is attenuation constant, β is phase constant, ω_0 is plasma frequency, e is electron charge, and m_e is electron mass.

3. Experimental Results and Discussion

Typical electron concentration profiles ahead of the shock wave in the gases under study (argon and xenon) are presented in Figure 3 at M = 14.5, p_1 = 4.3 torr.

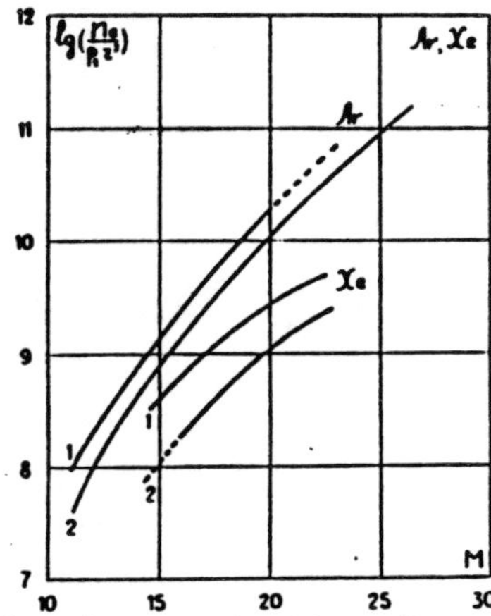

Fig. 4. Variation of $n_e/p_1 r^2$ (cm^{-5}/torr) with Mach number in argon and xenon at x = 10 (curve 1) and x = 20 cm (curve 2).

Figure 4 illustrates experimental data on the variation of the electron concentration referred to as initial pressure and squared shock tube channel radius r versus Mach number at two distances from the shock front in argon and xenon.

Experimental results show that in both argon and xenon for the entire range of Mach numbers considered, the precursor electron concentration at M = const decreases monotonously with the distance from the shock wavefront as seen in Figures 3 and 4 for M = 14.5. At a fixed distance from the shock front concentration n_e increases with Mach number.

Nevertheless, comparison of the experimental data on argon and xenon shows essential discrepancies in the charged particle profiles ahead of the shock waves in these gases. At the same Mach numbers, initial pressure, and distance from the shock wavefront electron concentration in argon is found to be higher than xenon in spite of the higher value of ionization energy of the argon atom (I_{Ar} = 15.76 eV, I_{Xe} = 12.13 eV). In addition, electron concentration dependence on Mach number in argon is stronger when compared with xenon. For instance, with a Mach number in argon increasing from M = 10 to M = 25 the n_e values correspondent to x = 10 cm and x = 20 cm rise by almost three orders of magnitude. Analogous variation of electron concentration ahead of the shock wave in xenon recorded in the range M = 15-23 is about 1.5 orders of magnitude.

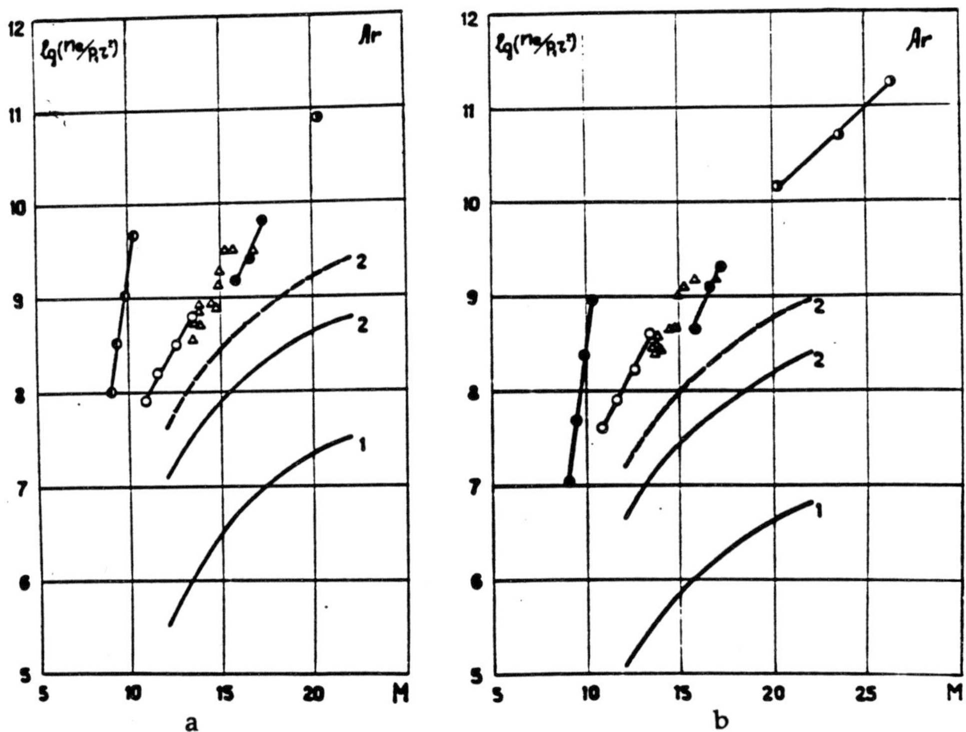

Fig. 5. Variation of $\lg(n_e/p_1 r^2)$ with Mach number in argon.

a) at x = 10 cm, b) at x = 20 cm. ◯ - data [13]; ◯ - [14]; ◯ - [15]; ◯ - [16]; △ - present results; 1 - calculation with only one highly excited state taken into account; 2 - calculation with the sum of Rydberg states taken into account. (n_e, cm^{-3}; p_1, torr; r, cm).

Figures 5 and 6 present comparisons of experimental and calculated data on electron concentration variation with the Mach number at two fixed distances from the shock wavefront in argon and xenon, respectively.

To calculate the upper n_e limit it was assumed that highly excited atoms formed due to absorption of the equilibrium flow region radiation were ionized instantaneously. Shock-heated gas radiation within absorption contours of Rydberg lines was treated as black body radiation, its absorption within the relaxation zone was ignored. The contributions of six Rydberg states [6, 21, 22] with excitation energy exceeding the energy of associative ionization were taken into account in ionization calculation. Correspondent results are shown by curve 2. Curve 1 is calculated under the assumption that only highly excited argon atoms at $4d[1\frac{1}{2}]_1^0$ state and xenon atoms at $5d^1[1\frac{1}{2}]_1^0$ state contribute to electron production. Solid curves show the results obtained with resonance broadening of the Rydberg line contours taken into account, dashed curves — with Van der Waals broadening mechanism.

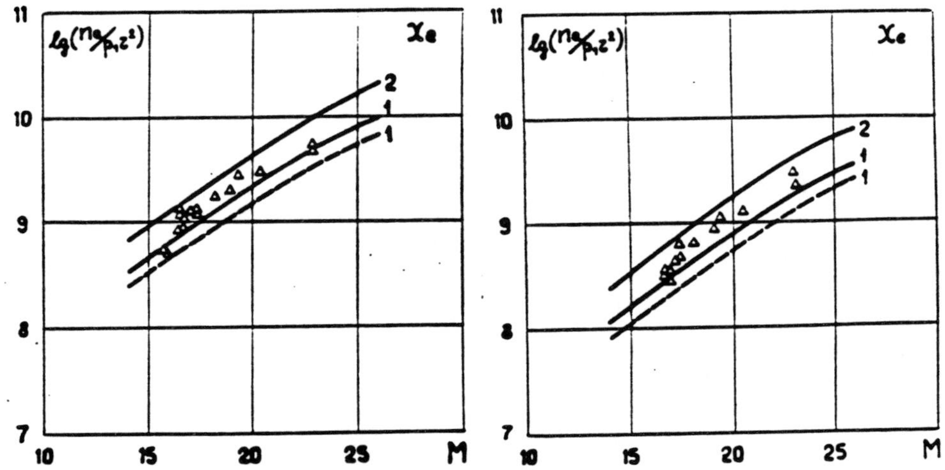

Fig. 6. Variation of $\lg(n_e/p_1 r^2)$ with Mach number in xenon.

a) at x - 10 cm, b) at $x = 20$ cm. $\triangle$ - present experimental results; 1 - calculation with only one highly excited state taken into account; 2 - calculation with the sum of Rydberg states taken into account. (n_e, cm^{-3}; p_1, torr; r, cm).

Experimental results show a monotonous increase of the precursor electron concentration in the gases under study with Mach number at $x = $ const. Experimental data for xenon are in good agreement with calculation results obtained with resonance mechanism of spectral line broadening taken into account. Taking resonance broadening for the sum of states in argon into account does not bring theoretical and experimental results into agreement. For example, at M = 15 experimental results exceed calculation by an order of magnitude and the discrepancy seems to increase at higher Mach numbers. This discrepancy diminishes when Van der Waals mechanism for spectral line broadening is taken into account.

As for an excess of measured n_e values in argon over n_e values in xenon at M = const, we must consider the effect as an anomaly in reference to conventional ideas about kinetics of elementary processes in monoatomic gas precursors. Indeed, it cannot be explained by the inclusion of additional electron production mechanisms, for instance, photoionization processes or excited atom ionization by electron impact. Taking into account photoionization of easily ionized impurities including molecular ones does not help either. The latter is substantiated by good agreement of all experimental results presented in Figure 5. These results were obtained on the facilities with different (from 10^{-2} to 10^{-6} torr) preliminary pumping out the low-pressure chamber of the shock tube and with the use of gases with different impurity concentration. Only the data [13] stand out against other results that confirm strong n_e dependence on Mach number.

The difference between precursor formation in argon and xenon is caused by the different atomic state structures of these gases. For instance, the energy gap between the $4d[1\frac{1}{2}]_1^0$ state and the nearest perturbing $4f[2\frac{1}{2}]_2$ state in argon is

$$\Delta E\left(4d[1\tfrac{1}{2}]_1^0 - 4f[2\tfrac{1}{2}]_2\right) = 0.05 \text{ eV}$$

while in xenon we have, according to [12]

$$\Delta E\left(5d'[1\tfrac{1}{2}]_1^0 - 6p'[\tfrac{1}{2}]_0\right) = 0.47 \text{ eV}.$$

Because of the closeness of the perturbing states Rydberg line contours in argon are influenced to a greater extent by Van der Waals than resonance broadening. In xenon under our experimental conditions the relative contributions of the above-mentioned mechanisms to Rydberg line broadening are comparable.

Van der Waals constants and the values of line-halfwidth of absorbing lines for the Van der Waals broadening mechanism were calculated according to [23, 24]. At $p_1 = 4$ torr they are:

$$C_6\left(4d[1\tfrac{1}{2}]_1^0, \text{Ar}\right) = 0.79 \cdot 10^{-29} \text{ cm}^6\text{s}^{-1}$$

$$\Delta\nu_W\left(4d[1\tfrac{1}{2}]_1^0, \text{Ar}\right) = 0.24 \cdot 10^9 \text{ s}^{-1}$$

$$C_6\left(5d'[1\tfrac{1}{2}]_1^0, \text{Xe}\right) = 0.70 \cdot 10^{-30} \text{ cm}^6\text{s}^{-1}$$

$$\Delta\nu_w\left(5d'[1\tfrac{1}{2}]_1^0, \text{Xe}\right) = 0.63 \cdot 10^8 \text{ s}^{-1}$$

The values of half-width of the same lines due to resonance excitation energy transfer [24] are

$$\Delta\nu_{us}\left(4d[1\tfrac{1}{2}]_1^0, \text{Ar}\right) = 0.87 \cdot 10^7 \text{ s}^{-1}$$

$$\Delta\nu_{us}\left[5d'[1\tfrac{1}{2}]^0_1,\ \mathrm{Xe}\right] = 0.10 \cdot 10^9\ \mathrm{s}^{-1}$$

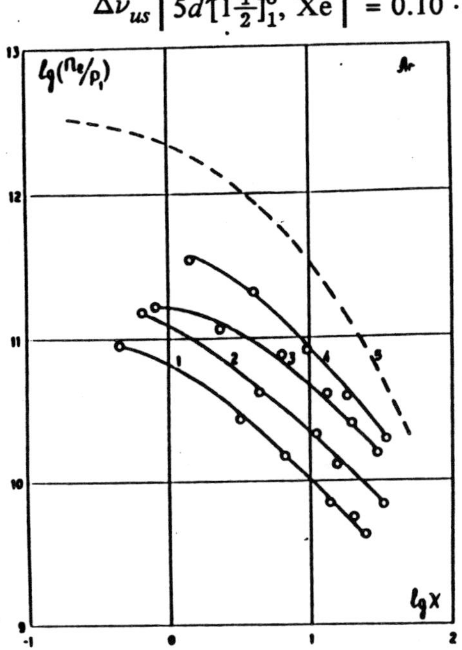

Fig. 7. Variation of $\lg(n_e/p_1)$ with lg x in argon.

1 - M = 13.5, p_1 = 5; 2 - M = 13.9, p_1 = 4; 3 - M = 14.9, p_1 = 3; 4 - M = 16.8, p_1 = 2; 5 - calculation accounting for geometrical factor. (n_e - cm^{-3}; p_1 - torr; r - cm).

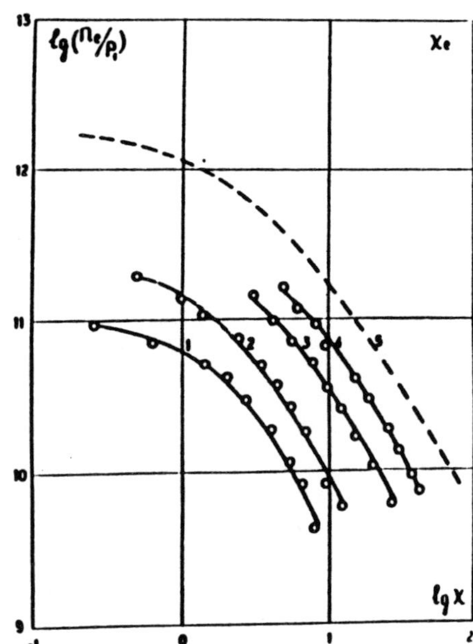

Fig. 8. Variation of $\lg(n_e/p_1)$ with lg x in xenon at p_1 = 4.3 torr.

1 - M = 14.5; 2 - M = 16.0; 3 - M = 17.3; 4 - M = 19.3; 5 - calculation accounting for geometrical factor (n_e - cm^{-3}, x - cm).

Relatively large Van der Waal broadening of the Rydberg lines in argon increases the rate of highly excited state population due to absorption of shock-heated gas radiation and, consequently, results in a higher degree of ionization in argon precursors.

High spatial resolution of the method used for plasma diagnostics allowed discovery of the difference in n_e variation in the precursors of the gases under study at a length up to 40 cm ahead of the shock wave.

Figures 7 and 8 present the data on electron concentration variation with x for some values of Mach number in argon and xenon. It is seen especially in the curves correspondent to small Mach numbers, that the shock wavefront approaching the tangent of slope of the curves $\lg(n_e/p_1) = f(\lg x)$ appreciably decreases (by absolute value) from its asymptotic value.

Note that in xenon at M ~ 14 electron concentration in the precursor was recorded at a distance of 0.5 cm from the shock thanks to the use of a two-wire transmission line in the microwave interferometer.

The reason for tangent of slope of the curves $\lg(n_e/p_1) = f(\lg x)$ variation is simple. Its decrease is caused by variation of radiative flux from shock heated gas along a semi-infinite cylindrical channel due to the geometrical factor:

$$P(x) = \int\limits_{\infty}^{x} \frac{1}{\sqrt{x'}} \left[1 - \left(1 + \left(\frac{x'}{r} \right)^{-2} \right)^{-3/4} \right] dx'$$

of Expression (5). The function $\lg(cP(x)) = f(\lg x)$ shown in Figure 8 with dashed curve is appropriate to the experimental curves in xenon (c is an arbitrary constant). The tangent of its slope is close to the calculated value near the shock wavefront and approaches an asymptotic value equal to -1.5 when the distance from the shock front increases.

An analysis of the curves for argon allows us to demonstrate the geometrical factor effect on electron concentration ahead of the shock wave as well. However, in this case there is only qualitative agreement. In contrast to xenon a rather slow decrease of electron concentration with a distance from the shock front is recorded in argon precursors, the tangent of slope is essentially less than the calculated value and is equal to -1 at distances up to 40 cm that significantly exceed the channel diameter. In all probability this discrepancy cannot be explained by omission of some kinetic processes of atom ionization in the precursors. It is known (see, for example, [5, 9]) that photoionization of the excited atoms is accompanied with more rapid n_e variation. One probable reason for the slow decrease of electron concentration in the precursors is absorption in the relaxation zone. Indeed, within the whole range of Mach numbers considered the length of relaxation zone in argon at M = const exceeds the relaxation length in xenon by several times [27, 28].

Taking the relaxation zone effect into account, the rate of production of the excited atoms at R-states is expressed instead of (6) by the following formula [9]

$$BR = \frac{n_R^B(T_{eq})}{3\theta_R \sqrt{\pi k_0^R r (16\frac{l}{r} + y)}} \left[1 - \left(\frac{y + \frac{l}{r}}{\sqrt{1 + (y + \frac{l}{r})^2}} \right)^{3/2} \right] \tag{14}$$

where l is relaxation zone length, $y \equiv x/r$.

Figure 9 shows tg γ variation with a distance from the shock wavefront in argon corresponding to experimental and theoretical results on electron concentration. Relaxation zone effect was taken into account in calculating the rate of excited atom production ahead of the shock wave. These results are shown by the solid lines ($l \neq 0$). Dashed lines depict the results obtained, the above-mentioned effect being ignored ($l = 0$.

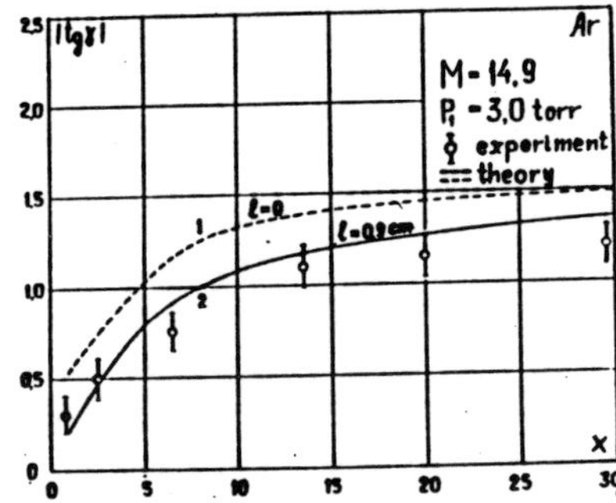

Fig. 9. Variation of tg γ absolute value with a distance from the shock wavefront in argon. Curves 1 and 2 are calculated at the following relaxation zone lengths:

$1 - l = 0$, $2 - l = 0.9$ cm.

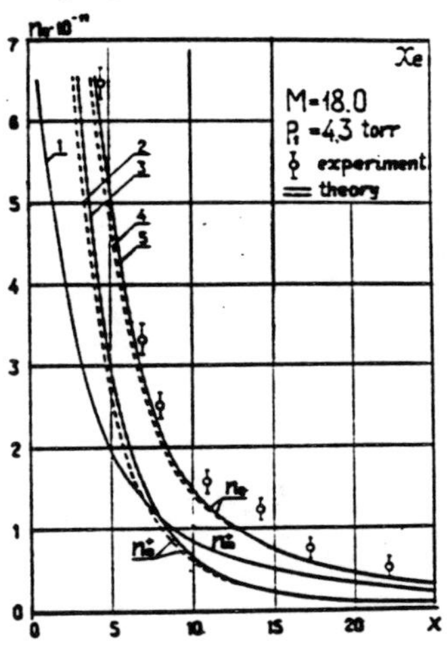

Fig. 10. Concentration of atomic ions n_a^+, molecular ions n_m^+, and electrons n_e (cm^{-3}) versus x (cm) in xenon.

1 - n_m^+; 2, 3 - n_a^+; 4, 5 - n_e. Solid curves - photoionization of excited states is taken into account; dashed curves - photoionization is ignored.

Results show that taking into account the relaxation zone effect in argon allows the discrepancy between experimental and theoretical data on tg γ values to be diminished. On the other hand, absorption of radiation in the relaxation zone results in reduction of the n_e absolute values. If this effect is ignored the calculated curves for argon lie below the experimental ones. Therefore we must seek an additional ionization mechanism.

As for xenon the relaxation zone effect is small because of small l values ($l = 0.1$ cm at M = 18, $p_1 = 4.3$ torr).

Let us consider electron production by two excited atom collisions (8). The rate constant for this process is equal to $1.1 \cdot 10^{-9}$ cm^{3}s^{-1} in argon [25] and is equal to $1.5 \cdot 10^{-9}$ cm^{3}s^{-1} in xenon [26].

Figure 10 presents atomic and molecular ions and electron concentrations calculated with the Molnar-Hornbeck associative ionization process (2) with ionization by two excited atom collisions (8) taken into account.

It is seen that the contribution of both processes are comparable near the shock front in xenon. As the distance from the shock front increases the relative contribution of reaction (8) diminishes and the Molnar-Hornbeck associative ionization process including highly excited atoms becomes dominant.

Experiments in argon do not show appreciable process (8) effect on electron concentration.

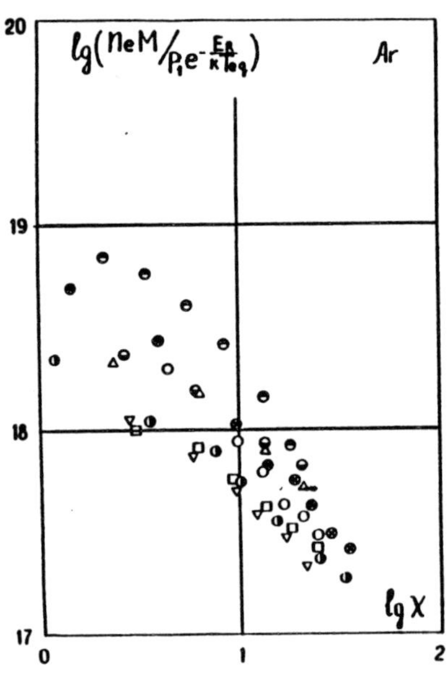

Fig. 11. Variation of $\lg(n_e M/p_1 \exp(-E_R/k\,T_{eq})$ with $\lg x$ in argon. $\ominus$ - M = 15.2, p_1 = 2; $\ominus$ - M = 14.9, p_1 = 2; O - M = 13.9, p_1 = 4; $\otimes$ - M = 16.8, p_1 = 2; OM = 15.8, p_1 = 4; $\triangle$M = 14.9, p_1 = 3; ∇M= 13.6, p_1 = 4.3; $\square$M= 13.4, p_1 = 5 (n_e - cm^{-3}; p_1 - torr, x - cm).

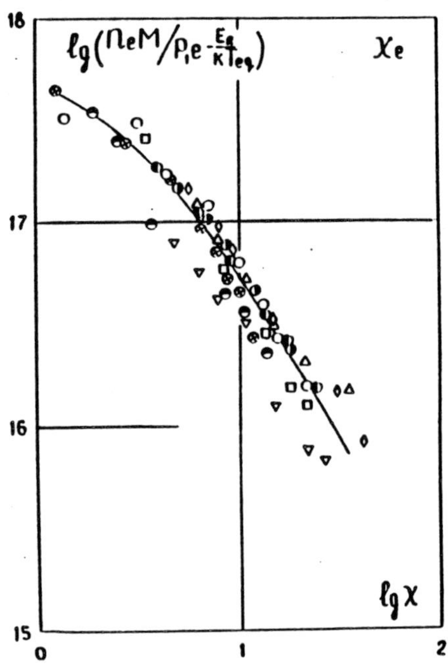

Fig. 12. Variation of $\lg(n_e M/p_1 \exp(-E_R/k\,T_{eq})$ with $\lg x$ in xenon.

OM = 16.8, p_1 = 4.3; $\otimes$M = 15.6, p_1 = 4.3; $\ominus$M = 15.9, p_1 = 4.3; $\lozenge$M = 19.3, p_1 = 4.3; OM = 18.9, p_1 = 4.3; OM = 18.0, p_1 = 4.3; $\square$M= 17.3, p_1 = 4.3; OM = 14.5, p_1 = 4.3; ∇M= 20.2, p_1 = 6; $\triangle$M = 22.9, p_1 = 2 (n_e - cm^{-3}; p_1 - torr; x - cm).

Finally, we note that the experimental data obtained on electron concentration distributions substantiate the conclusion that for a wide range of Mach numbers and initial pressures the dominant process of charged particle production in the precursors is instantaneous highly excited atom ionization by means of the Molnar-Hornbeck process. This conclusion is confirmed, particularly, by parameter $\dfrac{n_e M}{p_1 \exp(-E_R/k\,T_{eq})}$ (see (7)) invariancy as regards to initial pressure and Mach number.

All experimental data obtained at various initial pressures and Mach numbers are presented in Figures 11 and 12 in the form of $\dfrac{n_e M}{p_1 \exp(-E_R/kT_{eq})}$ dependence on lg x for argon and xenon, respectively. Here excitation energy of the Rydberg state taking part in the associative ionization process, E_R = 14.41 eV for argon and E_R = 11.16 eV for xenon. The data for xenon at Mach numbers M = 14.5-23 and initial pressure 2-6 torr within experimental accuracy lie on the single curve. In argon (M = 13.5-17, p_1 = 2-5 torr) there is a significant (up to an order of magnitude) scatter of experimental data at lg x < 1. However, a trend to bring these data closer together is seen at large lg x values.

The results presented with Expression (7) show that for the Mach number range considered the single value of the more general parameter

$$q = \frac{n_e M x^{-3/2}\,T_a^{3/2}}{p_1 r^2 \exp(-E_R/k\,T_{eq})}$$

governs electron concentration variation in xenon ahead of the shock wave in the cylindrical tube of arbitrary radius r, at $x > 2r$ and different values of x, M, p_1, T_a. According to experimental data this parameter value in xenon is equal to $0.31 \cdot 10^{21}$ deg$^{3/2}$/torr cm$^{7/2}$. The value calculated with only one excited state taken into account is $0.20 \cdot 10^{21}$. Taking into account the sum of Rydberg states mentioned above we obtain $0.34 \cdot 10^{21}$ [29].

4. Conclusions

The data on electron concentration distribution in argon and xenon precursors ahead of intensive shock waves are obtained.

The atomic structure effect on the precursor formation in these gases is substantiated. It is found that under the conditions of our experiments the contours of Rydberg absorbing lines are broadened according to a Van der Waal mechanism in argon and a resonance mechanism in xenon.

The dominance of two excited atom collisions in the inert gas ionization near the shock wavefront is shown.

References

1. L. M. Biberman, A. Kh. Mnatsakanian, and I. I. Yakubov, *Usp. Fiz. Nauk*, vol. 102, no. 3, 1970, pp. 431-464 (in Russian).

2. L. M. Biberman and B. A. Veklenko, *Zh. Eksper. Teor. Fiz.*, vol. 37, no. 1(7), 1959, pp. 164-169 (in Russian).

3. A. I. Lagarkov and I. T. Yakubov, *Optika i Spektr.*, vol. 14, no. 2, 1963, pp. 199-207.

4. G. K. Tumakaev, T. V. Zhikhareva, and V. R. Lasovskaya, *Zh. Tekhn. Fiz.*, vol. 41, no. 9, 1971, pp. 1986-1995 (In Russian).

5. A. E. Buluishev, N. G. Preobrazhenskii, A. E. Suvorov, and V. I. Yakovlev, *Fiz. Gazodinam.*, no. 6, 1976, pp. 19-21 (in Russian).

6. M. G. Vasiliev, T. V. Zhikhareva, and G. K. Tumakaev, *Zh. Tekhn. Fiz.*, vol. 49, no. 3, 1979, pp. 541-553 (in Russian).

7. M. Omura and L. L. Presley, *AIAA J.*, vol. 7, no. 12, 1969, pp. 2363-2365.

8. V. A. Gorelov and L. A. Kildushova, *Izv. Akad. Nauk SSSR, Mekh. Zhidk. i Gaza*, no. 2, 1971, pp. 147-151 (in Russian).

9. T. V., Zhikhareva, M. G. Vasiliev, and Yu. G. Morozov, *Zh. Tekhn. Fiz.*, vol. 45, no. 3, 1975, pp. 568-578 (in Russian).

10. M. G. Vasiliev and T. V. Zhikhareva, *Zh. Tekhn. Fiz.*, vol. 46, no. 6, 1976, pp. 1262-1270 (in Russian).

11. S. E. Frish, *Optical Atom Spectra*, Moscow: Fiz.-Mat. Izd. 1963 (in Russian).

12. A. R. Stringanov and N. S. Sventitskii, *Tables of Spectral Lines of Neutral and Ionized Atoms*, Moscow: Atomizdat, 1966 (in Russian).

13. L. B. Holmes and H. D. Weymann, *Phys. Fluids.*, vol. 12, no. 6, 1969, pp. 1200-1210.

14. S. Lederman and D. S. Wilson, *AIAA J.*, vol. 5, no. 1, 1967, pp. 70-77.

15. G. Kamimoro and K. Teshima, *Trans. Jpn. Soc. Aero-Space Sci.*, vol. 15, no. 29, 1972, pp. 124-140.

16. M. Pinegre, *Thes. Univ. Ruen*, France, 1976 (in French).

17. V. G. Maslennikov, A. K. Myakishev, E. V. Serova, and D. G. Zavarin, *All-Union Symp. on Aerophys. Res. Methods*, Novosibirsk, 1976, p. 25 (in Russian).

18. W. Makios, *RSI*, vol. 38, no. 3, 1967, pp. 352-354.

19. D. G. Zavarin, V. V. Rozhdestbenskii, and G. K. Tumakaev, In: *Diagnostics of Low Temperature Plasma*, Moscow: Nauka, 1979, pp. 154-158 (in Russian).

20. V. E. Golant, *Microwave Methods in Plasma Research*, Moscow: Nauka, 1968 (in Russian).

21. E. Eggarter, *J. Chem. Phys.*, vol. 62, no. 3, 1975, pp. 833-847.

22. C. M. Lee and K. T. Lu, *Phys. Rev. A.*, vol. 8, no. 3, 1973, pp. 1241–1257.

23. B. M. Smirnov, *Atomic Collisions and Elementary Processes in Plasma*, Moscow: Atomizdat, 1968 (in Russian).

24. I. I. Sobelman, *Introduction to Atom Spectra Theory*, Moscow: Fiz.-Mat. Izd., 1963 (in Russian).

25. V. B. Borisov, V. S. Egorov, and N. A. Ashurbekov, Thes.: *VI All-Union Conf. on Physics of Low Temp. Plasma*, vol. 1, Leningrad, 1983, p. 20 (in Russian).

26. N. B. Kolokolov and O. G. Toropov, *VI. All-Union Conf. on Physics of Low Temp. Plasma*, vol. 1, Leningrad, 1983, p. 23 (in Russian).

27. H. Petschek and S. Byron, *Ann. of Phys.*, vol. 1, no. 3, 1957, pp. 270–315.

28. J. A. Smith, *Phys. Fluids*, vol. 11, no. 10, 1968, pp. 2150–2161.

29. T. V. Zhikhareva, D. G. Zavarin, and G. K. Tumakaev, *Zh. Tekhn. Fiz.*, vol. 54, no. 3, 1984, pp. 477–485 (in Russian).

EXPERIMENTAL INVESTIGATION OF THE AERODYNAMIC CHARACTERISTICS AND GEOMETRIC PARAMETERS OF FLOWS ABOUT BLUNTED BODIES IN GASES WITH VARIOUS MOLECULAR STRUCTURES

A. P. Bedin, G. I. Mishin, M. V. Chistyakova

Abstract: Interest in gas flow with various molecular structures about blunted bodies arose in connection with the entry problem for Earth and other planets of the Solar system.

Reentry vehicle configurations significantly differ from each other due to different planet atmospheric parameters (pressure, density, composition), entry conditions (velocity, entry angle), loading, and a number of other factors. In some cases (dense atmosphere, high entry speed) vehicle configuration essentially changes during flight. This leads, naturally, to the necessity to investigate the influence of the shape of blunted bodies on the flow field and aerodynamic characteristics.

In addition, the difference in planetary atmosphere composition and other parameters demand investigation of the influence of the specific heat ratio on the aerodynamic characteristics of blunted bodies.

This paper presents the results of experimental investigations of two problems for cone-segment bodies and spheres. The experiments were carried out on pressure-tight ballistic ranges at the A. F. Ioffe Physical Technical Institute [1, 2] in argon (monoatomic gas, $c = C_p/C_v = 1.67$), air (two-atomic gas, $c = 1.4$), carbon dioxide (three-atomic linear molecule, $c = 1.29$), and freon-12 (multi-atomic gas, $c = 1.14$). The experiments were performed in air at atmospheric pressure and a temperature of 290 K; in other gases the pressure has been chosen to provide a Reynolds number independent of the type of gas for all Mach numbers. The cone-segment model parameters are presented in Figure 1. The investigated sphere diameters were equal to $d = 10$ mm. the Mach number was varied from 0.5 to 10, the incident angle from 0 to 40°. The Reynolds number $Re_{\infty,d}$ varied proportionally to the Mach number from 2.5 K 10^5 at M = 0.5 to $5K10^6$ at M = 10, and body diameter $d = 21$ mm.

1. Static and Unsteady Aerodynamic Characteristic Determination Technique

Determination of aerodynamic characteristics from trajectory measurements, carried out in ballistic experiments, is based on the use of the differential relation between aerodynamic force and moment coefficients and registered parameters of the model motion. Since in the present study the models were forced to oscillate in the vertical plane with no circular pitching the simplified equations of plane motion were used to process the trajectory measurements.

$$\frac{2m}{\rho s v^2}\frac{d^2x}{dt^2} = -C_x, \quad \frac{2m}{\rho s}\cdot\frac{d^2y}{dx^2} = C_y,$$

$$\frac{2I}{\rho s l}\frac{d^2\vartheta}{dx^2} - \left[\frac{I}{ml}C_x + lm\frac{\bar\omega_z}{z}\right]\frac{d\vartheta}{dx} = m_z. \tag{1}$$

Here x, y are longitudinal and transverse coordinates; ϑ is the pitching angle; t is time; C_x, C_y, m_z, $m_{\bar\omega_z}$ are the drag, lift, static moment, and damping moment coefficients, respectively; m, I are the mass and inertia moment of the model; l is character length; and ρ is the medium density under experimental conditions.

Equations show that the problem of static aerodynamic characteristic determination by the trajectory measurement data reduces mainly to calculation of derivatives contained in the equations. Direct differentiation of the experimental dependences can lead to significant errors in the derivative values and, as a result in the aerodynamic characteristics. Therefore, to diminish the effect of coordinate and time measurement errors on the aerodynamic characteristics experimentally obtained dependences $x = x(t)$, $y = y(x)$, and $\vartheta = \vartheta(x)$ had been preliminarily smoothed before differentiating by the least squares method using the following approximating functions:

$$x = F_k(t) = \sum_{\nu=1}^{k} a_\nu f_\nu(t),$$

$$y = G_k(x) = \sum_{\nu=1}^{k} b_\nu g_\nu(x), \tag{2}$$

$$\vartheta = H_k(x) = \sum_{\nu=1}^{k} c_\nu h_\nu(x),$$

where $f_\nu(t)$, $g_\nu(x)$, and $h_\nu(x)$ are given, the coefficients a_ν, b_ν, and c_ν are unknown and must be determined from the set of conditional equations:

$$x_i = \sum_{\nu=1}^{k} a_\nu f_\nu(t_i), \quad y_i = \sum_{\nu=1}^{k} b_\nu g_\nu(x_i),$$

$$\vartheta_i = \sum_{\nu=1}^{k} c_\nu h_\nu(x_i), \quad i = 1, 2, ..., n, \tag{3}$$

obtained as a result of substitution of experimental values of x_i, y_i, ϑ_i, t_i into (2) (here n is the number of registration planes, $n > k$). Solution of the set (3) was carried out by the least squares method on a computer. The program was designed to provide the possibility of simultaneous computation of several approximations of each function $x = x(t)$, $y = y(x)$, and $\vartheta = \vartheta(x)$; k approximations were chosen for further analysis, namely, $F_k(t)$, $G_k(x)$, and $H_k(x)$, satisfying the condition of absence of the systematic error:

$$\sigma_{x_k}^2 = \frac{\sum\limits_{i=1}^{n} \left[x_i - \sum\limits_{\nu=1}^{k} a_\nu f_\nu(t_i) \right]^2}{n - k} = \sigma_x^2,$$

$$\sigma_{y_k} = \frac{\sum\limits_{i=1}^{n} \left[y_i - \sum\limits_{\nu=1}^{k} b_\nu g_\nu(x_i) \right]^2}{n - k} = \sigma_y^2, \tag{4}$$

$$\sigma_{\vartheta_k}^2 = \frac{\sum\limits_{i=1}^{n} \left[\vartheta_i - \sum\limits_{\nu=1}^{k} c_\nu h_\nu(x_i) \right]^2}{n - k} = \sigma_\vartheta^2.$$

Here σ_x, σ_y, and σ_ϑ are the mean square errors of coordinate measurements, depending on a number of factors; for the A. F. Ioffe PTI installations they are: $0.02 \text{ mm} \leqslant \sigma_x$, $\sigma_y \leqslant 0.2 \text{ mm}$, $0.03° \leqslant \sigma_\vartheta \leqslant 0.3°$. Such a wide range of the

42   *Bedin et al.*

mean square error variation makes the choice of an approximation satisfying
(5) difficult. Therefore, the following rule is used if $\sigma_{x,y,\vartheta_{k+1}} = \sigma_{x,y,\vartheta_k}$, then
k-approximation satisfies Conditions (4). Differentiating the chosen func-
tions $x = F_k(t)$, $y = G_k(x)$, $\vartheta = H_k(x)$ with respect to time or longitudinal
coordinate and substituting the results into (1) it is easy to obtain the parametric dependence of aerodynamic characters on the incidence angle and the
mach number M.

The model trajectories can essentially differ from each other due to
the difference in aerodynamic and inertial model characteristics, medium
sizes, and parameters, etc. This, naturally, leads to variety in the trajectory
measurement processing techniques and applied approximative functions.
Some of these techniques are described below.

Polynomials of the form

$$x, y, \vartheta = \sum_0^k a_\nu t^\nu \tag{5}$$

or

$$t, y, \vartheta = \sum_0^k a_\nu x^\nu \tag{6}$$

can be used to determine the drag coefficient C_{x_0}, to process the trajectory
measurements for models, flying at balance incidence angles, to process the
oscillating model flight measurements at $\lambda \gg L$, where λ is the model oscillation wavelength, L is the working section length of the ballistic range; to
process experiments with statically unstable models. This method of applying for aerodynamic characteristics maximum accuracy is achieved in the
middle of the working section. To estimate the principal part of the aerodynamic coefficients mean square errors caused by model coordinate measurement errors, the following formulae can be used:

$$\sigma_{c_x c_y} = \frac{24m}{\rho S L^2} \sigma_{xy} \sqrt{\frac{5(n-1)^3}{n(n+1)(n^2-4)}},$$

$$\sigma_{m_z} = \frac{24m}{\rho S l L^2} \sigma_\vartheta \sqrt{\frac{5(n-1)^3}{n(n+1)(n^2-4)}}. \tag{7}$$

These formulas suggest that the testing cross-sections are arranged uniformly along the working section. So far as the method (at the A. F. Ioffe PTI
installation $\sigma_{C_x,C_y} = 0.002$) can be obtained in a single experiment for a

single incidence angle only, then to obtain the dependences of these coefficients on the incidence angles to the same accuracy, it is necessary to carry out a series of experiments.

Processing of measurement results for oscillating motion of the model can be performed with the use of the following functions:

$$x = \sum_{\nu=0}^{3} a_\nu t^\nu + \sum_{\nu=1}^{k} b_\nu \sin\nu\omega t + \sum_{\nu=1}^{k} B_\nu \cos\nu\omega t,$$

$$y_1\vartheta = c_1 + c_2 x + \sum_{\nu=1}^{k} g_\nu \sin\frac{2\pi\nu}{\lambda}x + \sum_{\nu=1}^{k} G_\nu \cos\frac{2\pi\nu}{\lambda}x. \tag{8}$$

Experience shows that for trajectory data being processed the terms containing cosines can be excluded from y and ϑ without noticeable loss of accuracy. The best conditions for aerodynamic characteristic determination using trigonometric functions correspond to $L = (0.5\text{-}1)\ \lambda$. Components of the aerodynamic coefficient errors caused by a coordinate measurement error appear to be close to the following estimates:

$$\sigma_{c_y} = \frac{8\pi^2 m}{\rho S L^2}\,\sigma_y \sqrt{\frac{2}{n}\sum_{\nu=1}^{k} \nu^4 \sin^2\frac{2\pi}{\lambda}\nu x},$$

$$\sigma_{m_z} = \frac{8\pi^2 I}{\rho S l L^2}\,\sigma_\vartheta \sqrt{\frac{2}{n}\sum_{\nu=1}^{k} \nu^4 \sin^2\frac{2\pi}{\lambda}\nu x}. \tag{9}$$

It was suggested here that $C_i = G_i = 0$. Hence, the maximum aerodynamic coefficient determination error by this method corresponds to the amplitude value of the incidence angle. The value of this error is proportional to $\sqrt{1+3^4+5^4+...+K^4}$, where K is an odd number, i.e., the accuracy drops with the increase of the number of harmonics retained in the series expansions (8). Thus at the A. F. Ioffe PTI installation $K = 3$ corresponds to $\sigma_{c_y} = 10 \cdot \sigma_{m_z} = 0.003 + 0.03$. Hence, in technique application we must take care to prevent the incidence angle amplitude from exceeding its critical value α_K for which the conditions (4) for K equal to, say 3, are still valid. If the incidence angle in the experiment exceeds α_K so that (4) is valid only at $K > 3$, then to obtain the accuracy correspondent to $K = 3$ the series of experiments at $\alpha < \alpha_K$ must be carried out and the results compared to those at $\alpha > \alpha_K$. The comparison allows determination of error function for aerodynamic characteristics found in the experiment at $\alpha > \alpha_K$. Excluding this function we can obtain more accurate values of aerodynamic coefficients in a wider range of incidence angles.

The method considered shows high accuracy of determination of the first several harmonics in Expression (8) for y and ϑ as well as the coefficients at t^ν. This feature can be utilized to obtain nonlinear aerodynamic characteristics with higher accuracy using the experimental data on oscillating motion of the model. To this effect a series of experiments is carried out at a given regime in which the models are forced to oscillate with various incidence angle amplitudes. Trajectory measurements of each experiment are processed using Relation (8). This results in determination of unknown coefficients A_0, B_1, and G_1 in the expansions

$$C_x = A_0 + {}_{A1}t + \sum_{\nu=1}^{k} A_\nu \cos\frac{4\pi\nu}{T},$$

$$C_y = \sum_{\nu=1}^{k} B_\nu \sin\frac{2\pi\nu}{\lambda}x, \tag{10}$$

$$M_z = \sum_{\nu=1}^{k} G_\nu \sin\frac{2\pi\nu}{\lambda}x,$$

and then using the whole series data the dependences $A_0 = A_0(\alpha_0)$, $B_1 = B_1(\alpha_0)$, $G_1 = G_1(\alpha_0)$ are found; here α_0 is the incidence angle amplitude. Processing of these dependences is performed by the least squares method in accordance with the relations

$$A_0 = a_0 + a_2\frac{\alpha_0^2}{2} + \frac{3}{8}a_4\alpha_0^4 + ...,$$

$$B_1 = b_1\alpha_0 + \frac{3}{4}b_3\alpha_0^3 + \frac{5}{8}b_5\alpha_0^5 + ..., \tag{11}$$

$$G_1 = g_1\alpha_0 + \frac{3}{4}g_3\alpha_0^3 + \frac{5}{8}g_5\alpha_0^5 + ...$$

Coefficients a_i, b_i, and g_i which must be determined from experiment are identical to coefficients of the power series expansions in the incidence angle for the aerodynamic characteristics:

$$C_x = a_0 + a_2\alpha^2 + a_4\alpha^4 + ...,$$

$$C_y = b_1\alpha + b_3\alpha^3 + b_5\alpha^5 + ..., \tag{12}$$

$$M_z = g_1\alpha + g_3\alpha^3 + g_5\alpha^5 + ...,$$

because Relations (11) were obtained from (12) and (10) under the assumption

$$\alpha = \alpha_0 \sin \frac{2\pi}{T} t,$$

$$\alpha = \alpha_0 \sin \frac{2\pi}{\lambda} x, \tag{13}$$

that is valid for linear or weak nonlinear dependence of the moment coefficient on the incidence angle. If this coefficient is essentially nonlinear and there are high harmonics in the incidence angle approximations (13) then to deduce calculation formulas similar to (11) we must substitute expansion

$$\alpha = \alpha_1 \sin \frac{2\pi}{\lambda} x + \alpha_3 \sin 3\frac{2\pi}{\lambda} x + \ldots \tag{14}$$

into (12) and equate the free terms or coefficients of the first several harmonics in (10) and (12). The formulas obtained are cumbersome and are not presented here. The values of the coefficients a_i, b_i, and g_i found in experimental dependences A_0, B_1, and $G_1 = f(\alpha_0)$ processing are substituted into (12) to give dependence of aerodynamic coefficients on an incidence angle at a given Mach number M. Another method for aerodynamic characteristic calculation is possible that reduces dependences on Mach number at $\alpha =$ const. to the determination of A_0, B_1, and G_1. The set of these dependences serves as initial data to obtain the a_i, b_i, and g_i coefficients and then the aerodynamic characteristics. Estimates showed that nonlinear character determination accuracy obtained by this method is comparable with linear aerodynamic coefficients and C_{x_0} accuracy.

Thus the static aerodynamic characteristic error at ballistic range can vary significantly depending on experimental method, result processing techniques, and model and medium parameters. Therefore, the right choice of methods and parameters plays an important role in testing program design. These experiments were organized so that the accuracy of the model aerodynamic characteristic determination (Figure 1) at an incidence angle about 40° was in the range $\sigma_{C_x} = \sigma_{C_y} = 10\,\sigma_{m_z} = 0.005 - 0.02$.

Reporting the technique for unsteady aerodynamic characteristic determination will proceed from the pitching equation in the form:

$$\frac{d^2\alpha}{dx^2} + P\frac{d\alpha}{dx} + Q\alpha = 0 \tag{15}$$

where

$$P = -\frac{\rho s}{2m}(\dot{C}_x - C_y^\alpha) - \frac{\rho s d^2}{2I}(m_z^{\bar\omega}z + m_z^{\dot\alpha}); \quad Q = -\frac{\rho s d}{2I}m_z^\alpha.$$

If $p = p(x)$, $Q = Q(x)$,. then Equation (15) can be treated as a linear equation having the solution

$$\alpha = \alpha_0 e^{-\frac{1}{2}\int p dx} \cdot f(x/\lambda, \varphi_0),\tag{16}$$

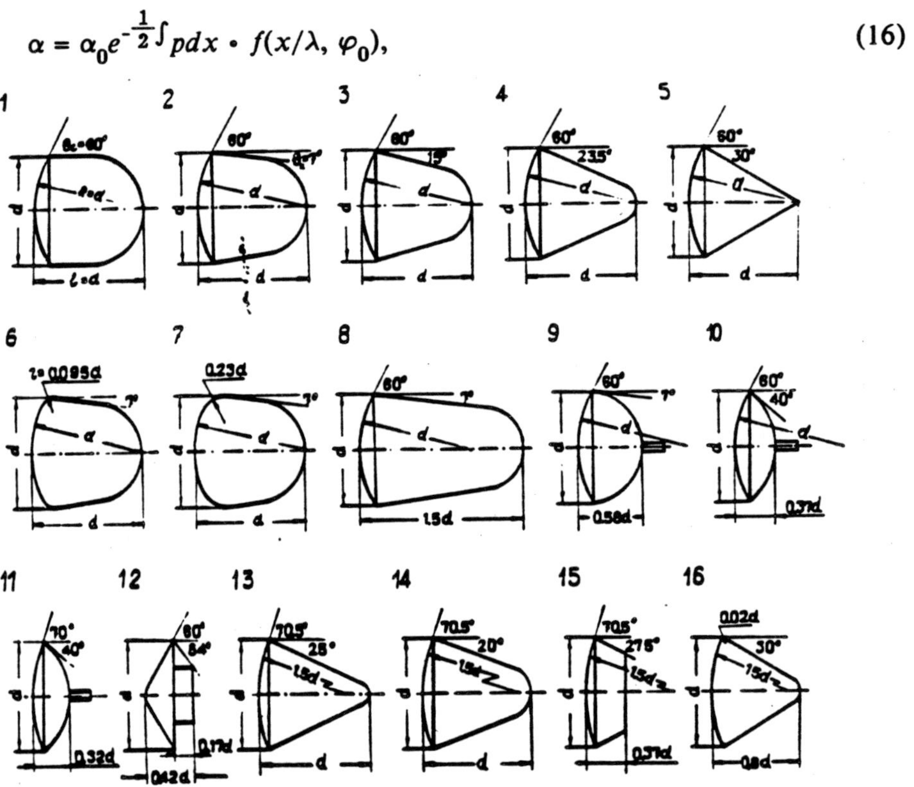

Fig. 1. The models under study. d (mm): 1-9 — 21, 10-12 — 28, 13-16 — 26.

where $f(x/\lambda, \varphi_0)$ is the periodic function with wavelength $\lambda = \lambda(\alpha_0)$ and limits of variation $f = \pm 1$, φ_0 is the oscillation phase. Hence, choosing the phase φ_0 so that at $x = 0$ and $x = \lambda$ the function f will be equal to 1 and taking a logarithm of the ratio of incidence angle amplitudes correspondent to chosen points α_1 and α_3 we will have:

$$\int_0^\lambda p dx = 2\ln\frac{\alpha_1}{\alpha_3}\tag{17}$$

whence taking into account expression for p from (15) we will obtain:

$$\overline{m_z^{\overline{\omega}_z}} = \int_0^\lambda \frac{(m_z^{\overline{\omega}_z} + m_z^\alpha)dx}{\lambda} = -\left[\frac{1}{\lambda}\ln\frac{\alpha_1}{\alpha_3} + \frac{4}{k}(\overline{C}_x - \overline{C}_y^\alpha)\right]k \cdot r_u^2 \tag{18}$$

where

$$k = \frac{16m}{\rho s}; \quad r_u^2 = \frac{I}{md^2}; \quad \overline{C}_x = \frac{\int_0^\lambda C_x dx}{\lambda}, \quad \overline{C}_y^\alpha = \frac{\int_0^\lambda C_y^\alpha dx}{\lambda}.$$

This relation allows us to determine the wavelength averaged value of the damping moment coefficient on the basis of the experiment. The dependence of $\overline{m_z^{\omega_z}} = f(\alpha)$ can easily be obtained by using the experimental data carried out at a given Mach number and various amplitudes. Processing this dependence by means of relations:

$$\overline{m_z^{\overline{\omega}_z}} = a_0 + \frac{1}{2}a_2\alpha_0^2 + \frac{3}{8}a_4\alpha_0^4 + \frac{5}{16}a_6\alpha_0^6 + \dots \tag{19}$$

(the goal of the processing is to find the unknown cords a_i) deduced from (18) under the assumption that

$$m_z^{\overline{\omega}_z} + m_z^\alpha = a_0 + a_2\alpha^2 + a_4\alpha^4 + \dots \tag{20}$$

and

$$\alpha = \alpha_0\cos\frac{2\pi}{\lambda}x \tag{21}$$

we can obtain the dependence $m_z^{\overline{\omega}_z} + m_z^\alpha = f(\alpha)$.

The accuracy of the damping moment coefficient determination by this method depends on the ratio of the incidence angle measurement error to its amplitude value, therefore, a number of experiments must be carried out at small angles of α.

In testing data processing it was found that this method gave good results at a monotonous dependence of $m_z^{\omega_z}$ on the incidence angle. If the character of the dependence steeply changed starting from a certain incidence angle the results appeared to be inadequate because of poor convergence of the power series (19). In such cases the experimental data was processed by means of the technique reported in Ref. [3].

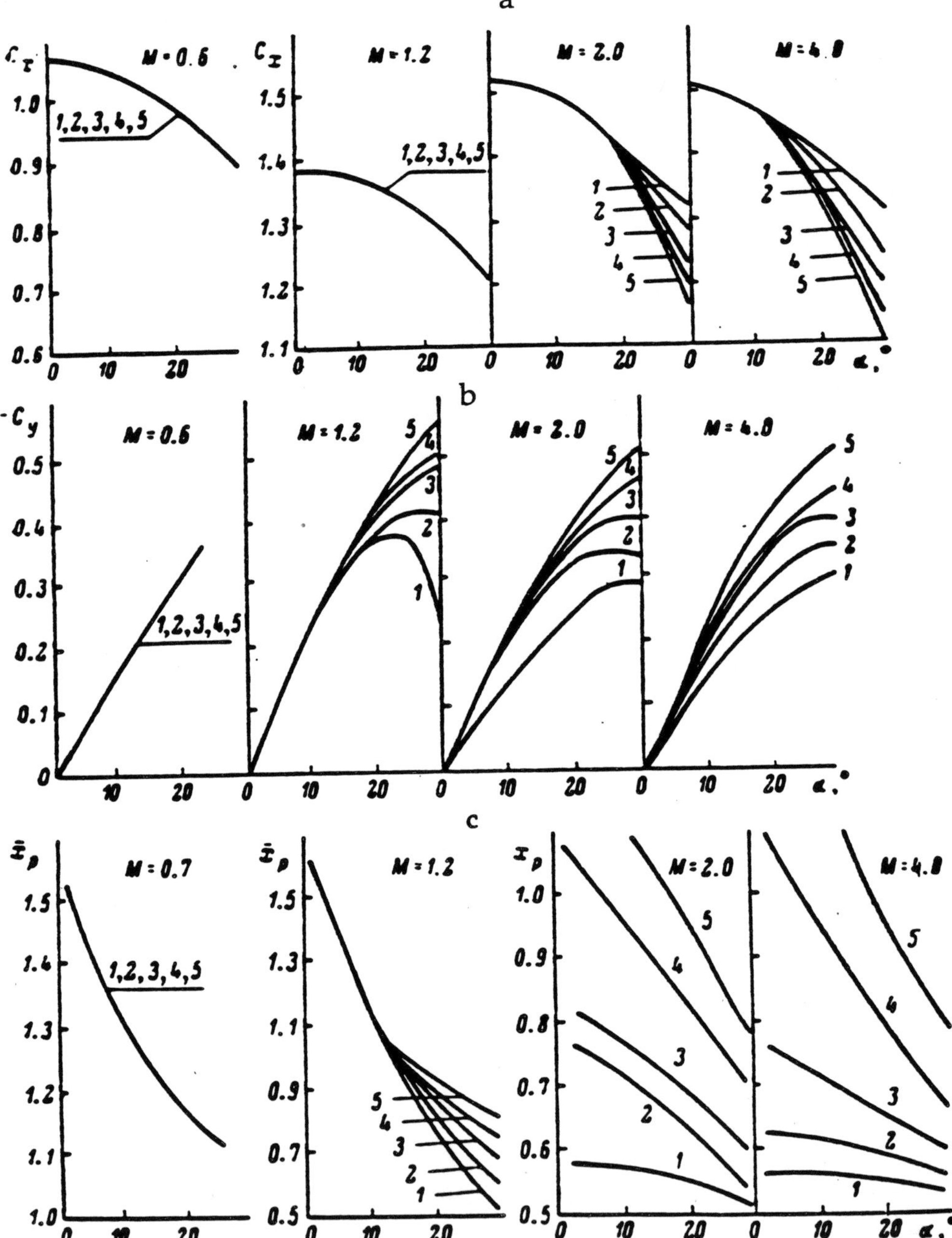

Fig. 2. Dependences of drag (a), lift (b), coefficients, and pressure center coordinate (c) on the incidence angle for models with $\theta_c = 60°$, $\bar{7} = 1$, $\bar{r} = 0$.

2. Experimental Results and Discussion

2.1. Static Aerodynamic Characteristics

The data obtained in the experiments performed in air for models 1-11 allowed analysis of the cone-segment body geometry effect on the static aerodynamic characteristics (see also Ref. [4]).

The inverse cone angle is θ_k. Investigation of the effect of this parameter on the aerodynamic characteristics has been carried out for models having the same segment of the angle $\theta_c = 60°$, the same aspect ratio of $\overline{l} = l/d = 1$, and various cone angles of $\theta_k = 0, 7, 15, 23.5,$ and $30°$ (the models 1-5, Figure 1). Measurement results of the drag and lift coefficients as well as the pressure center coordinates $\overline{x}_p = x_p/d$ as a function of the incidence angle at Mach numbers equal to 0.6, 1.2, and 2.4 are presented in Figure 2 for the models mentioned above. As can be seen from the data presented, the value of the inverse cone angle does not affect the aerodynamic characteristics of the segment body at M = 0.6. This fact can be explained as follows: at M < 1 the lateral surface of the body is situated completely in the separation zone, the pressure in this zone is independent of θ_k, which can be judged by the independence of the frontal segment departure angle on the inverse cone angle. At supersonic velocities the cone angle growth leads to the drag coefficient decrease and to increase of the lift coefficient modulus. The pressure center moves downstream. Such effects are caused by decrease of the local incidence angle of the lateral surface and the correspondent diminishing of its contribution to aerodynamic coefficients. Variation of C_x and C_y with θ_k leads to aerodynamic quality $K = C_y/C_x$ changes (see Figure 3) which represents K obtained by the authors at the ballistic range, at the wind tunnel [5], and by the calculation method [6] (the models have $\theta_k = 7°$ and $30°$ at M = 4). The figure shows that at $\alpha = 35°$ and $\theta_k = 30°$ model quality is 1.5 times greater than that of a model with $\theta_k = 7°$. There is close agreement between experimental data and theoretical calculations.

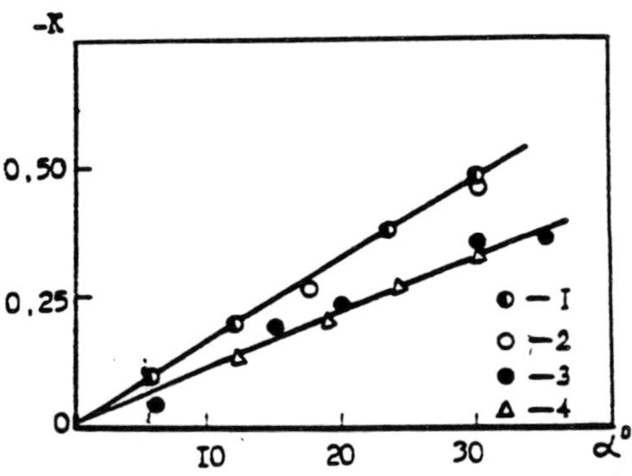

Fig. 3. Dependence of aerodynamic quality on the incidence angle for models with $\theta_c = 60°$, $\overline{l} = 1$, $\overline{r} = 0$.

θ_k^0:

1 — 30, 2 — 30 [5], 3 — 7, 4 — 7 [6].

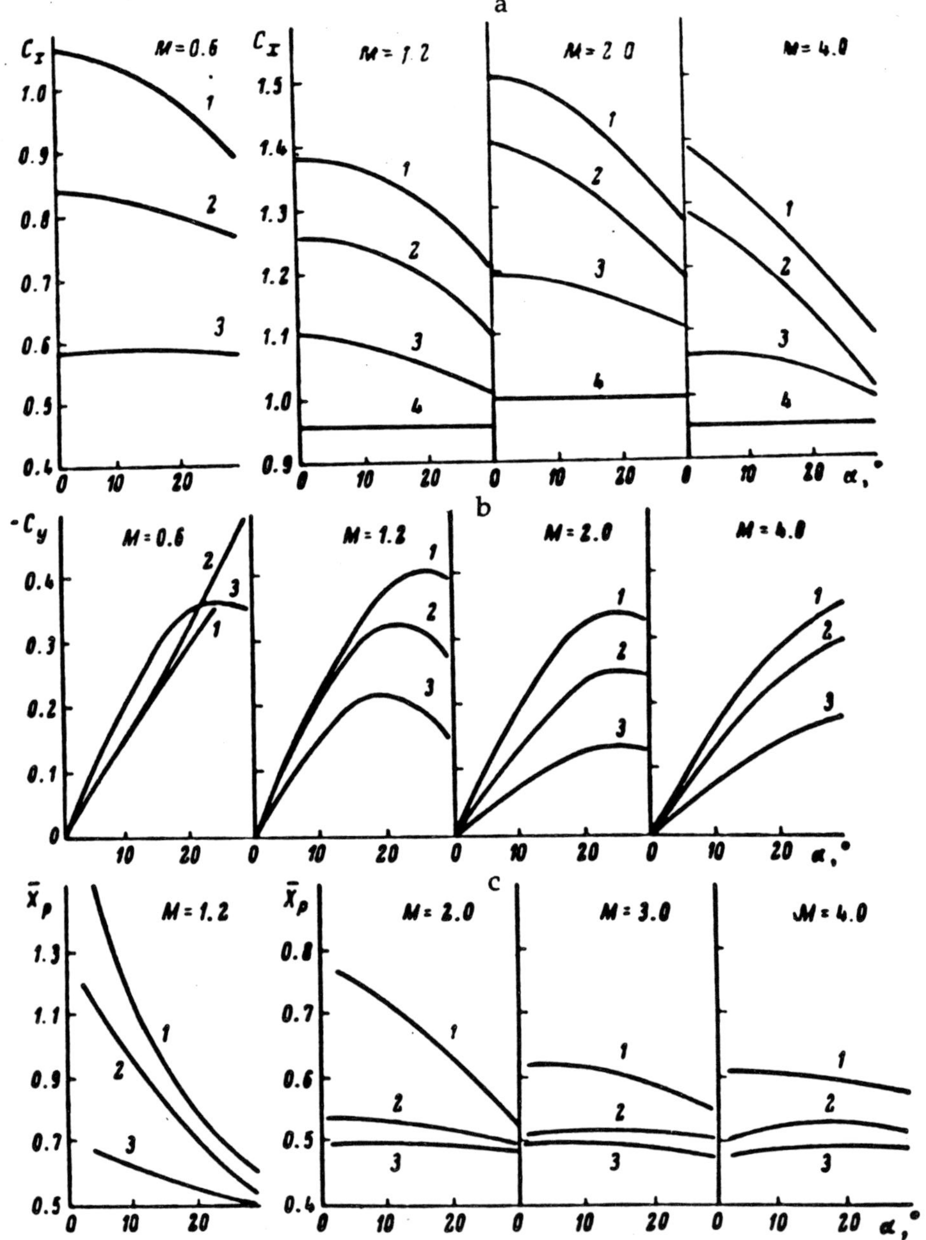

Fig. 4. Dependence of drag (a), lift (b), coefficients and pressure center coordinate (c) on the incidence angle for models with $\bar{R} = 1$, $\bar{7} = 1$, $\theta_k = 7°$. $\bar{r}$: 1 — 0, 2 — 0.095, 3 — 0.23.

The edge rounding radius of the cone-segment body is r. The effect of this parameter on the aerodynamic characteristics is presented in Figure 1 for models 2, 6, and 7 having the same segment radius $R = d$, the same inverse cone angle $\theta_k = 7°$, and the same aspect ratio $l/d = 1$. The dependences $C_x = C_x(\alpha)$, $C_y = C_y(\alpha)$, $\bar{x}_p = \bar{x}_p(\alpha)$ obtained experimentally at M = 0.6, 1.2, 2.4 and $\bar{r} = r/d = 0.095, 0.25$ are presented in Figure 4. The figure shows that an increase of $\bar{r}$ results in the drag coefficient decrease approaching this coefficient to that of the sphere in the whole Mach number range under study. The sphere drag coefficient in Figure 4 corresponds to the data of Ref. [7]. The lift coefficient decrease approaches zero correspondent to the sphere when the rounding radius increases. The aerodynamic quality steeply decreases. The pressure center moves upstream.

The model aspect ratio is $l = l/d$. Influence of this parameter on the aerodynamic characteristics was investigated for models 2, 8, 9 (Figure 1). The models were of the same cone angles $\theta_k = 7°$ and the same segments of $\theta_c = 60°$, the aspect ratios were $\bar{l} = 0.58, 1$, and 1.5. Figure 5 represents the experimental aerodynamic characteristics of the models. At subsonic velocities as well as at M > 1 and small incidence angles the drag coefficient is practically independent of the aspect ratio. This occurs due to the fact that the lateral surface of the body is situated in the separation zone or in the aerodynamic "shadow." The effect of aspect ratio on the drag coefficient at M > 1 comes into play at $\alpha > 15°$, and for the model with $\bar{l} = 1.5$ the dependence $C_x = C_x(\alpha)$ appears to be nonmonotonic because of the effect of the lateral surface on C_x when coming out of the "shadow" region. As the Mach number increases the lateral surface influence on the aerodynamic characteristics decreases and the frontal segment influence on these characteristics also decreases resulting in the nonmonotonic behavior of C_x at a large incidence angle for model 9. According to Figure 5 the aspect ratio does not effect the lift coefficient at M < 1 (due to the "shaded" lateral surface) and at M > 1 this effect is essential (in this case C_y already $\alpha \sim 20°$ provides >0 for the model with $\bar{l} = 1.5$). The lift coefficient modulus decreases with $\bar{l}$ growth caused by the positive lateral surface lift coefficient increase due to enlarging of this surface. The model aerodynamic quality drops as the aspect ratio increases. The static moment at M > 1 increases with aspect ratio because of the model elements situated behind the mass center.

The frontal segment bluntness parameter is θ_c. The testing data for models 10 and 11 allow consideration of the nature of the effect on aerodynamic characteristics. As in previous experiments the models differ from each other by a single parameter θ_c of 60° for model 10 and 70° for model 11. Figure 6 represents the aerodynamic characteristics determined in experiments. Considering this figure we can determine that the drag coefficient and the lift coefficient modulus increase with the pressure on the segment. The value of this pressure approaches the stagnation point. When θ_c grows the moment coefficient decreases because the pressure on the upper and lower sides of the segment are becoming equal.

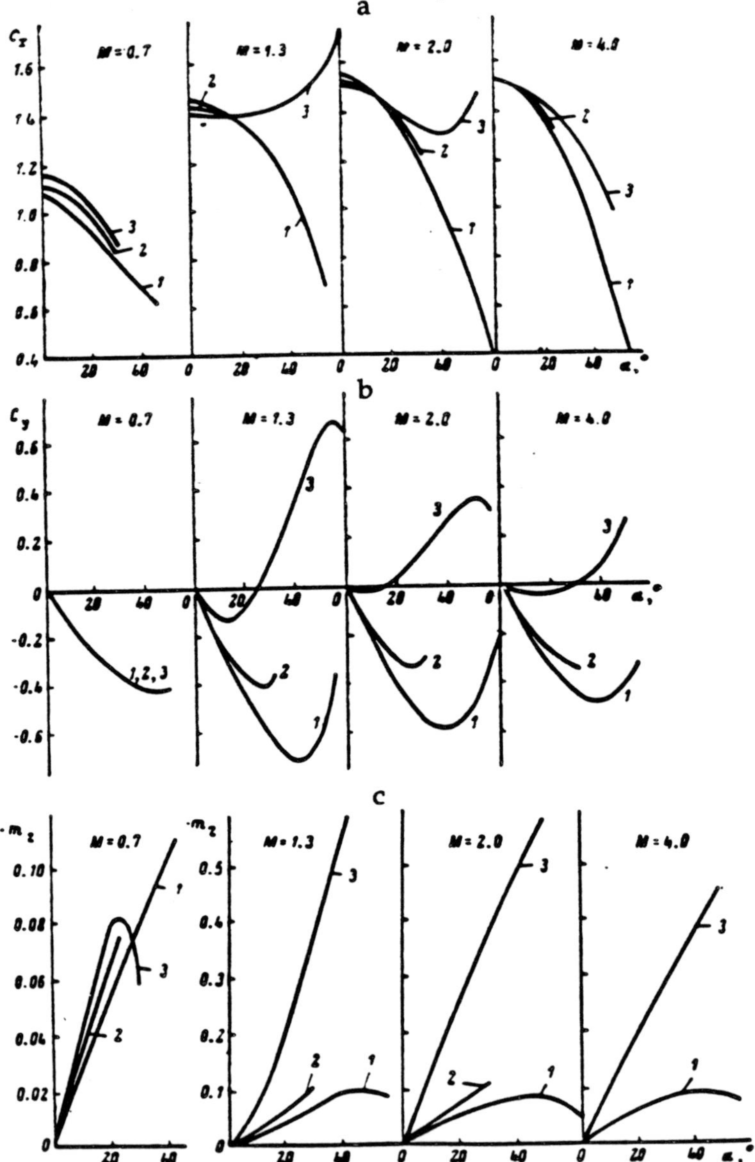

Fig. 5. Dependences of drag (a), lift (b), and moment (c) coefficients for models with $\theta_c = 60°$, $\theta_k = 7°$, $\bar{r} = 0$, $\bar{x}_T = 0.15$.
$\bar{l}$: $1 - 0.58$, $2 - 1$, $3 - 1.5$.

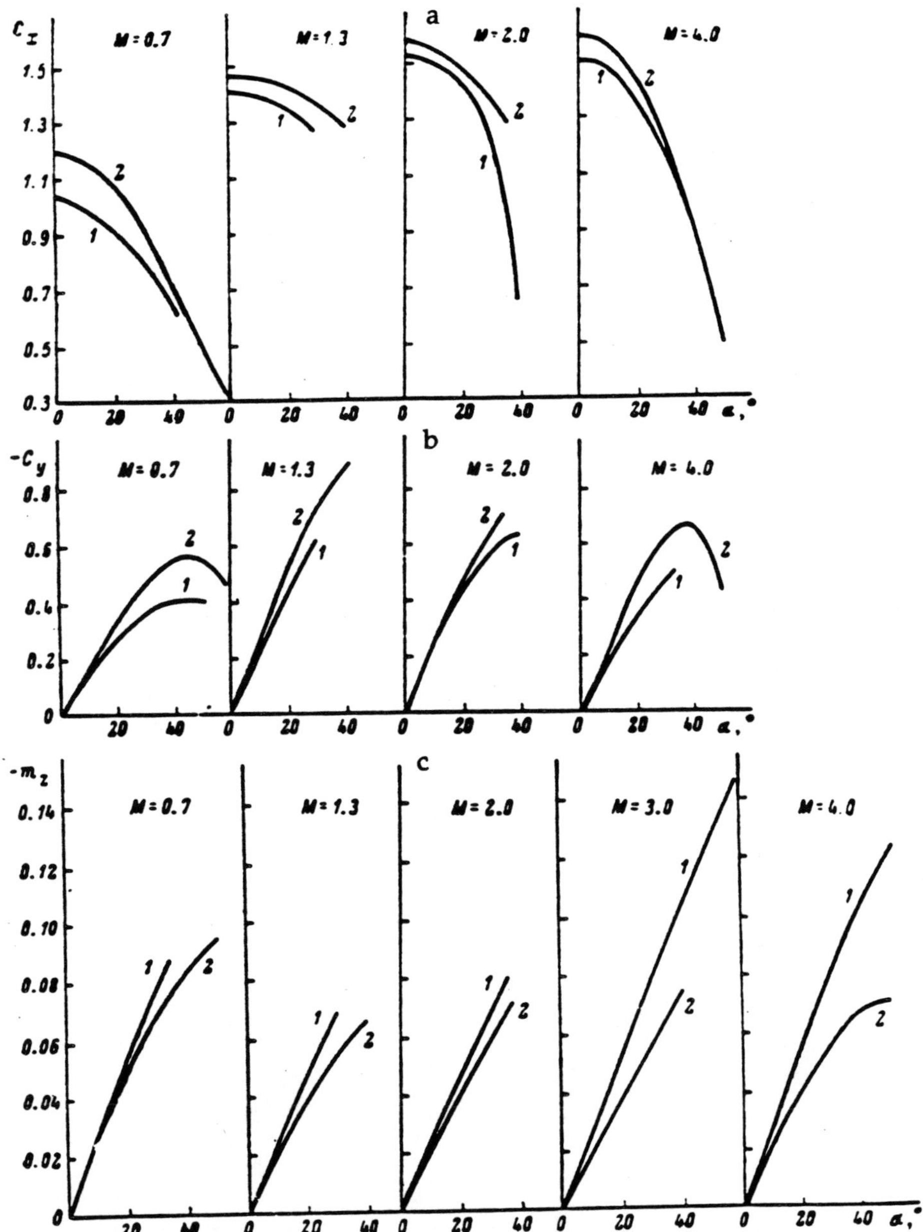

Fig. 6. Dependences of drag (a), lift (b), and moment (c) coefficients for models with $\theta_k = 40°$, $\bar{r} = 0$, $\bar{x}_T = 0.15$.
$1 — \theta_c = 60°$, $\bar{l} = 0.3$; $2 — \theta_c = 70°$, $\bar{l} = 0.32$.

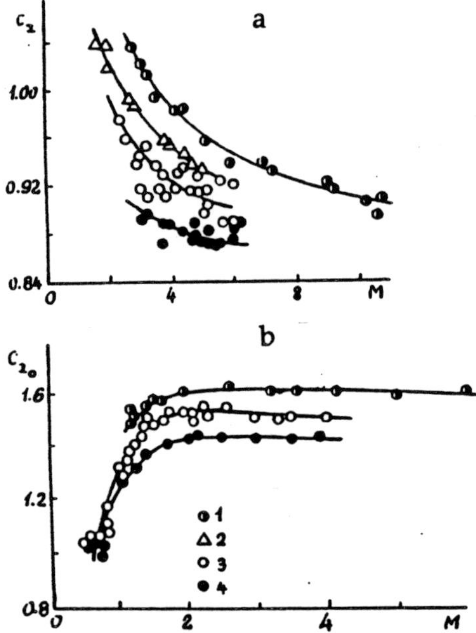

Fig. 7. Dependences of drag coefficients of sphere (a) and cone-segment body (b) on Mach number in various gases.

γ: 1 — 1.14, 2 — 1.29, 3 — 1.41, 4 — 1.67.

The study of the effect of the specific heat ratio on static aerodynamic characteristics was carried out for sphere and cone-segment body (model 2) (see also Ref. [8]). If the drag coefficient of a sphere and cone-segment body in argon, air, carbon dioxide, and freon-12 are measured (Figure 7) we can judge the nature of the effect of this parameter. Figure 7a shows the dependence of the sphere drag coefficient on the Mach number, Figure 7b presents the dependence $C_{x_0} = C_{x_0}(M)$ for the cone-segment model 2. The specific heat ratio increase from $\gamma = 1.14$ to 1.67 produces a 10-15% variation of the drag coefficient (see figure). Representing the results of the measurements in the form $C_{x_0}/P_0' = f(M)$ where $P_0' = 2P_0'/\rho V^2$, P_0' is the stagnation pressure, it is possible to correlate the results for various gases. Such a correlation principle follows modified Newton's formula for drag at hypersonic speed [9]. Correlations $C_x/P_0' = f(M)$ for a blunt body obtained by processing the present experimental data and the data of Refs. [7, 10] are presented in Figure 8. The stagnation pressure was calculated using the perfect gas theory. Figure 8 presents the results for models with $\gamma \leqslant 1$ and $\theta_k \geqslant 7°$. Since the model base geometry does not effect the drag coefficient C_{x_0} at considered values of γ and θ_k the experimental points correspondent to various models being plotted on graph are indistinguishable. The figure demonstrates satisfactory correlation of the data on blunted body drag coefficients in the whole range of Mach numbers in various gases. It provides the recalculation formula for drag coefficients in various gases $C_{x0_2} = (C_{x0}/P_0')_1 \cdot P_{0_2}'$. Figure 9 shows that at nonzero incidence angles the character of the drag coefficient dependence on the specific heat ratio does not change at least for model 2.

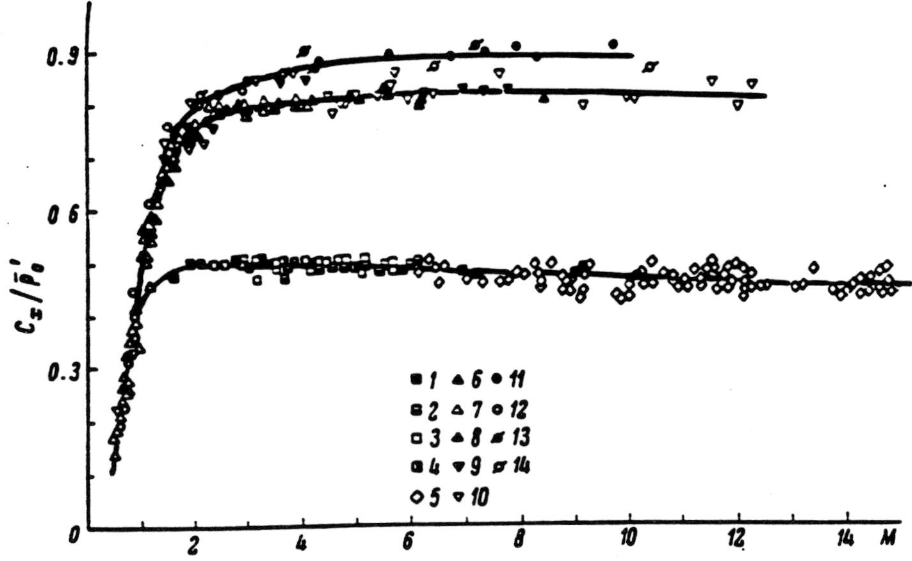

Fig. 8. Correlation of the results of blunted body drag coefficient measurements in various gases.

Sphere γ: 1 - 1.14, 2 - 1.29, 3 - 1.41, 4 - 1.67, 5 - 1.41 [7]; Segment $\theta_c = 60°$, γ: 6 - 1.14, 7 - 1.41, 8 - 1.67; Blunted cone $\theta_c = 60°$, γ: 9 - 1.16 [10], 10 - 1.41 [10]; Segment $\theta_c = 70°$, γ: 11 - 1.14, 12 - 1.41; Blunted cone $\theta_c = 70°$, γ: 13 - 1.16 [10], 14 - 1.41 [10].

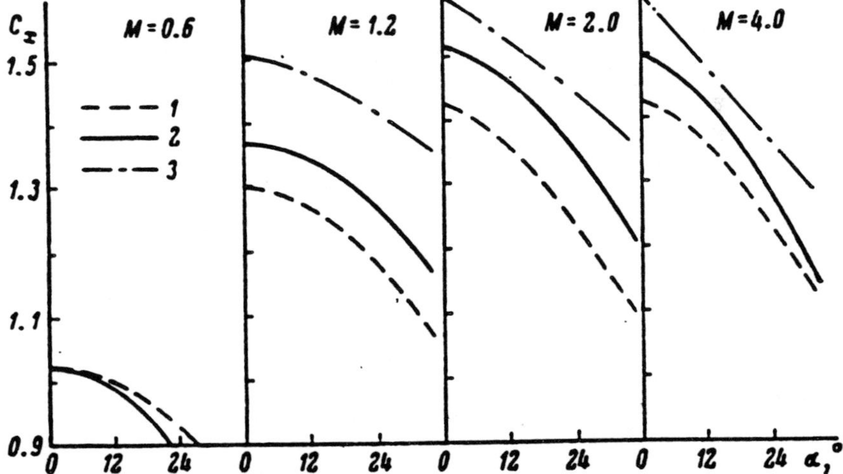

Fig. 9. Dependence of drag coefficient of model with $\theta_c = 60°$, $\theta_k = 7°$, $\overline{l} = 1$ on the incidence angle in various gases.

γ: 1 - 1.67, 2 - 1.41, 3 - 1.14.

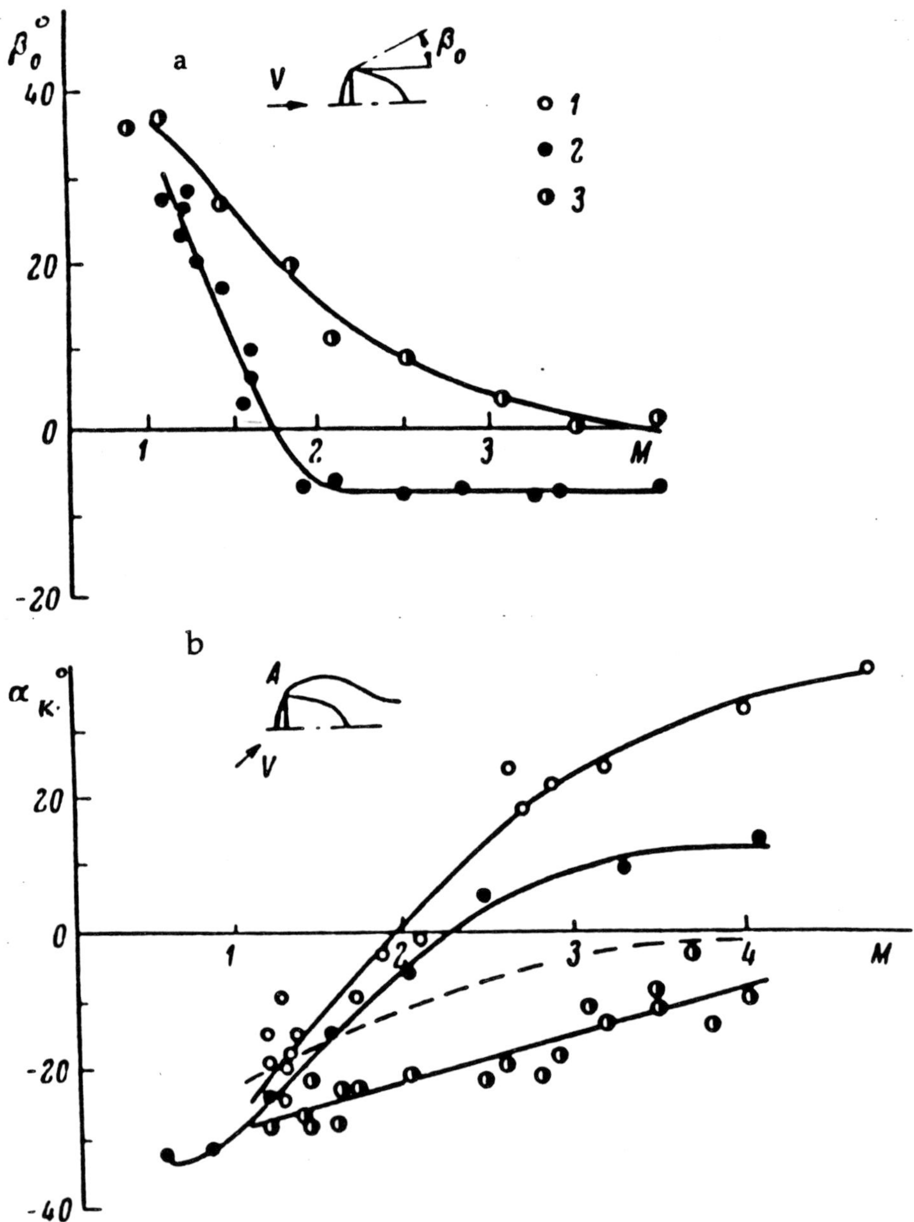

Fig. 10. Dependences of flow departure angle (a) and critical incidence angle (b) on Mach number for models with $\theta_c = 60°$, $\theta_k = 7°$, $7 = 1$.
γ: 1 - 1.14, 2 - 1.41, 3 - 1.67.

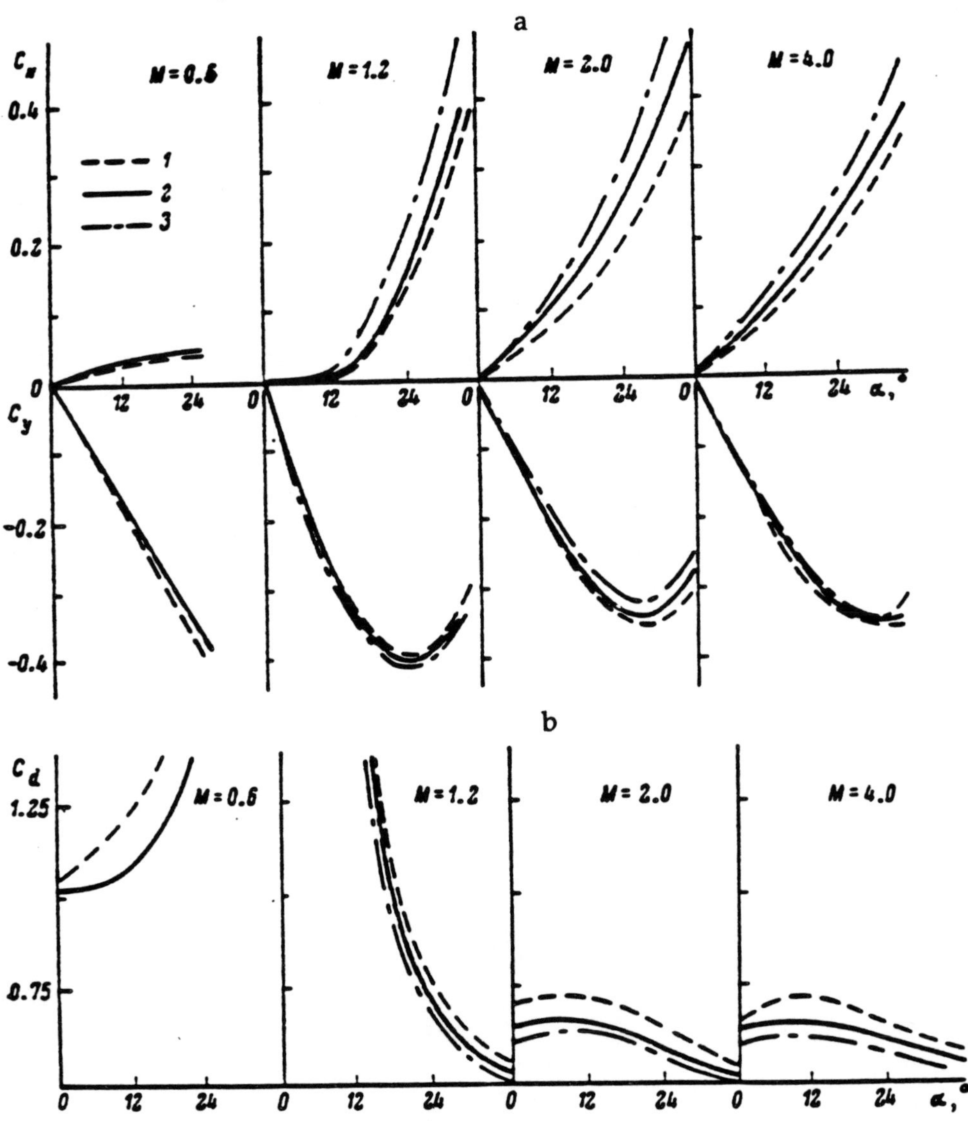

Fig. 11. Dependences of normal force C_N, lift C_y coefficients (a) and pressure center coordinate (b) on the incidence angle for model with $\theta_c = 60°$, $\theta_k = 7°$, $\overline{l} = 1$.

γ: 1 - 1.67, 2 - 1.41, 3 - 1.14.

The specific feature of the flow about the segment body is the expansion of the domain of nonseparated flow along the lateral surface with increase of the adiabatic exponent by reason of the turn angle growth (Figure 10a). The figure represents the results of the flow turn angle β measurements at zero incidence angle for $\gamma = 1.14$ and $\gamma = 1.41$. The nature of the dependence $\beta = \beta(\gamma)$ does not change at $\alpha \neq 0$. This phenomenon results in later flow separation from the lateral surface of the model. In the medium with a lower specific heat ratio (Figure 10b presents the dependences of the critical incidence angle α_k on Mach number M for various γ). At $\alpha < \alpha_k$ the flow about the lateral surface of the model (its upper half) occurs without separation (except nearest the edge). In air and in freon a completely separated flow takes place if the incidence angle exceeds its critical value. In argon complete separation occurs at an incidence angle α_1 noticeably greater than α_k. The dependence $\alpha_1 = \alpha_1(m)$ is shown in Figure 10b by the dashed line. The differences in the character of the flow about the lateral surface of the model lead to peculiarities of the lift coefficient dependence γ. At M = 1, 2, and $\alpha < 25°$ the lateral surface is located within the separation zone and does not practically influence the aerodynamic characteristics. In this case as the adiabatic exponent decreases, the absolute value of the coefficient C_y increases (Figure 11a). At M $\geqslant$ 2 the character of the dependence $C_y(\gamma)$ is reversed. C_y decreases with γ. This effect can be explained by the increase of the positive component of the lateral surface lift coefficient due to surface pressure growth and the expansion of the nonseparated flow domain. The same reason along with the pressure growth at front surface causes the increase of C_N when γ decreases. The pressure center moves upstream (Fig. 11b). The model aerodynamic quality at M > 1 increases with γ independently on the flow about the lateral surface.

2.2. Unsteady Aerodynamic Characteristics

The unsteady aerodynamic characteristics were determined for models 9-15 with trim $\bar{x}_T = x_T/d = 0.258$, 0.171, 0.141, 0.244, 0.319, 0.364, and 0.26, respectively. In addition, the center of mass of model 15 was displaced by $\bar{y}_T = 0.02$ from the symmetry axis. Experiments were carried out both in air and in freon-12. The dependences of the damping moment coefficient $m_z^{\bar{\omega}_z} + m_z^{\alpha}$ on the incidence angle determined experimentally are plotted on graphs (figure 12) for several values of M. Examination of the graphs reveals that the models under study occasionally turn out to be dynamically unstable. As a rule the model instability was observed either at small incidence angles (the first instability region) or at $\alpha > 30°$ (the second region). The mass center moving towards the body base of both regions expand and

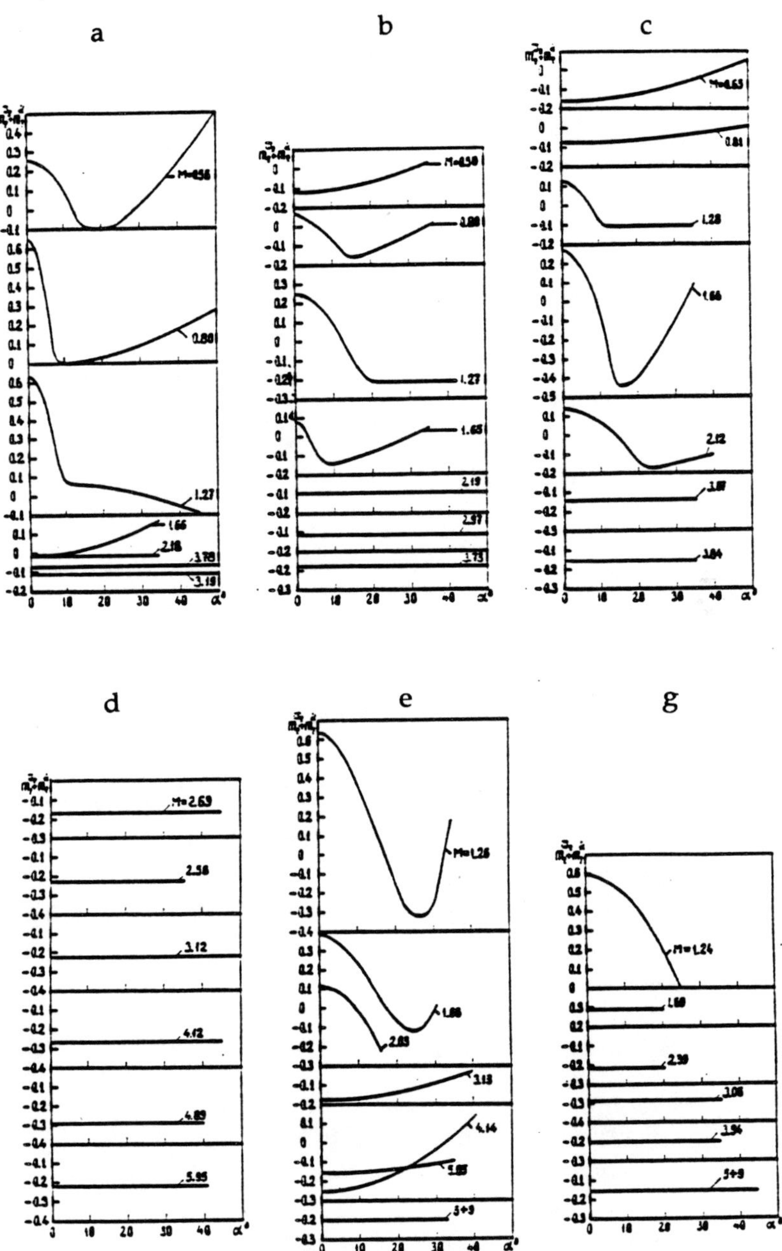

Fig. 12. Dependence of damping moment coefficients for models 9–14 on the incidence angle.

Air: a-d — 9, 10, 11, 12, at M = 2.69. Freon-12: d-f — 12 (except M = 2.69), 14, 15.

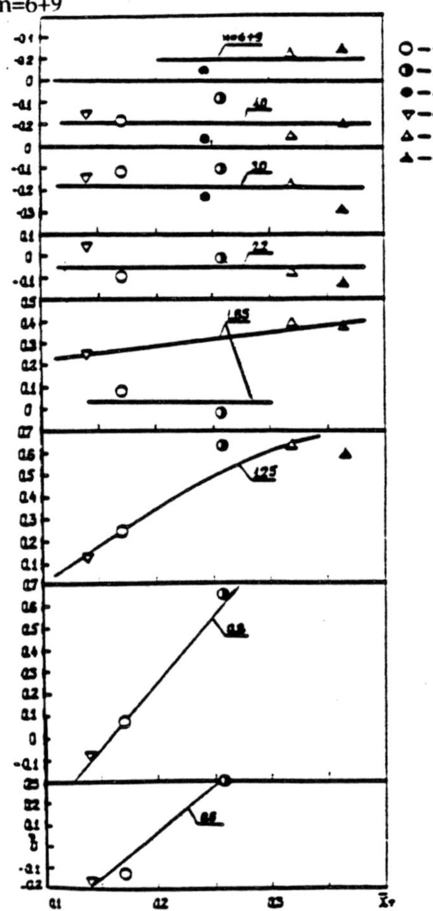

Fig. 13. Dependence of damping moment coefficient on mass center location for models 9-14 at incidence angle $\alpha = 0$.

Air: θ_c^0: 1, 2 - 60; 4 - 70. Freon: 3 - 60; 5, 6 - 70.5.

sometimes coalesce. As the Mach number increases the first instability region collapses, disappearing completely at $M > 2.5$, while the second one displaces towards the greater incidence angle values. Small center of mass displacement does not practically effect the damping moment coefficient. The bases of the models are aerodynamically shaded at small incidence angles and therefore cannot influence the coefficient $m_z^{\bar{\omega}_z} + m_z^{\alpha}$. This fact has been confirmed by comparison of this coefficient measurement at zero incidence angle for various models (see dependences $m_z^{\bar{\omega}_z} + m_z^{\alpha}$ on the coordinate $\bar{x}_T$ (Figure 13)). The data analysis obtained allows us to establish, in addition, that the damping moment coefficient is practically independent of γ. Blunting of the model has a noticeable effect on the damping moment coefficient only at $M \sim 1.6$. The dependences $m_z^{\bar{\omega}_z} + m_z^{\alpha} = f(x_T)$ were used to calculate the damping force $C_N^{\bar{\omega}_z} + C_N^{\alpha}$ and the pressure center coordinate C_d. Calculation results according to

formula for the damping moment coefficient recalculation for trim variation [11, 12] at $\alpha = 0\text{-}20°$ are presented in Figures 14 and 15. It is seen that at $\alpha = 0$ and $M \approx 1$ the coefficient of damping forces reaches its maximum. At $M > 2$ this coefficient is close to zero. At $M = 1.4\text{-}2$ the coefficient $C_N^{\bar{\omega}_z} + C_N^{\alpha}$ depends on the model bluntness radius $\bar{R} = R/d$. The minimum of coefficient $C_d = 0.1$ corresponds to $M = 1.3$ and $\alpha = 0$. Since all models used to have a trim of $\bar{x}_T > 0.1$, dynamic instability arose.

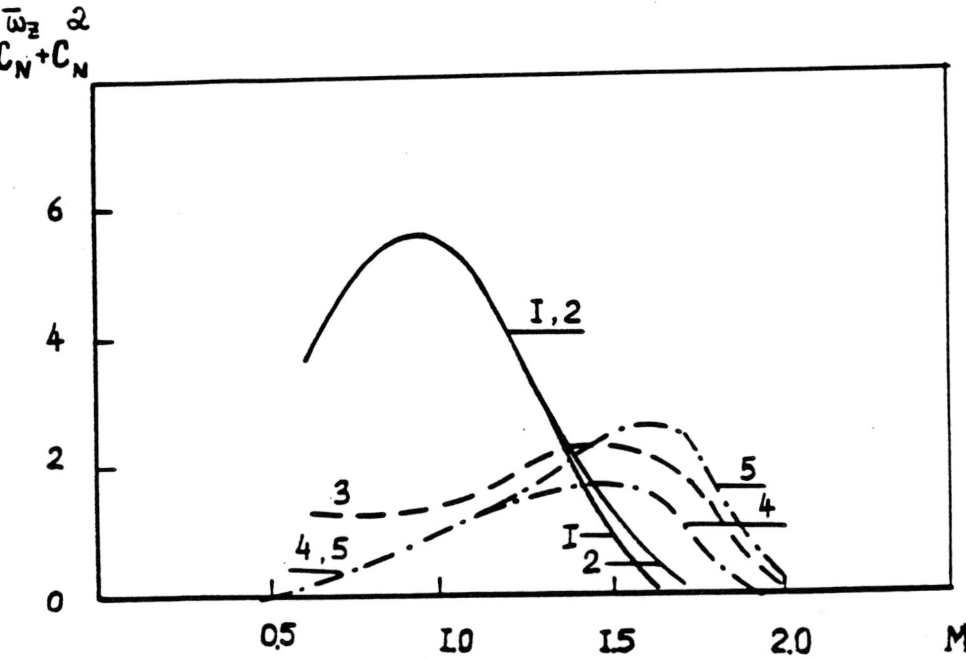

Fig. 14. Dependence of damping force coefficient on Mach number at $\overline{x}_T$ = 0.1–0.25.

α^0: 1, 2 - 0; 3 - 10; 4, 5 - 20. $\overline{R}$: 1,4 - 1; 2, 5 - 1.5; 3 - 1-1.5.

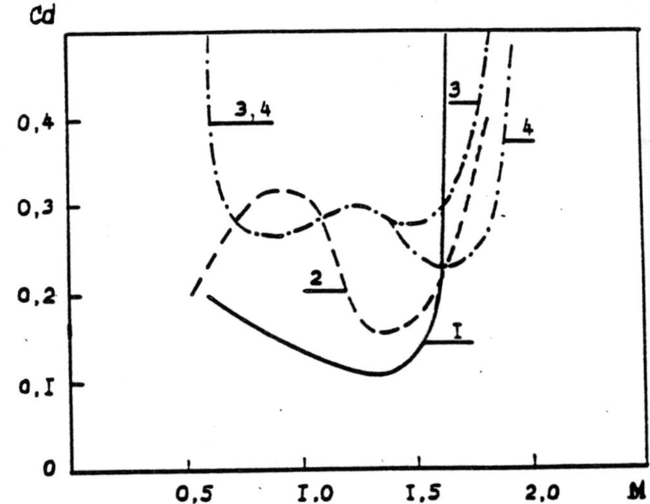

Fig. 15. Dependence of damping force pressure center location on Mach number at $\overline{x}_T$ = 0.1–0.25.

α^0: 1 - 0; 2 - 10; 3, 4 - 20. $\overline{R}$: 1, 2 - 1-1.5; 3 - 1.5; 4 - 1.

2.3. Geometrical Flow Parameters

The nature of the adiabatic exponent effect on the flow turn angle β for fixed boundary layer separation points *semper idem*: if β increases then γ decreases. To substantiate this fact we can examine Figure 10a relating to model 2 (see also Ref. [8]). Figure 16 relating to cone-segment model 16 (see also Ref. [13]) and in addition the data of Ref. [8] relating to hemisphere-cylinder. In all these cases the incidence angle was zero, however, the same dependence $\beta(\gamma)$ was observed at a nonzero incidence angle. Thus, the revealed dependence $\beta(\gamma)$ is universal. The dependence of β on the adiabatic

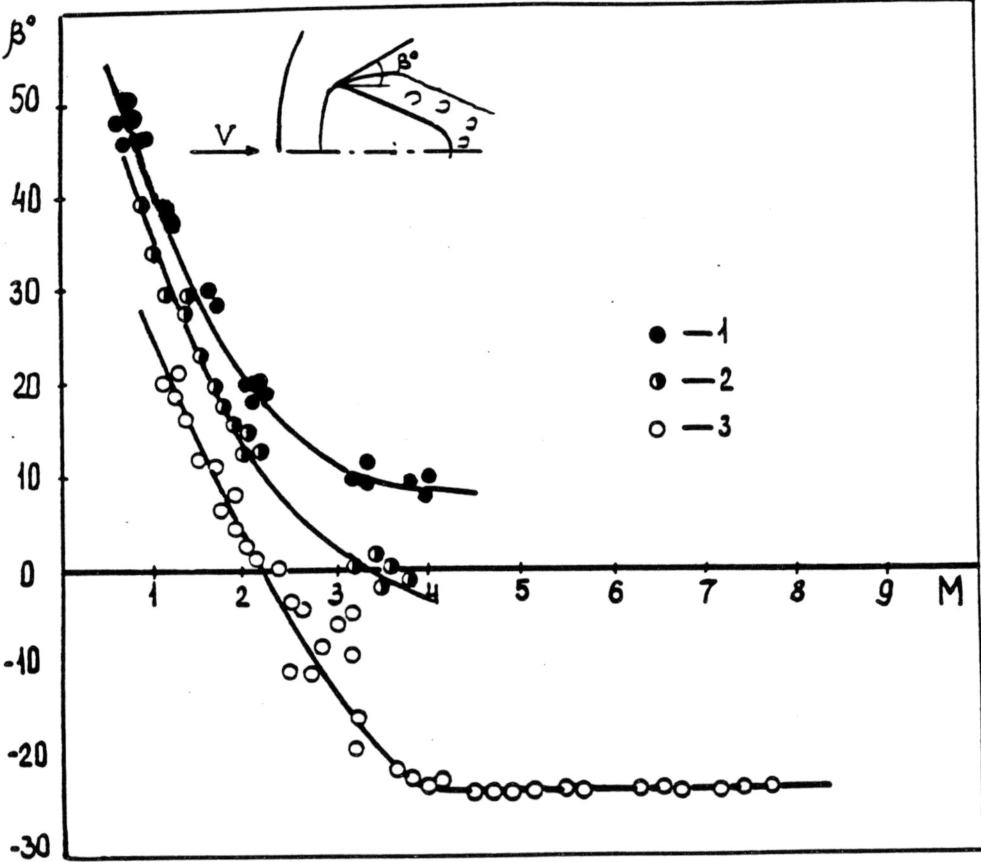

Fig. 16. Dependence of flow departure angle β from the bow segment of model 16 at zero angle of attack on Mach number.

γ: 1 - 1.67, 2 - 1.41, 3 - 1.14.

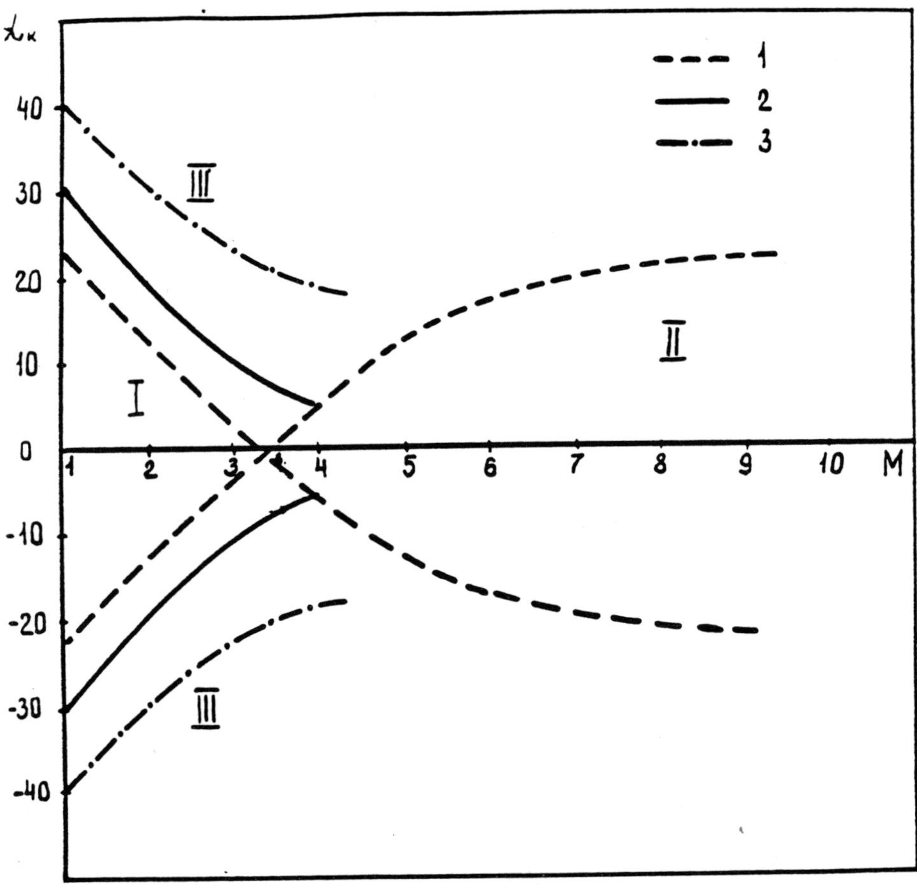

Fig. 17. Dependence of critical incidence angle on Mach number.
γ: 1 - 1.14, 2 - 1.41, 3 - 1.67.

exponent also determines the dependence of α_k on this parameter, α_k being the critical incidence angle correspondent to transition from a nonseparated to a separated flow or vice versa as can be seen from the data of Figures 10b or 17. Examination of the figures indicates that three regimes of the flow about the lateral surface are possible, namely, entirely separated boundary layer (domain 1), partly separated boundary layer (domain 3), and nonseparated boundary layer (domain 2). The extent of each domain in the M - α

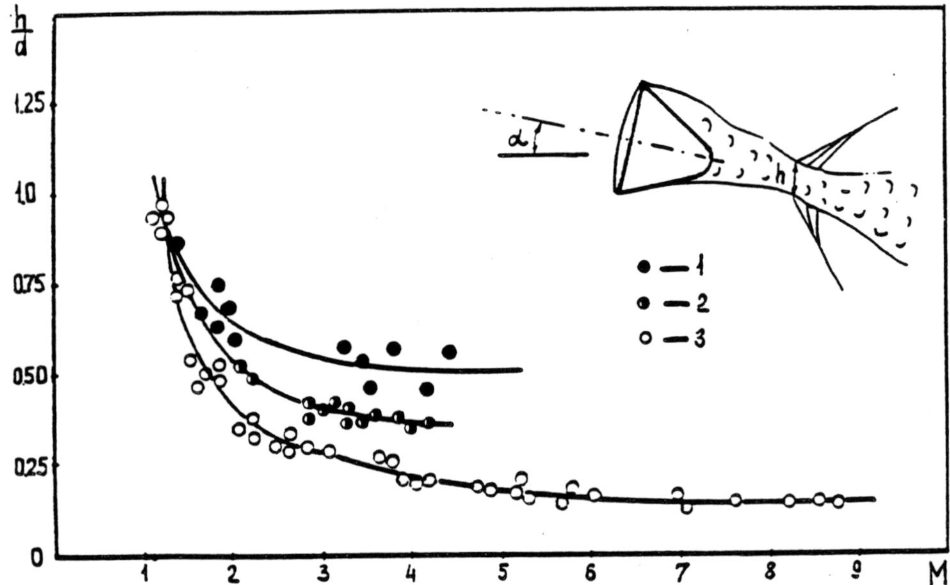

Fig. 18. Dependence of the wake throat width behind model 16 on the Mach number.

γ: 1 - 1.67, 2 - 1.41, 3 - 1.14.

plane depends on the value of γ. In particular, as γ decreases the second domain expands while the first one collapses. The character of flow departure angle dependence on γ determines the character of the near wake parameters, namely, wake throat width h and the distance from the wake throat to the base of the model L_c on γ as well (see Figures 18, 19). From the data presented it follows that both h and L_c decrease with γ. However, decrease of the throat width with γ was not observed for some models. Thus, for model 12 (Figure 1) $h \neq h(\gamma)$ not only at zero incidence angle but at $\alpha \sim 40°$ as well (see Figure 20). Nevertheless, the distance from the throat to the model base decreases with γ. This can be explained by the fact that the boundary layer separation for model 12 always occurs at the front segment while in the case of model 16 it also occurs at the base segment, depending on γ and M. The flow departure from the segment of lesser diameter naturally diminishes the wake width. The wake throat location for model 12 depends essentially on the adiabatic exponent, especially at small incidence angle values as can be seen in Figure 21.

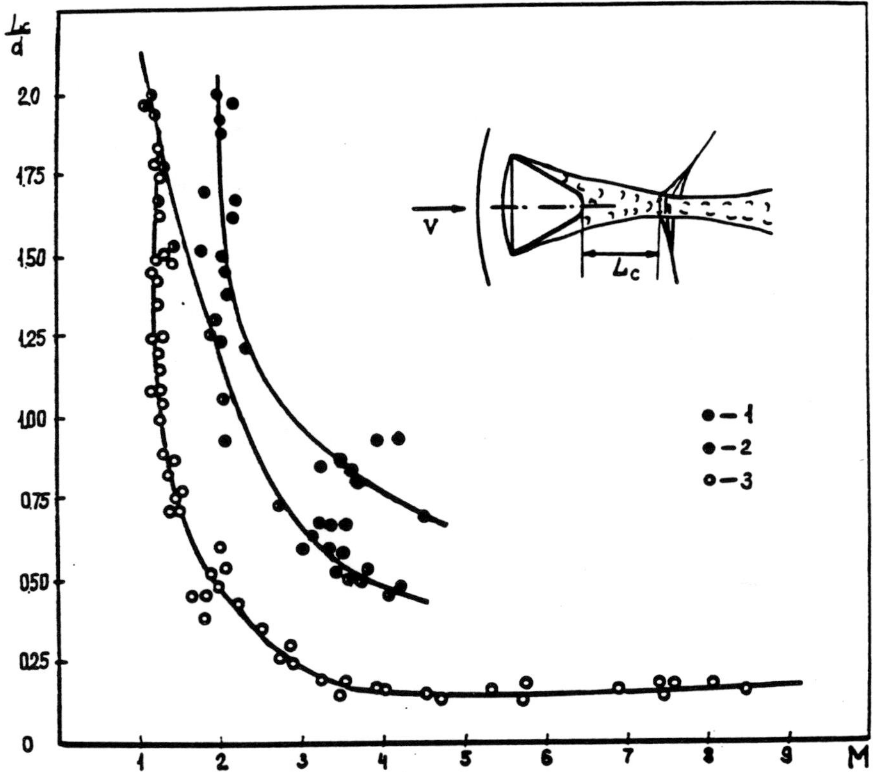

Fig. 19. Wake throat location behind model 16 versus Mach number.
γ: 1 - 1.67, 2 - 1.41, 3 - 1.14.

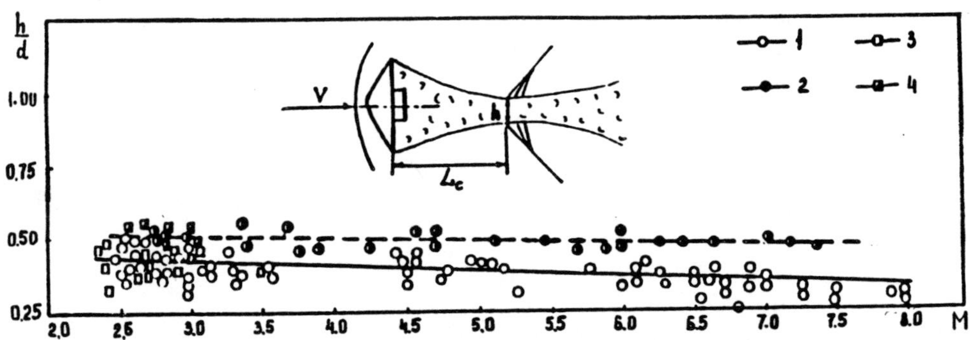

Fig. 20. Dependence of the wake throat width behind model 12 on Mach number.
α^0: 1, 3 - 0-10, 2, 4 - 30-50. γ: 1, 2 - 1.14; 3, 4 - 1.41.

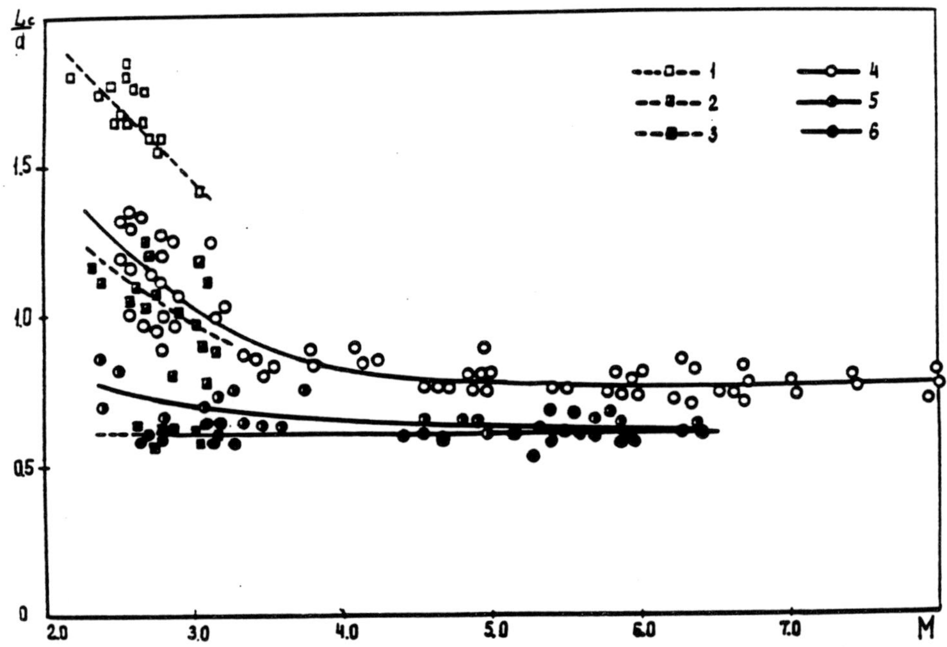

Fig. 21. Wake throat location behind model 12 versus Mach number.
α^0: 1, 4 - 0; 2, 5 - 20; 3, 6 - 40. γ: 1-3 - 1.41, 4-6 - 1.14.

References

1. Yu. A. Dunaev and G. I. Mishin, *Izv. AN SSSR, Otdelenie Tekhn. nauk, Mekh. i Mash.*, no. 2, 1959, pp. 188-190 (in Russian).

2. G. I. Mishin and N. P. Mende, In: *Aerophysical Studies of Supersonic Flows*, Moscow: Nauka, 1967, pp. 163-168.

3. A. P. Bedin, In: *Physical-Gasdynamical Ballistic Studies*, Leningrad: Nauka, 1980, pp. 48-56.

4. A. P. Bedin, G. I. Mishin, and M. V. Chistyakova, *Ibid.*, pp. 9-24.

5. V. G. Artonkin and K. P. Petrov, *Trudy TsAGI*, no. 1361, 1971, 25 pp.

6. V. B. Minostsev and G. F. Telenin, *Nauchnye Trudy Inst. Mekh. MGU*, no. 5, 1970, pp. 4-19 (in Russian).

7. A. P. Krasilschikov and V. P. Podobin, *Izv. AN SSSR, Mekhn. Zh. i Gaza*, no. 4, 1968, pp. 190-193 (in Russian).

8. A. P. Bedin, V. P. Melesheko, G. I. Mishin, et al., Preprint: *Report on 24th Congress of International Astronautical Federation*, Baku, 1973, 15 pp.

9.   V. V. Lunev, *Hypersonic Aerodynamics*, Moscow, 1975, 327 pp.

10.   M. V. Krumins, In: *19th Congress on International Astronautical Federation*, Report no. 138, New York, 1968.

11.   S. M. Belotserkovskii, *Izv. AN SSSR, Otdelenie Tekhn. Nauk*, no. 7, 1956, pp. 53-70 (in Russian).

12.   N. V. Krasnov, V. N. Koshevoi, A. N. Danilov, et al., *Aerodynamics of Rockets*, Moscow, 1968, 772 pp.

13.   A. P. Bedin, G. I. Mishin, *Zh. Tekhn. Fiz.*, vol. 55, no. 4, 1985, pp. 719-722 (in Russian).

ABNORMAL RELAXATION AND INSTABILITY OF SHOCK WAVES IN GASES

A. P. Bedin, G. I. Mishin, A. P. Ryazin,
G. E. Skvortsov, N. I. Yushenkova

Abstract: **Results of theoretical and experimental investigations of peculiar regimes of shock wave propagation in gases**

1. Introduction

More than thirty years ago the question of the possible instability of a simple contour shock wave (SW) was stated by theoreticians (D'yakov, 1954); in experimental studies indications of the specific features of some shock wave propagation regimes in a number of gases have emerged.

It should be considered that shock wave instability (SWI) was discussed Ref. [1] which was the first special study of shock wave instability regimes in argon and carbon oxide. The general nature of a very strong shock wave ($V \sim 10$ km/s) instability was determined by Gordeev [2]. He investigated a number of gases and revealed the universal pattern of the shock wave instability. The nature of the instability found was not discussed. Previous analysis ignoring relaxation processes failed to answer this question.

In order to describe the nature of the shock wave instability the authors of the present paper carried out a detailed investigation. The results obtained are presented in Refs. [3-13]. In studying shock wave instability a number of peculiarities of the flow, microscopic parameters, and radiation have been discovered. Their analysis has led to revelation of rather common phenomenon, namely, abnormal relaxation (AR). The properties discovered are inherent to the whole class of gases and are the effects of abnormal relaxation. The most impressive effect of abnormal relaxation is a relaxation instability of the shock wave (RISW). Together with experimental study of the abnormal relaxation an initial theory of abnormal relaxation and the relaxational instability of the shock wave has been proposed and conditions for relaxational instability of the shock wave in xenon have been determined using the formulated kinetic model. Some results obtained by the authors have been confirmed by other investigators (see, for example, [14]). Undoubtedly, the study of abnormal relaxation and its numerous effects will continue.

The experiment was performed for a representative series of gases: Ar, Kr, Xe, O_2, N_2, CO_2, CCl_2F_2, CF_4, and CF_6 in a wide range of conditions, namely, initial pressure of 1 torr at 2 atm; shock wave velocity 0.6-6.5 km/s. Three experimental installations were used: ballistic range (A. F. Ioffe Physical Technical Institute), shock tube (Moscow State University), low-pressure tunnel (Institute of Chemical Physics of the Academy of Sciences of the USSR). Registration was provided by means of shadow and interferometric visualization, spectroscopy, and electron concentration measurements. The shock wave velocity decrement (drag) and heat flux on the body surface were also measured. The information obtained is estimated by hundreds of experiments (thousands of measurements), only a few of them, the most demonstrative, were published. It is reasonable to group the experimental data in the following manner. The main portion consists of visualized flow fields, namely, flows about flying bodies and shock wave flows with relaxation zones (up to test flow region). These photographs represent approaching the abnormal regime, its development during velocity increase, and further evolution of perturbations. The other group of data consists of the flow parameter measurements, namely, drag of a flying body, heat flux on the body surface (low-pressure tunnel), as well as resistance of shock wave propagation. All these parameters change sharply in the vicinity of the abnormal relaxation regime. The third group contains the radiation data, namely, intensity distribution in the shock wave flow field and spectroscopic data on the characteristic zone behind the shock wave. Specific features of radiation: nonmonotonous luminescence behavior along the shock wave path, unusual spectral lines and bands as well, result from the peculiarity of the kinetic processes in the abnormal relaxation. There are data on the mixing and admixture influence on abnormal relaxation development.

On the basis of the data obtained a series of abnormalities, the abnormal relaxation effects, was discovered. The group of effects related to distribution of the flow uniformity and formation of unsteady perturbations under steady conditions, was most completely brought to light. These effects are shock wave bifurcation, the "bubbles" on the shock wavefront, "internal" shock waves in the relaxation zone, strong distortion and destruction of the shock wave, as well as resonance destruction and specific turbulence induced behind the flying body. The second group is represented by abnormal drag and heat flux on the body surface variation in the abnormal relaxation regime. Among the radiation-related effects are the nonmonotonous regular and irregular radiation distribution, high radiation density right behind the shock wavefront, so-called "flash." Such a radical effect as essential direct and inverse influence of a small amount of admixture was also discovered.

The effects indicated above display the phenomenon of abnormal relaxation and can be explained in the framework of this idea.

What is the abnormal relaxation phenomenon? First, the name of the phenomenon describes the physical situation in which the essential excitation of internal freedom provides, in contrast to commonly known discharging perturbations, an increase of various disturbances. The general mechanism of the abnormal relaxation is as follows. At a certain shock wave velocity, threshold excitation of high levels of particles occurs. The excited particles, possessing a noticeable share of shock wave kinetic energy, as a result of peculiar kinetics, form highly excited particles which transfer their energy into thermal and radiation energy. These ideas supplied by particle excitation level data and a simple kinetic model of abnormal relaxation permit us to interpret the experimental data.

2. Discussion of Experimental Data

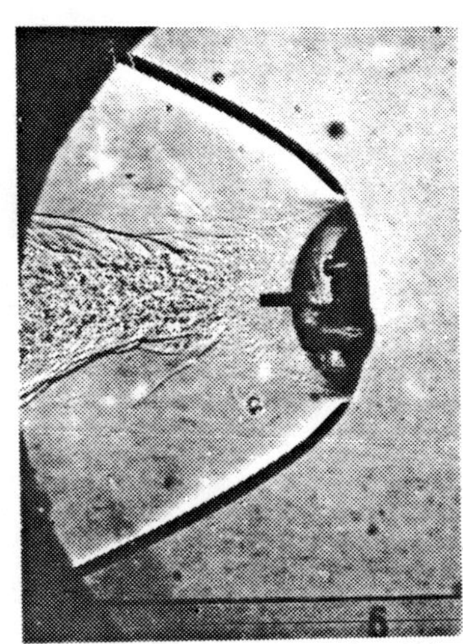

Fig. 1. Initial stage of RISW in CCl_2F_2

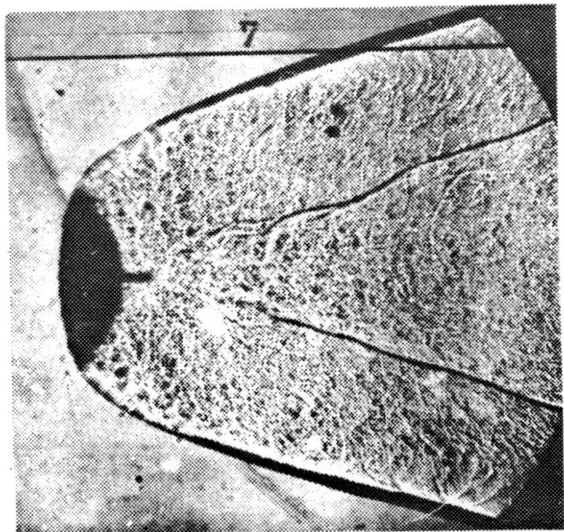

Fig. 2. Developing RISW in CCl_2F_2.

Figures 1-10 represent the first group effects. Figure 1 shows the flow about the segmental body (freon-12, M = 3.9, P_0 = 0.038 MPa), single disturbance of shock wave. Figure 2 corresponds to freon-13, M = 6.8, P_0 =

0.038 MPa. The shock wavefront ripple and induced turbulence in the wake are seen. Figure 3 shows neutral curves in freon-12 for spheres of 27 mm in diameter (1-3, 12), 20 mm in diameter (7-9, 11), and 10 mm in diameter (4-7,10); curves 3, 6, 9 correspond to disturbed shock layer flow and shock wavefront; curves 2, 5, 8 correspond to disturbed shock layer flow and stable shock wavefront; curves 1, 4, 7 correspond to stable flow. Figure 4 represents the interferogram of the flow field behind the shock wave in Ar at $V = 5.3$, $P_0 = 25$ torr, a two-front structure is seen. Figure 5 represents a spectrogram of the flow behind the shock wave in Kr at M = 14.7, P_0

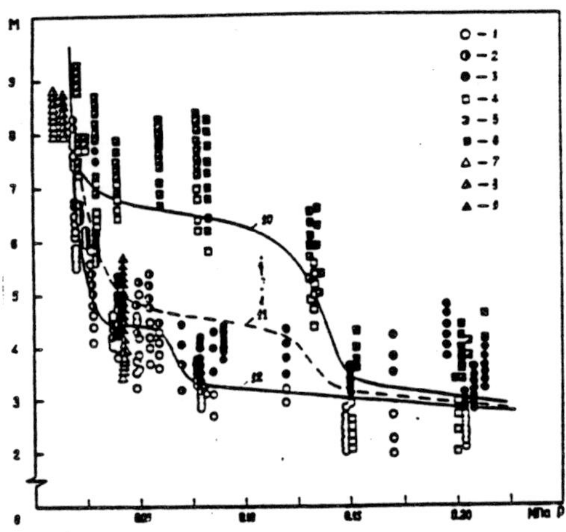

Fig. 3. Neutral curves in CCl_2F_2.

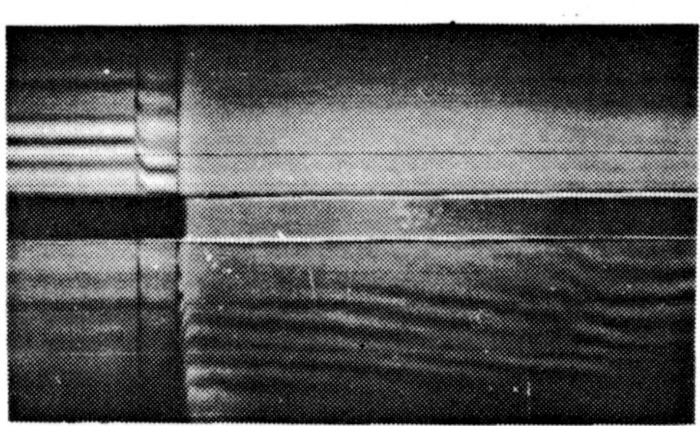

Fig. 4. Interferogram of the flow behind the shock wave in Ar.

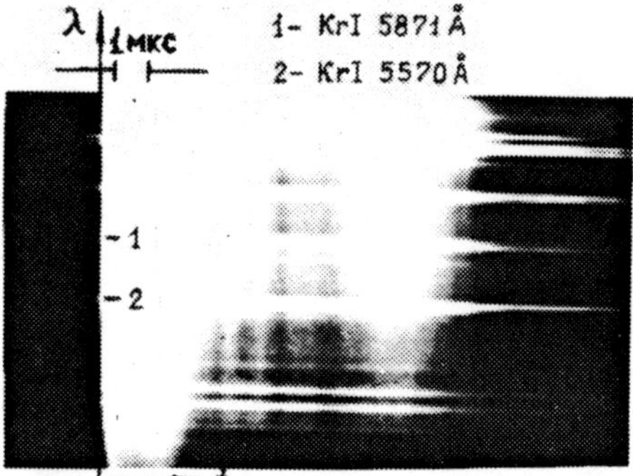

Fig. 5. Flow behind the shock wave Kr.

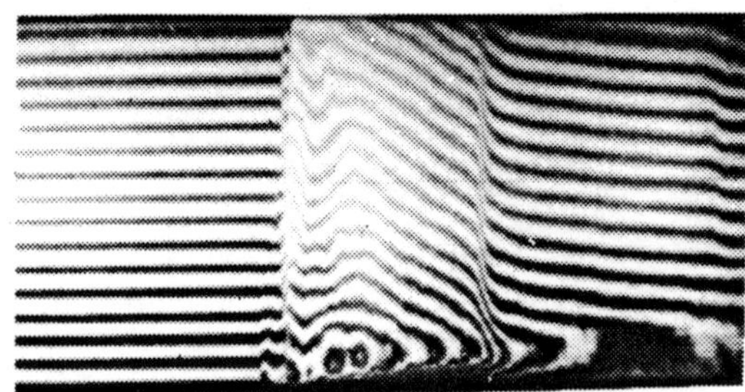

Fig. 6. Flow behind the shock wave in Ar.

Fig. 7. Interferogram of the flow behind the shock wave in Xe.

Fig. 8. Flow behind the shock wave in 0.5 Ar + 0.5 Xe.

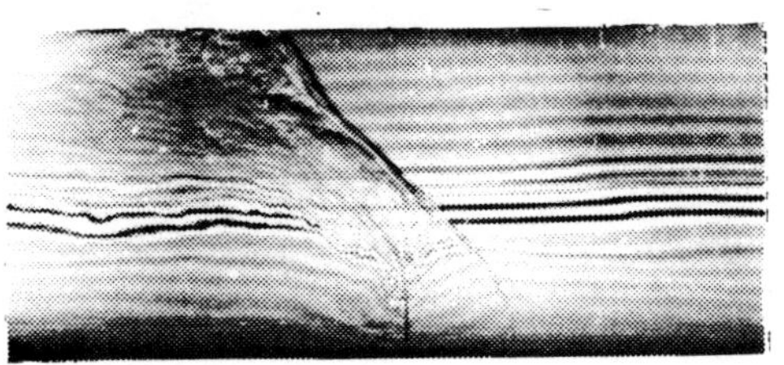

Fig. 10. Flow behind the shock wave in CO_2.

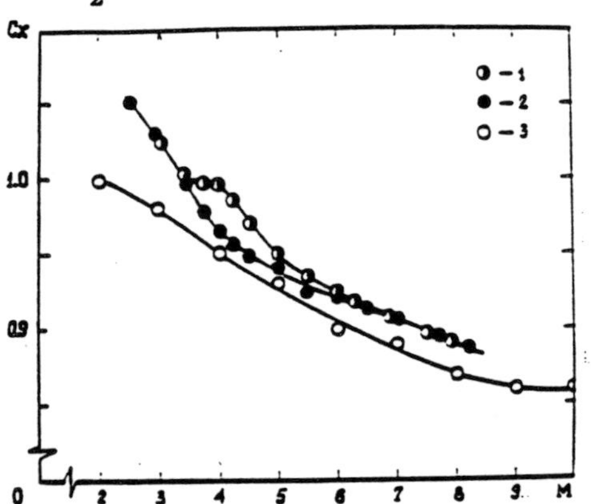

Fig. 11. Sphere drag coefficient vs. the Mach number.

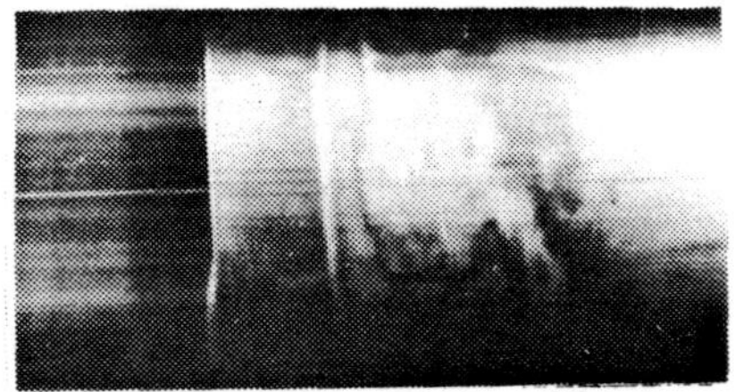

Fig. 9. Flow behind the shock wave N_2.

= 100 torr. Radiation is nonequilibrium, 2 mm shock front bifurcation is observed. Figure 6 shows an interferogram of the flow behind the shock wave in Ar at M = 19.6, P_0 = 25 torr. Shock wavefront bifurcation, nonmonotonic profile of n_e, its irregular oscillation and anticascade are seen (the photograph was taken in 18/05/78). Figure 7 represents the interferogram of the flow behind the shock wave in Xe at M = 13.6, P_0 = 50 torr. Triangular disturbance near the wall, nonmonotonic profile and oscillation of n_e are observed (the photograph was taken in 02/03/77). Figure 8 shows the interferogram (upper part of the figure) and the radiation intensity (lower part of the figure) of the flow behind the shock wave with a gas mixture of 0.5 Ar and 0.5 Xe, V = 6.5, P_0 = 10.7 torr. Shock wavefront is completely destroyed. Figure 9 represents the flow behind the shock wave in N_2 at V = 6.8, P_0 = 10 torr. Disturbance near the wall, shock wave bifurcation and internal shock wave (ISW) are seen. Figure 10 represents the flow behind the shock wave in CO_2 at V = 4.4, P_0 = 10 torr. Shock wave destruction (resonance regime of shock wave instability) is observed.

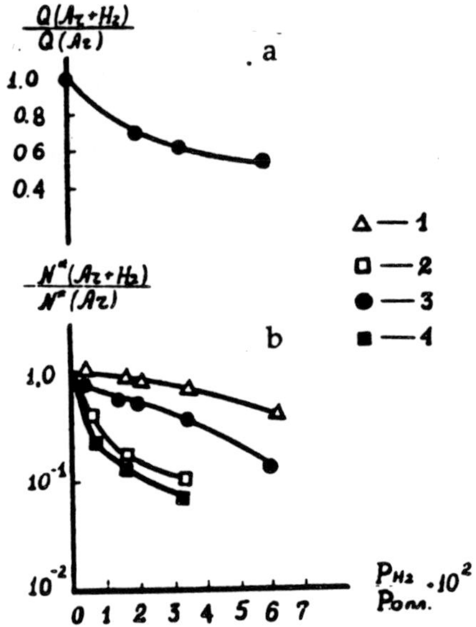

Fig. 12. H_2 admixture influence on the heat flux (a), and on Ar excited atom concentration in the shock wave (b).

The second group of effects is illustrated by Figures 11 and 12. Figure 11 represents drag coefficient dependence on the Mach number for a sphere. If the gas is freon-12; curves 1 and 2 correspond to Re = $1.5 \cdot 10^7$ and Re = $1.2 \cdot 10^6$, respectively; for air, curve 3 corresponds to Re = $2 \cdot 10^6$. The influence of H_2 admixture on heat flux (curve a) and shock excited Ar atoms (curve b) is presented in Figure 12. Curves 1, 2, 3, and 4 correspond to 4p, 4d, and 4d, 5p, and 5p, 5d, respectively.

Radiation effects are represented in Figures 13, 14, and 15. Radiation distribution for the mixture 0.9 Ar + 0.1 Xe behind the shock wave is illustrated in Figure 13. Wavelength λ = 5451 Å, M = 17, 22.6, and 23.1, P_0 = 10 torr. Figure 14 represents the spectrum of Xe in the shock wave for abnormal relaxation in a low-pressure tunnel. The pressure $P_0 \approx$ 1 torr, 1-2 and 3-4 indicate intervals common to the spectrum obtained in the shock tube (Figure 15). Figure 15 shows the spectrum of Xe in a shock wave obtained in the shock tube, M = 31, P_0 = 3 torr. XeII lines originate as a result of reaction with Xe^{+*} and Xe^m. Lines 1-2 and 3-4 indicate intervals common to the spectrum obtained in the low-pressure tunnel (Figure 14).

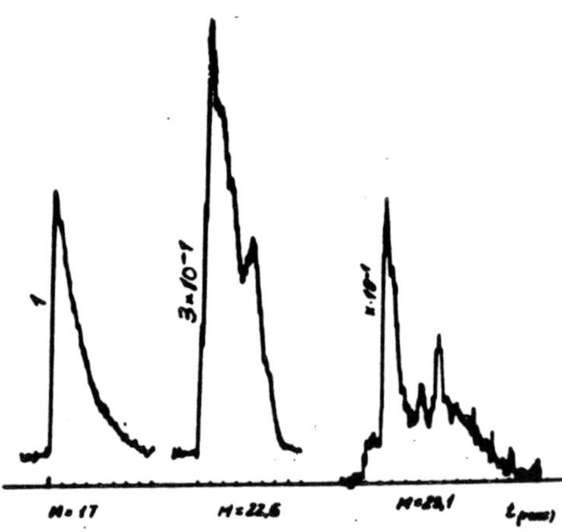

Fig. 13. Radiation distribution behind the shock wave in 0.9 Ar + 0.1 Xe.

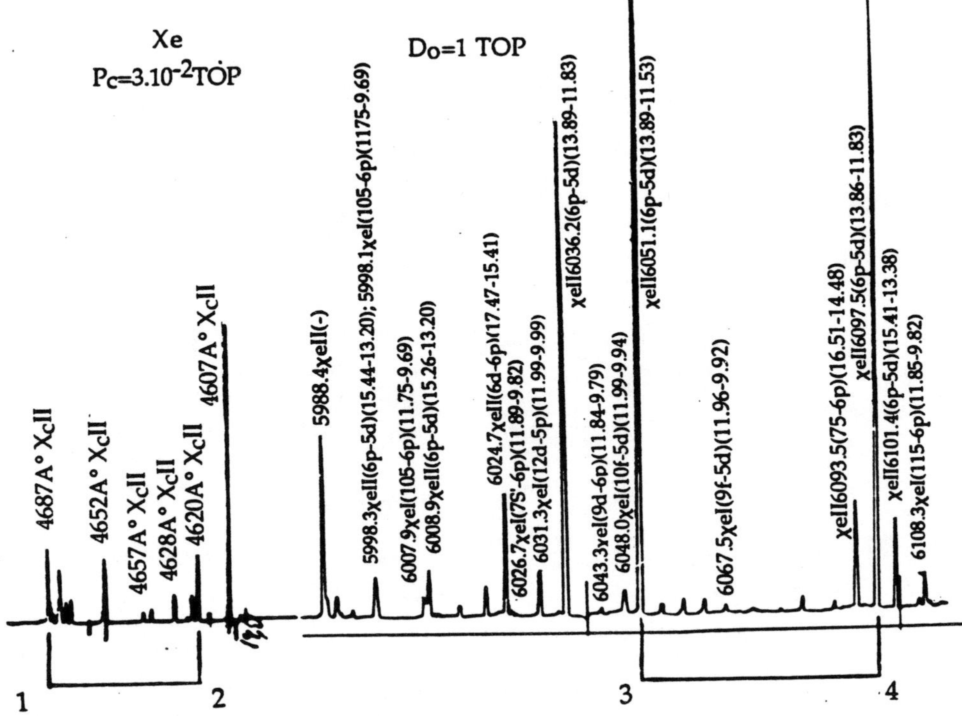

Fig. 14. Xe spectrum in the shock wave for the abnormal relaxation regime obtained in a low-pressure tunnel.

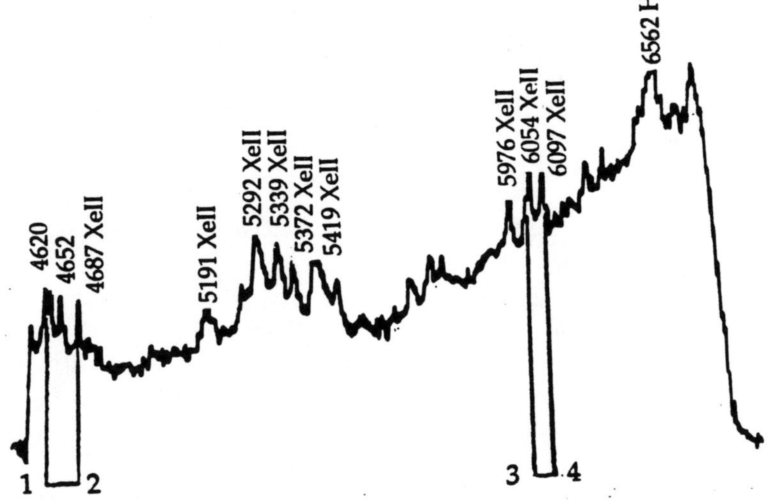

Fig. 15. Xe spectrum in the shock wave for the abnormal relaxation regime obtained in the shock tube.

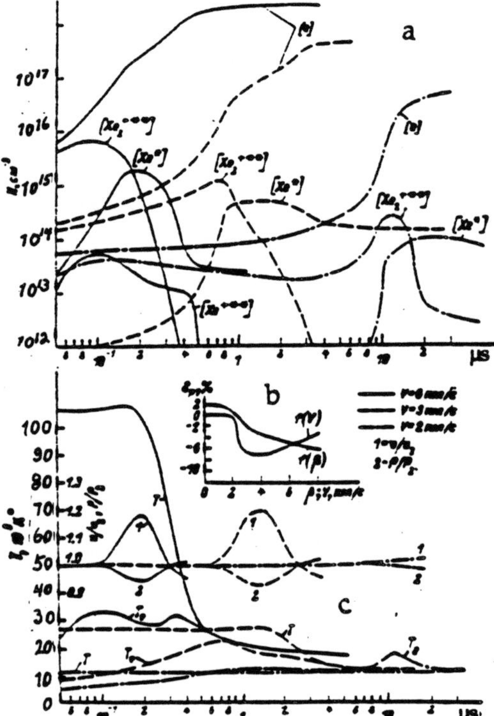

Fig. 16. a) Component concentration distribution behind the shock wave in Xe; b) translational and electron temperature distribution; c) depth of the pressure "cavity" dependence on the kinetic parameter β and the shock wave velocity.

Using the Ar kinetic model for Xe calculations we can obtain parameter distribution behind the shock wave. Figure 16a represents component concentration distributions behind the shock wave in Xe at various shock wave velocities. ($P_0 = 5$ torr), Figure 16b represents distribution of translation and electron temperatures behind the shock wave, Figure 16c represents dependence of the pressure "cavity" depth on kinetic parameter $\beta = K_g/(K_{20}\cdot[\text{Xe}])$ and shock wave velocity, here K_g and K_{20} are the constants of reaction rates for $\text{Xe}_2^+ + h\nu \to \text{Xe}^{+**}$ and $\text{Xe}_2^+ + \text{Xe} = \text{Xe}^+ + 2\text{Xe}$, respectively.

Results of theoretical analysis of abnormal relaxation and relaxational instability of the shock wave are presented in [4, 8-13]. The theory includes: general discussion of the nature of abnormal relaxation and relaxational instability of the shock wave, qualitative criteria, as well as gasdynamic description using a simple relaxation model; formulation of an adequate microscopic model of abnormal relaxation; analysis of the show stability and other effects of the abnormal relaxation in the framework of relaxation gasdynamics.

General ideas of the nature of abnormal relaxation mentioned above are supplied by concepts of highly nonequilibrium processes, namely, quality of boundary, abnormality, and alternation of nonequilibrium. These concepts directly result in threshold formula, applied from the very beginning of the investigations to determine the abnormal relaxation regime:

$$mV^2 = \alpha \mathcal{E}_a, \tag{1}$$

here m is the mass of the gas particles before the shock wave, V is the shock wave velocity, $\mathcal{E}_a$ is the particle excitation energy, $\alpha(\rho_0, T_0)$ is the coefficient dependent on conditions before the shock wave, ρ_0 and T_0 are the unperturbed gas density and temperature, respectively. Coefficient α was chosen to be equal to 1 for averaged experimental conditions ($P_0 = 10$ torr, $T_0 = 290$ K). For $\mathcal{E}_a$ appropriate energies were chosen, namely, metastable excitation, dissociation, or ionization energies. Concretized as in Formula (1) the threshold regimes of abnormal relaxation for all the gases under study were determined, the dependence of α on P_0 was also obtained (T_0 being constant).

Let us demonstrate use of Formula (1) for various gases and velocities at $P_0 = 10$ torr. Consequences of the formula provide partial interpretation of the abnormal relaxation effects. First, choosing for $\mathcal{E}_a$ dissociation or ionization energies, we can obtain the shock wave velocities leading to significant change of the flow field. For calculation convenience appropriate units of measure are used in (1):

$$\mu V_c^2 = 96 \, \alpha \mathcal{E}_a,$$

μ is atomic unit of mass, $[V_c] = $ km/s, $[\mathcal{E}_a] = $ eV. At $\alpha = 1$, $\mathcal{E}_a = \mathcal{E}_{\text{dis,ion}}$ we can learn that critical velocities for Xe, Ar, O_2, N_2, CO, CO_2, and NO are equal to 3, 6.3, 4, 5.9, 5.3, 3.4, and 1.8, respectively. These values agree with experimental data (the last velocity was determined by dissociation energy equal to 1.67 for the process $2NO_2 = N_2 + 2O_2$, the velocity obtained agrees with the result of Zaslonko). Formula (1) predicts a bifurcation regime, if the lowest excitation energy of atom or molecule is taken as $\mathcal{E}_a$. For Xe, Ar, and N_2 the threshold velocities are equal to 2.4, 5.2, and 4.9, respectively. The first and second values approach the values obtained by A. P. Rayzinov (Figure 4). Formula (1) defines a multiplicity of the abnormal relaxation and shock wave instability regimes accounting for the sequence of the characteristic excitation or dissociation (ionization) particle energies, as well as nonequilibrium alternation law. Regimes correspondent to successive dissociation or ionization are predicted in the simplest way (for example, for Xe, V_{c2} is equal to 3.9 and for CO_2, to 5.3). If the meanings of the $\mathcal{E}_a$ sequence are nearly equal to each other, the abnormal relaxation intensity decrease between the nearest velocity thresholds is hard to distinguish. This is valid for freons and even for Xe at $V > V_{c1}$. An illustrative example of the abnormal relaxation and relaxational instability of the shock wave is provided by CO_2. Accounting for possible ways of dissociation accompanied by splinter excitation, we can obtain abnormal relaxation thresholds:

$$V_{c2} = 4.55, \quad \mathcal{E}_{a2} = \mathcal{E}_d \, (CO_2 \rightarrow CO + O(2^1 S_o)) = 9.61;$$

$$V_{c3} = 5.4, \quad \mathcal{E}_{a3} = \mathcal{E}_d \; (CO_2 \to CO(A^1\Pi) + O) = 13.4;$$

$$V_{c4} = 6.1, \quad \dot{\mathcal{E}}_{a4} = \mathcal{E}_d \; (CO_2 \to C + O + O) = 16.6;$$

$$V_{c5} = 6.7, \quad \mathcal{E}_{a5} = \mathcal{E}_{a4} + \mathcal{E}(O(2^1S_o));$$

$$V_{c6} = 7.3, \quad \mathcal{E}_{a6} = \mathcal{E}_{a5} + \mathcal{E}(O(2^1S_o)), \text{ etc.}$$

The first and fourth regimes of the abnormal relaxation displayed as the shock wave instability were discovered by the authors of Ref. [1]; the second and third regimes were demonstrated in Ref. [9]. The strongest and most sharply resonant shock wave instability corresponds to the second regime V_c = 4.4 ± 0.1 (Figure 10). Within the threshold velocity intervals for CO_2 the shock wave is smooth and there are no "visible" density nonuniformities.

The second section of the relaxational instability of the shock wave theory consists in applying the relaxation gasdynamics supplied by the simplest model of relaxation, to obtain the instability conditions [4, 9]. The governing physical condition, namely, intensive transfer of excitation energy, a thermal one, provides observed subcritical behavior of gasdynamic parameters. The flow instability in the relaxation zone has to follow pressure and density "cavities." The regions of regular decrease and increase of the gas density in the relaxation zone are observed in our experiments in the shock tube (by means of interferogram of the flow, see Figures 6, 7). A stability analysis of small dynamical perturbation of a steady nonuniform flow gives a number of instability conditions, in particular a condition of formation of sonic perturbations in the relaxation zone [9].

The third section of the abnormal relaxation theory includes the elaboration of the kinetic model of the phenomenon.

The simplest kinetic scheme taking into account "internal" features of the abnormal relaxation phenomenon is as follows [4]:

$$1. \; A_2 + A_2 \to A_2 + A_2^x;$$

$$2. \; A_2^* + A_2 \to A_2^* + A_2; \tag{2}$$

$$3. \; A_2^* + A_2^* \to A + A + A_2 + \Delta E;$$

where A_2 is a two atomic molecule, A_2^x, A^* are excited and strongly excited molecules, $2\mathcal{E}(A_2^*) > \mathcal{E}d$, ΔE is an energy released during dissociative disactivation. The scheme (2) reflects three specific features of the abnormal relaxation mechanism, namely: 1) rise of considerable concentration of excited particles; 2) formation of superequilibrium number of highly excited particles with sufficient lifetime; 3) collisions of these particles with each other, resulting in dissociation of one of the particles with complete disactivation. Such a scheme is adequate to separate stages of the process, however it is too simplified. For instance, formation of highly excited particles, which can intensively transfer the excitation energy into a thermal one, is performed for noble gases by $A^x + A^+ \rightarrow A_2^+ + e$ (or for nitrogen by $N_2^* + N_2^* \rightarrow N_4^+ + e$); disactivation process accompanied by heating of the "splinters" is dissociative neutralization: $A_2^+ + e \rightarrow A + A + \Delta E$. The kinetic scheme containing the reactions with highly excited particles is introduced for the first time, and correspondent coefficients of the reaction rates is adequate to peculiarities revealed in the abnormal relaxation and the shock wave in xenon. This scheme is presented in [12].

In this model highly nonequilibrium excitation arises as a result of absorption of radiation from the equilibrium zone and formation of a molecular ion excited into self-ionization state $A_2^+ + h\nu \rightarrow A_2^{+*}$, $\mathcal{E}(A_2^{+*}) \approx 23$. Introduction of such a process is based on the observation of correspondent lines in the xenon spectrum (Figures 14, 15).

The fourth section of the theory is analysis of the stability and abnormal relaxation effects on the basis of relaxation gasdynamics, supplied with an appropriate kinetic model (previously used relaxation models do not predict the shock wave instability). Such an analysis for xenon was carried out in [13]. It was found that the flow behind the shock wave was steady during shock wave velocity increase. When the pressure "cavity" formed, the stability analysis for small dynamical perturbations was carried out. For a certain depth of the "cavity" the flow was shown to be unstable and correspondent velocity was defined as critical. This velocity value was found to agree with experimental data (Figure 16). Accounting for excitation states, the nonequilibrium percussive adiabat for the bottom of the "cavity" was calculated. It displayed characteristic Z-form.

3. Conclusion

The study revealed abnormal phenomenon in the shocked flows under certain conditions, explained the main features of these flows at the abnormal relaxation regime and, in particular, shock wave instability. It is worth noting that numerous intriguing questions arose as a result of the study. The answers to these questions can be obtained only in complex experimental–theoretical investigation.

References

1. R. Griffiths, R. Sanderman, and H. Hornung, *J. Phys. D: Appl. Phys.*, vol. 9, 1976, p. 1681.
2. V. E. Gordeev, *Dokl. AN SSSR*, vol. 239, 1987, p. 117 (in Russian).
3. A. S. Barishnikov, et al., *Pisma v Zh. Tekhn. Fiz.*, vol. 5, 1979, p. 281 (in Russian).
4. A. S. Barishnikov and G. E. Skvortsov, *Zh. Tekhn. Fiz.*, vol. 49, no. 11, 1979, p. 2483 (in Russian).
5. A. P. Ryazin and B. Ferkhat, *Vestnik MGU*, no. 1, 1979, *Vestnik MGU, Fiz.-Astron.*, no. 2, 1980 (in Russian).
6. A. P. Ryazin, *Pisma v Zh. Tekhn. Fiz.*, vol. 6, no. 9, 1980 (in Russian).
7. N. I. Yuschenkova, *Pisma v Zh. Tekhn. Fiz.*, vol. 6, no. 21, 1980 (in Russian).
8. A. P. Bedin, G. I. Mishin, and G. E. Skvortsov, *Pisma v Zh. Tekhn. Fiz.*, vol. 7, no. 10, 1981 (in Russian).
9. G. I. Mishin et al. *Zh. Tekhn. Fiz.*, no. 11, 1981, p. 2315 (in Russian).
10. A. S. Prostnev, N. I. Yuschenkova, *Pisma v Zh. Tekhn. Fiz.*, vol. 8, no. 9, 1982 (in Russian).
11. G. E. Skvortsov, *Pisma v Zh. Tekhn. Fiz.*, vol. 9, no. 12, 1983, p. 744 (in Russian).
12. N. I. Yuschenkova, G. I. Mishin, and O. V. Roschin, *Pisma v Zh. Techn. Fiz.*, vol. 11, no. 9, 1985 (in Russian).
13. V. G. Grigor'ev, G. I. Mishin, N. I. Yuschenkova, and O. V. Roschin. *Pisma v Zh. Tekhn. Fiz.*, vol. 12, no. 4, 1986, p. 228 (in Russian).
14. A. Houwing, T. McIntyre, P. Taloni, R. Saudeman, *J. Fluid Mech.*, vol. 170, 1986, p. 319.

SONIC AND SHOCK WAVE PROPAGATION IN WEAKLY IONIZED PLASMA

G. I. Mishin

Abstract: This article presents the state-of-the-art of research on sonic and shock wave propagation in weakly-ionized plasma. A theory which accounts for the double electrical layer on the shock-wave front is formulated.

1. Introduction

Studying the dynamical characteristics of low-temperature plasma is of great scientific importance. In practice, it is related to the necessity to perfect flow and acoustic lasers and sonic and shock wave (SW) application in plasma diagnostics [1-3].

1. One of the original studies in this field was done by Blum [4], who discovered that the electrically modulated corona discharge generated sonic waves and external sonic field affected the discharge.

Important results were obtained when glow discharge of direct current in argon and krypton at a pressure of 13-38 torr was modulated [5]. The wavy behavior of the discharge column corresponding to frequencies of radial and azimuthal acoustic modes arising in plasma was noted.

Similar effects were observed in Ref. [6, 7], which indicate electron density oscillations with a frequency equal to or greater than plasma oscillation frequency caused in a weakly ionized plasma sonic perturbations of the same frequency.

In Ref. [8] argon plasma acoustic instability resulting from discharge current modulation was theoretically considered and experimentally investigated.

Investigation of sonic oscillation damping in argon decaying plasma using oscillations preliminarily excited by means of an external voltage at a frequency of 5-10 kHz applied to discharge tube electrodes was carried out in Ref. [9].

A number of papers have studied the effect of acoustic fields on the weakly ionized plasma. The acoustic fields were generated by external sonic and ultrasonic sources. In Ref. [10], for example, periodic alternating radiation of a plasma column correspondent to generating sonic field frequency was observed. In Ref. [3] modulation of ion density in the gas discharge developing under the sonic field was seen.

Consideration of the results of parallel investigations [11] allowed us to establish that the neutral component density oscillations arose and propagated ni the plasma boundary layer, where electron oscillations of Tonks-Dattner-type resonance were possible [12, 13].

2. In plasma generated by means of power enough ultrahigh frequency electric pulses or direct current, a number of investigators observed low frequency sonic waves (1-10 kHz) [14-22] appearing during discharge as a result of the joint effect of electrodynamical pressure and local heating pressure pulses.

Berlande, et al. [14] were among the pioneers of these studies. They discovered electron concentration oscillations in experiments on the pulse generated cryogenic helium plasma.

In Ref. [16] pulsed input of additional high-frequency energy was shown to give rise to acoustic oscillations in glow direct current discharge in xenon.

Study of longitudinal and radial sonic waves generated in neon at a pressure of 5-15 torr under current pulse action [17] showed that their frequency was related to translational gas temperature, average molecular weight (therefore, dissociation degree), and discharge tube geometry.

In Ref. [18] excitation of acoustic waves in CO_2 by low voltage electron pulses was reported. In has been suggested that the high efficiency of this process is caused by CO_2 molecule dissociation.

Oscillations of the discharge column in stationary pulsed tubes [19, 21] were used to estimate the discharge temperature.

Ref. [20] is devoted to a detailed experimental study of radiation intensity oscillations in helium plasma generated in a glass vessel in electrodeless high-frequency discharge (8 MHz). The vessel walls were cooled by liquid helium. Using the measurements of standing sonic wave periods, the translational atom temperature of plasma was determined as a function of excitation pulse duration at various discharge currents and gas densities.

3. Interplay of the behavior of neutral and charged plasma components forces critical consideration of the possibility of plasma temperature determination by sound velocity measurement. This problem was stated clearly by Ingard and Gentle [23] who considered theoretically the properties of three eigenmodes of oscillations in plasma; namely, ordinary sonic wave, degenerated ion-sonic wave, and degenerated plasma oscillations. The energy transfer between the components was also taken into account.

Refs. [24-28] are devoted to direct experimental investigation of sound propagation in plasma.

Experiments [24] have shown that acoustic oscillations were amplified if the propagation directions of ultrasonic waves and electron drift coincided.

In Ref. [25] the sound phase velocity in argon plasma of direct current was measured as a function of electron density. A high velocity increase with the electron density was shown.

Increasing the amplitude and velocity of acoustic waves of frequencies of 3.8 kHz and 7 kHz occurring in wave propagation in plasma [26] is related to energy transfer from electrons to neutrals.

Sound velocity measurements in air at a frequency of 4-25 kHz and a pressure of 1 torr, as well as electron temperature measurements in glow direct current discharge with isolated electrodes [27] showed that in this case the ion-sonic wave velocity was governed by electron temperature and noticeably exceeded the acoustic mode velocity.

Ishida et al. [28] developing the study from Ref. [25], established the sound velocity dependence on the frequency of acoustic oscillations in gas-discharge argon plasma at a pressure of 2 torr. A frequency decrease from 7 to 5 kHz is accompanied by a steep (more than double) plasma sound velocity increase.

Asanaliev et al. [29] show a twofold decrease of the drag of the sphere moving at Mach number 0.05 and Reynolds number 4-7 in argon plasma as compared with the drag of the sphere moving at the same parameters in neutral argon. It can be considered as an indirect confirmation of the correspondent increase of weak perturbation propagation velocity.

4. Refs. [30-40], presented chronologically, are devoted to a theoretical discussion of the acoustic effects observed in weakly ionized plasma.

In studies [30-32] ion-sonic perturbation and neutral-acoustic wave correlation is discussed. It is shown that acoustic wave instability caused by significantly exceeding the electron temperature over a heavy particle can be the reason for spontaneous sonic oscillations and acoustic wave amplification.

Ingard [33] presents a detailed analysis of the rise of acoustic perturbations in weakly ionized plasma developing as a result of electron-heavy particle energy transfer. At certain conditions sonic wave amplification and velocity must increase. The model suggested is used to explain sonic wave development in moving strata and acoustic modulation of plasma post-luminescence. An expression for the rate of the electron-heavy particle energy transport is obtained which is used in a number of studies.

Frequency dependence of electrostatic energy density in continuous or slightly decaying longitudinal plasma oscillations as well as the correlations of kinetic, electrostatic and thermal energy densities are discussed in Ref. [34].

Theoretical investigation of acoustic waves in weakly ionized plasma [35] showed that at sound frequency $\omega \ll \omega_s = \omega_p(nT/n_iT_e)$ (where ω_p is the collision frequency for neutral-charged particle interaction, n, n_i are the densities of neutral and charged components, T, T_e are the neutral and electron temperatures, respectively) oscillations of neutral particles, ions, and electrons coincided in phases and amplitudes. But at $\omega \gtrsim \omega_s$ there is a noticeable difference in their phases leading to electric field perturbation by the sonic wave which results in a decrease of the sonic wave amplification as compared with the case of sinphase oscillations of electrons and heavy particles. For this reason the neutral gas temperature increase leads to acoustic wave amplification only at $\omega \ll \omega_s$.

Aubrecht [36] deduced a dispersion equation for sonic waves. Weakly ionized plasma in an electric field of glow discharge at a pressure of 1 torr or greater was considered. It was mentioned that at certain conditions acoustic oscillations became unstable, their amplitude and velocity increased. Computation results were in good agreement with experimental data [15].

Authors of Refs. [37, 39] considered a single acoustic pulse (weak shock wave) in a weakly ionized gas. Calculations showed that such a pulse gave rise to ion density and electric field perturbation, propagating before the pulse front and decreasing as exponent function at a distance of $(T_e/T)l$ where l is a mean free path in neutral gas.

In Ref. [38] the time variation of the equilibrium temperature of a plasma neutral component, heated by electrons, was calculated. It was found that the dependence obtained failed to give acoustic instability in weakly ionized plasma. This conclusion contradicts all previously published theoretical and experimental results and can refer, to some extent, to standing sonic waves only.

Tokhaya [40] considers the problem of acoustic wave generation in weakly ionized plasma located in an external alternating electric field. It is shown that, in this case, sonic oscillations arise whose frequency is equal to that of the external field. Conditions for the sonic wave increase are determined and an expression for maximum amplitude increment is obtained.

In electrical molecular gas discharge vibrational energy exceeds translation energy due to inelastic electron scattering on molecules. Therefore, in an acoustic wave moving in the plasma of molecular gas along with energy transfer from electrons to neutrals, vibrational energy transfers into translational energy. This occurs in wave nodes, where according to adiabatic law, the gas density and temperature increase and vibrational de-excitation leads to an increase of sound velocity and amplitude [41].

Thus, there are reasons to consider the phenomena of acoustic oscillation generation and sonic wave amplification in a weakly ionized plasma as theoretically explained.

5. The main features of shock waves in a weakly ionized plasma were discovered experimentally in Refs. [42-50]. Results of theoretical investigations of the shock wave properties in low-temperature plasma are presented in Refs. [51-65].

The essential differences of the shock wave properties in weakly ionized nonisothermal plasma and neutral gases are related to the presence of a charged component in plasma as well as significant difference of electron and other particle masses and temperatures [37, 51-56].

Chutov [42] showed experimentally that the shock wavefront width significantly exceeded the mean free path of the plasma particles. At the shock wavefront an electrical voltage jump to several volts was seen. The electric field vector was found to be directed along the shock wave propagation. Additional ionization occurred even at the front of weak shock waves.

Investigations of the velocity and amplitude of shock waves in the plasma of longitudinal [43-45] and transverse [45, 48] glow discharges showed that the shock wave velocity essentially exceeded (by 1.5 or more times, depending on experimental conditions) the shock wave velocity in neutral gas at a plasma temperature. The plasma shock wave amplitude was far less than a neutral gas.

Weak dependence v_p on the velocity of a shock wave entering the plasma was observed.

It was determined experimentally that the discharge current density affected significantly the shock wave velocity in plasma v_p; namely, at $j \leqslant 1$ mA/cm^2 was approximately equal to the thermal velocity, at $j \leqslant 15$ mA/cm^2 it essentially exceeded thermal velocity, and at $j \leqslant 50$ mA/cm^2 the effect of superthermal increase of shock wave velocity in plasma diminished.

The shock wave velocity v_p is found to be independent of the electric field vector as well as plasma volume. The length of the plasma column was varied from 30 mm to 160 mm, and in some experiments was equal to 500 mm. This velocity is also independent of the type of gas (air, argon [45, 48], N_2, CO_2 [50]).

The investigations of shock wave propagation in the decaying air plasma of longitudinal glow discharge ($p = 30$ torr, $j = 30$ mA/cm^2) at the moments succeeding discharge turn-off [44] showed that the characteristic time of the shock wave velocity approaching thermal velocity was ~10^{-2} s, i.e., greater than that of recombination.

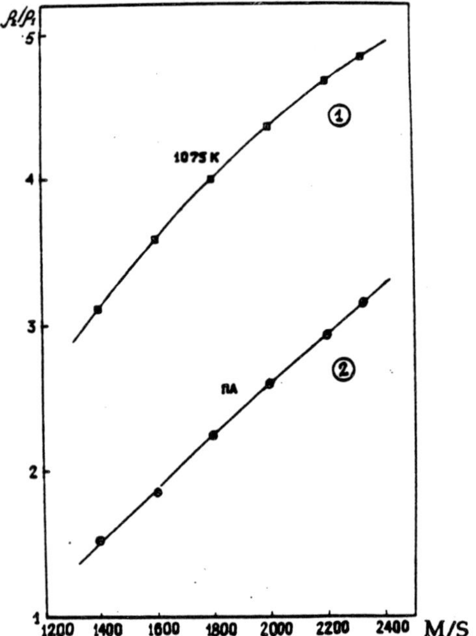

Fig. 1. 1) density ratio in the shock in air at T = 1075 K, 2) density ratio in the glow discharge air plasma (P = 30 torr, T = 1075 K), V is the shock wave velocity.

Investigations of the interaction of the shock wave and the decaying plasma of electrodeless pulse UHF discharge [46] showed that plasma generated by UHF radiation possessed the same abnormal properties with regard to the shock wave propagation as the electrode glow discharge plasma. These properties arose in UHF plasma at a radiation wavelength η = 10 cm during a short pulse of 40 μs and disappeared in a time approximately equal to that for glow discharge.

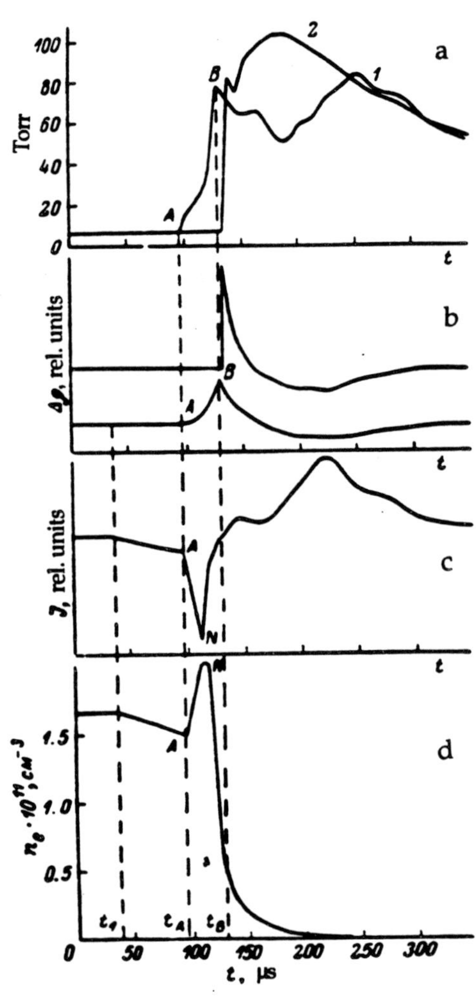

Fig. 2. Oscillograms of signals of piezoelectric transducer $P(t)$.

a - Schlieren-system; b - plasma luminescence $J(t)$; c - UHF interferometer $N_e(t)$; d - the shock wave in air plasma p = 6 torr, entry velocity of the shock wave V = 1000 m/s, V_p = 1500 m/s. 1 - air plasma, 2 - air without discharge.

In the plasma shock wave a significant decrease in the density ratio took place which, as could be seen in Figure 1, differed by approximately two times as compared to air at a plasma temperature for velocities $v_p \approx$ 1400-2325 m/s.

Figure 2 [50] shows pressure $p(t)$, density gradient $\Delta\rho(t)$, electron concentration $n_e(t)$, and plasma luminescence $J(t)$ time variation and correlation in the course of shock wave propagation through the horizontal glow discharge plasma (air, $p = 6$ torr). It can be seen that there is a precursor AB in front of the shock wave. Its length is about 30 mm at an average shock wave velocity in plasma of 1500 m/s (velocity of the shock wave entry into plasma is 100 m/s). The precursor at a low-pressure gas flow with smoothly increasing density, containing a peak of electron concentration and sinphase drop of luminescence signal. Variation of n_e and J is ~30% of their values in undisturbed plasma.

The precursor forms gradually, approaching its final constant value as the shock wave propagates in plasma. At the parameters mentioned above a stationary precursor forms at a distance of 50 mm from the discharge edge.

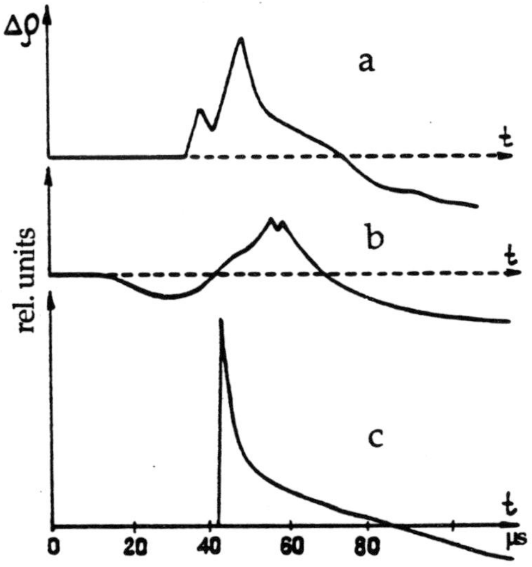

Fig. 3. a - Schlieren signal in the air plasma $p = 6$ torr, $V = 1040$ m/s; b - $V = 1130$ m/s; c - $V = 1050$ m/s; a - at a distance less than 50 mm; b - at a distance of 100 mm; c - control Schlieren signal, without discharge.

It is worth noting that in some cases in front of strong shock waves (~1000 m/s) in glow discharge plasma at a distance less than 50 mm from the shock wave entry the Schlieren signal has double-peak form (Figure 3a), and at a distance more than 50 mm the Schlieren signal shows a negative pulse of density gradient (Figure 3b), the latter indicates the heat precursor existence in front of the shock wave and related rarefaction zone predicted in Ref. [47].

A weak shock wave dissipation in a laser spark in air was shown in [66]. Phenomena accompanying shock wave propagation in low temperature carbon plasma are described in Refs. [67, 68]. These results confirm the previously discovered [43-45] effect that the maximum temperature ratio at the air-plasma boundary is insufficient for explaining the shock wave Mach number decrease in plasma.

In order to understand the mechanism of shock wave acceleration in plasma shock wave propagation in longitudinal glow discharge plasma at a pressure of 5 torr in the presence of a magnetic field was investigated [46]. It was shown experimentally that a magnetic field up to 400 Oe directed along the discharge axis parallel to shock wave propagation did not affect the shock wave velocity in plasma. At the same time, applying a transverse magnetic field causes the shock wave velocity to decrease to the "thermal" velocity of shock wave in plasma and the amplitude to increase to an adequate value.

Further increase of the magnetic field did not lead to shock wave velocity decrease and amplitude increase. The transverse magnetic field correspondent to the "thermal" shock wave velocity increased when the discharge current decreased, namely, magnetic fields of 110, 170 and 220 Oe correspond to electric current densities of 11, 6.5, and 3.5 mA/cm^2, respectively. The value of $\omega\tau$ was approximately equal to 0.4, i.e,. the shock wave velocity variation was seen even if plasma electrons were weakly magnetized.

In connection with the magnetic field influence on the shock wave dynamics the charged component behavior during shock wave propagation in plasma was investigated by means of double and single electric probes operated simultaneously with a piezoelectric transducer [49].

The medium under study was vertically directed gas-discharge air plasma at a pressure of 33 torr, temperature at the discharge center was 1400 K. The shock wave velocity near the discharge axis was 416 m/s without discharge and 1320 m/s in plasma.

The double probe current variation is proportional to the plasma specific conductivity variation. The plasma specific conductivity can be obtained from:

$$\sigma = a\frac{n_e T_e}{n\, Q_{ia}\sqrt{\bar{m}T}},$$

where Q_{ia} is the ion-neutral collision cross-section, m is the molecular mass, α is the coefficient of proportionality.

Measurements carried out in the plasma undisturbed by the shock wave showed that the double probe current and therefore σ varied across the discharge column from a minimum value at the periphery to a maximum value at the center.

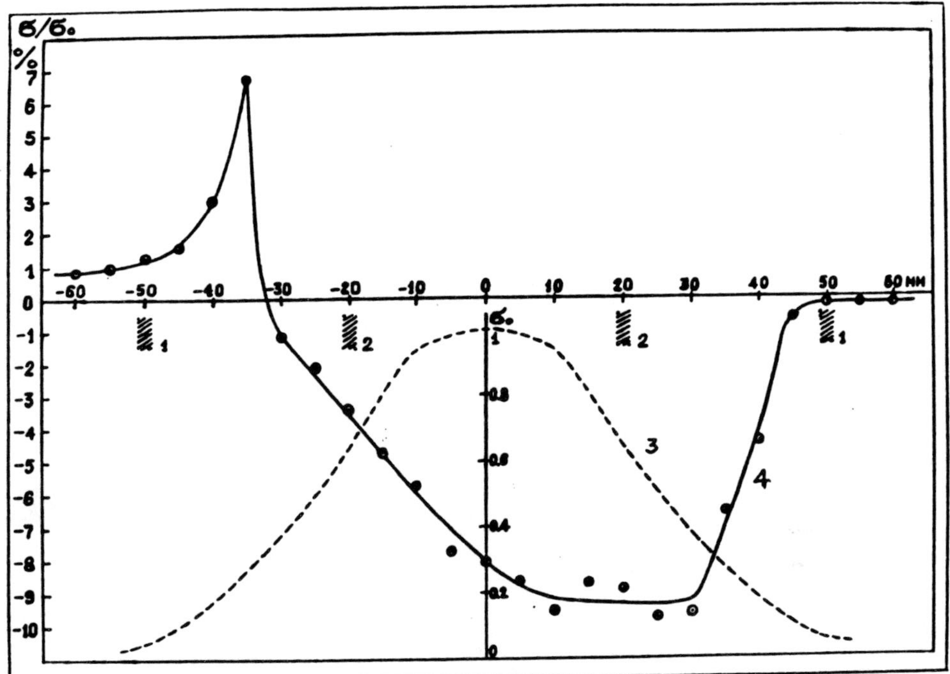

Fig. 4. Conductivity variation in the shock wave, propagating in vertical column of glow discharge in air.

1 - outer boundary of discharge; 2 - boundary of highly luminous zone of discharge; 3 - plasma conductivity variation across the discharge without shock wave; 4 - the same in the shocked plasma.

Interferogram investigation revealed a number of specific features of conductivity behavior observed in shocked plasma (Figure 4).

Shortly before the shock wave entered the plasma weakly luminescence zone the conductivity in the shock wave increased, although the shock wave was weak and the temperature increase could not provide additional ionization.

However, in the middle of this zone the situation changed qualitatively. Conductivity decreased, when the shock wave arrived at the plasma domain, where $n_e T_e$ had reached a large value. It should be noted that there is no conductivity jump on the shock wavefront in this case.

6. Several explanations were proposed for the shock wave velocity and the simultaneous amplitude variation inadequacy to plasma temperature.

One of the theories was based on the possibility of releasing the super-equilibrium energy obtained by the particles during plasma generation. Estimates showed the permissibility of this version [57].

A detailed theoretical study of the shock wave propagation in nitrogen under the conditions of nonequilibrium vibrational excitation was carried out in Refs. [59-61]. Further analysis will be based on the ideas contained in [59].

In the glow discharge in molecular gas, energy input in the electrons transfers mainly into vibrational excitation if the electric field is not too high. Because the gas temperature is low enough in discharge and the characteristic time of V-T relaxation is large, vibrational energy ~1 eV is accumulated that significantly exceeds its equilibrium value correspondent to gas temperature. Behind the shock wavefront because of temperature and concentration increase the characteristic time of the V-T process significantly decreases and, as a result, the degree of vibrational excitation in the shock wave diminishes to the thermal excitation level. Additional energy $\langle \varepsilon_{V1} \rangle$ - $\langle \varepsilon_{V2} \rangle$ transfers into translational and (partly) rotational degrees of freedom. Energy release in the shock layer leads to abnormal shock wave dynamics in plasma. Solution of the equations for the shock wave in two-atomic gas (nitrogen), taking into account energy input into discharge, and the mean vibrational energy $\langle \varepsilon_V \rangle$ correlation gives the Hugoniot adiabat for conditions adequate for the experiment (pressure, density, temperature, energy input, electric field).

Comparison of the theoretical analysis and experimental data showed that the V-T process in the shock wave failed to explain the effects observed, because:

1. calculations show the velocity "threshold" $v_p \geqslant 2.5 \cdot 10^5$ cm/s for this phenomenon, otherwise the relaxation zone becomes too wide, but experiments show that even weak shock waves display abnormal properties and there is no relaxation zone;

2. in the course of energy release in the shock layer the shock wavefront remains thin (an order of some mean free paths) while the shock wavefront is smeared in plasma;

3. the shock wave acceleration and amplitude decrease occur in monoatomic gases as well. In this case it is hard to assume the existence of a large enough reservoir of internal energy in front of the shock wave;

4. the experimentally obtained values of p_2/p_1 and ρ_2/ρ_1 are less than that correspondent to shock adiabat and Rayleigh law for the regimes under study;

5. the mechanism discussed above fails to explain the transverse magnetic field influence on the shock wave dynamics in plasma.

Ref. [58] presents the idea that a body flying in plasma at high velocity, thermoemission phenomenon must rise, causing electric current and magnetic fields near the body. These currents relate to temperature gradients in the body itself or in the shock heated gas at hypersonic flight,. Comparison of heating estimates, obtained in the framework of this mechanism, with experimental data indicates a necessity to develop a more effective mechanism for shock wave-plasma interaction.

The shock wave acceleration in various gases due to ohmic heat release in the shock layer, caused by the plasma conductivity increase due to the gas temperature increase behind the shock wave was considered in Ref. [64]. In this case shock wave acceleration must be accompanied by simultaneous amplitude increase at Mach numbers M ≳ 2. However, experiments show that shock wave acceleration in plasma occurs at lower Mach numbers and is accompanied by the amplitude decrease and, in addition, is observed after the discharge is switched off.

Based on the results presented above, a theory which accounts for the double electrical layer on the shock wavefront can be formulated.

A double electrical layer on the shock wavefront [51-54] arises due to electron density gradients in the shock wave and high electron mobility, resulting in electron diffusion relative to ions leading to electric potential jump on the shock wavefront.

The value of this jump $\delta\varphi$ depends on the shock wave intensity:

$$e\delta\varphi = kT_{eo} \ln \frac{\rho_2}{\rho_1}$$

Electron shock wave heating is a nonadiabatic process because electron energy in the double layer increases and ion energy decreases [63]. The electron temperature increase can be obtained from:

$$\frac{T_{eo}}{T_1} = \frac{1}{1 - (\gamma - 1)\ln(\rho_2/\rho_1)},$$

This formula shows that the mechanism considered explains a significant pumping of electron energy in the shock wave observed in experiments. Large electron velocity results in an increase of the electron concentration ahead of the shock wave. High electron heat conductivity as well as generation and dissipation of ion-sonic waves lead to neutral plasma component heating and thermal precursor formation [47]. The gas heating in front of the shock wave results in a small increase of pressure (Figure 2). At high electron temperature the precursor gas density is lower than that of the free stream (Figure 3b). THe plasma conductivity decrease can be related to the temperature T increase according to the conductivity formula.

As the precursor formation is accompanied by the energy transfer to neutrals, it forms gradually and the shock wave travels a certain distance in plasma during formation. The precursor forms while moving in plasma, it acquires a practically unchanged shape and then moves in a self-similar regime.

The suggested model, developed from Ref. [47], explains the plasma shock wave properties revealed experimentally [42-50]. The most important of them are:

- shock wave acceleration occurs simultaneously with amplitude decrease and essential shock front smearing:
- abnormal shock wave properties including monoatomic gases;
- weak shock waves must also display abnormal properties (the formation time (path) in this case will increase correspondingly);
- a magnetic field transverse to the shock wave propagation decelerates electrons ("magnetization") and prevents precursor formation;
- the precursor rise correlates with the behavior of plasma conductivity during shock wave propagation;
- the precursor forms in a domain where the conductivity increase is replaced by its decrease (Figure 4).

Estimates based on the experimental data on pressure and density ratios in the shock waves in argon showed that the specific energy transferred from the shock wave into precursor amounts to 20% of its value in front of the shock wave. The Mach number, correspondent to the pressure and density jump, is approximately equal to 1.2, while the ratio of the shock wave velocity to the thermal sound velocity is 2.7 times greater.

The precursor phenomena, in the sense of gasdynamics, are similar to the existence of the effective velocity of small perturbation propagation, exceeding the thermal sound velocity in weakly ionized nonequilibrium plasma.

This is the state-of-the-art of the problem.

References

1.   C. G. Suits, no. 1, 1941, p. 720.
2.   H. Edels and D. Whittaker, *Proc. Roy. Soc.*, vol. 240A, no. 1220, 1857, pp. 54-66.
3.   K. W. Gentle and U. Ingard, *Appl. Phys. Lett.*, vol. 5, 1964, pp. 105-106; K. W. Gentle, U. Ingard, and G. Bekefi, *Nature*, vol. 203, no. 4952, 1964, pp. 1369-1370.
4.   B. W. Blum, I. Dyer, and U. Ingard, *J. Acoust. Soc. Am.*, vol. 26, no. 1, 1954, p. 139.

5.   S. D. Strickler and A. B. Stewart, *Phys. Rev. Lett.*, vol. 11, no. 12, 1963, pp. 527-529.

6.   M. Sicha, V. Vesely, J. P. Novak, and V. Fuchs, *Third Czech. Conf. on Electronics and Vacuum Phys. Trans.*, Prague, Sept. 23-28, 1965, pp. 139-152; S. Subertova, *Third Czech. Conf. on Electronics and Vacuum Phys. Trans.*, pp. 167-173.

7.   M. Schulz and U. Ingard, *Phys. Fluids*, vol. 10, no. 5, 1967, pp. 1031-1036.

8.   G. M. Sessler, *Acoustic and Plasma Waves in Ionized Gases. Physical Acoustics (Principles and Methods)*, Ed. by W. P. Mason, vol. IV, Part B, 1968, pp. 123-213.

9.   I. V. Gerasimov, *Doklady AN Tad. SSR*, vol. 16, no. 9, 1973, pp. 21-23 (in Russian).

10.   K. Wojaczek, *Beitrge Plasmaphys.*, vol. 1, no. 3, 1960/61, pp. 127-136 (in German).

11.   I. V. Gerasimov and V. F. Nozdrev, *Doklady AN Tad. SSR*, vol. 14, no. 2, 1971, pp. 20-23 (in Russian).

12.   L. Tonka, vol. 37, no. 11, 1931, pp. 1458-1483.

13.   A. Dattner, *Phys. Rev. Lett.*, vol. 20, no. 6, 1963, pp. 205-206.

14.   J. Berlande, P. D. Golden, and L. Goldstein, *Appl. Phys. Lett.*, vol. 5, no. 3, 1964, pp. 51-52.

15.   S. Subertova, J. Kracik, and J. B. Slavik, *Acustica,* vol. 15, no. 2, 1965, pp. 104-109.

16.   E. Heyess, *Beitrage Plasmaphys.*, vol. 6, no. 5, 1966, pp. 377-387 (in German).

17.   Yu. G. Kozlov and A. M. Shukhtin, *Zh. Tekhn Fiz.*, vol. 38, no. 9, 1968, pp. 1465-1471 (in Russian).

18.   V. L. Exner and L. Pekarek, *Appl. Phys. Lett.*, vol. 14, no. 1, 1969, pp. 37-38.

19.   L. V. Babin, A. Ya. Balagurov, and M. A. Plyshevskii, *Teplofiz. vys. temp.*, vol. 7, no. 3, 1963 (in Russian).

20.   I. Ya. Fugol, G. P. Reznikov, *Sherchenko Zh. Exp. u Teor. Fiz.*, vol. 56, no. 5, 1969, pp. 1533-1545 (in Russian).

21.   I. V. Demenik, et al., *Teplofiz. vys. Temp.*, vol. 8, no. 2, 1970, pp. 443-444 (in Russian).

22.   M. Fitaire and T. D. Mantei, *Phys. Fluids*, vol. 15, no. 3, 1972, pp. 464-469.

23.   U. Ingard and K. W. Gentle, *Phys. Fluids*, vol. 8, no. 7, 1965, pp. 1396-1397.

24.   K. Yatsui, T. Kobayashi, and Y. Inuishi, *J. Phys. Soc. Jpn.*, vol. 24, no. 5, 1968, pp. 1186-1187 (in Japanese).

25.   Y. Ishida and T. Idehara, *J. Phys. Soc. Jpn.,* vol. 35, no. 6, 1973, pp. 1747-1752 (in Japanese).

26.   M. Hasegawa, *J. Phys. Soc. Jpn.*, vol. 37, no. 1, 1974, pp. 193-199 (in Japanese).

27.   A. P. Saxena and C. P. Shrivastava, *Indian J. of Phys.*, vol. 48, no. 2, 1974, pp. 311-317.

28.   Y. Ishida, T. Idehara, and H. Inada, *Jpn. J. of Appl. Phys.*, vol. 14, no. 10, 1975, pp. 1571-1574 (in Japanese).

29.   M. K. Asanaliev, V. S. Engelsht, E. P. Pakhomov, M. A. Samsonov, I. M. Yartsev, and Zh. Zh. Zheenbaev, *Proc. XV Int. Conf. on Phenomena in Ionized Gases*, 1981, Minsk, July 14-18 (in Russian).

30.   I. Alexeff and R. V. Neidigh, *Phys. Rev.*, vol. 129, no. 2, 1963, pp. 516-527.

31.   S. L. Kahalas and L. W. Parker, *Phys. Fluids*, vol. 8, no. 9, 1965, pp. 1731-1737 (in Russian).

32.   L. D. Tcending, *Zh. Tekhn. Fiz.*, vol. 8, 1965, pp. 1731-1737.

33.   U. Ingard, *Phys. Rev.*, vol. 145, no. 1, 1966, pp. 41-46.

34.   U. Ingard and W. M. Manheimer, *Phys. Fluids*, vol. 9, no. 8, 1966, pp. 1608-1610.

35.   U. Ingard and M. Schulz, *Phys. Rev.*, vol. 158, no. 1, 1967, pp. 106-112.

36.   L. Aubrecht, *Phys. Lett.*, vol. 27A, no. 8, 1968, pp. 526-527.

37.   U. Ingard and M. Schulz, *Phys. Fluids*, vol. 11, no. 3, 1968, pp. 688-689.

38.   P. Kaw, Acoustic Instability in Weakly Ionized Gases, *Phys. Rev.*, vol. 188, no. 1, 1969, pp. 506-508.

39.   M. Schulz and U. Ingard, *Phys. Fluids*, vol. 12, no. 6, 1969, pp. 1237-1245.

40.   Tokhaya, *Zh. Tekhn. Fiz.*, vol. 41, no. 3, 1971, pp. 497-499 (in Russian).

41.   E. Ya. Kogar and Y. N. Malnev, *Ibid.*, vol. 47, no. 3, 1977, pp. 653-656 (in Russian).

42.   Yu. I. Chutov, *Prikl. Mathem. Tekhn. Fiz.*, no. 1, 1979, pp. 124-130 (in Russian.

43.   A. I. Klimov, et al., *Pisma v Zh. Tekhn. Fiz.*, vol. 8, no. 7, 1982, pp. 439-443 (in Russian).

44.   A. I. Klimov, et al., *Ibid.*, vol. 8, no. 9, 1982, pp. 551-554 (in Russian).

45.   I. V. Basargin and G. I. Moshin, *Shock Wave Propagation in Plasma of Longitudinal and Transverse Glow Discharge*, Преprint FTI, no. 880, 1984.

46.   V. G. Gorshkov, et al. *Zn. Tekhn. Fiz.*, vol. 54, no. 5, 1984, pp. 995-998 (in Russian).

47.   G. I. Mishin, *Pisma v Zh. Tekhn. Fiz.*, vol. 11, no. 5, 1984, pp. 274-278 (in Russian).

48.   I. V. Basargin and G. I. Mishin, *Ibid.*, vol. 11, no. 4, 1985, pp. 209-215 (in Russian).

49.   I. V. Basargin and G. I. Mishin, *Ibid.*, vol. 11, no. 21, 1985, pp. 1297-1303 (in Russian).

50.   V. A. Gorshkov, et al., *Zh. Tekhn. Fiz.*, vol. 57, no. 10, 1987, pp. 1893-1898 (in Russian).

51.   V. D. Shafranov, *Ibid.*, vol. 32, no. 6, 1957, pp. 1453-1459 (in Russian).

52.   J. Jukes, *J. Fluid Mech.*, vol. 3, no. 3, 1957, p. 275.

53.   Ya. B. Zeldovich and Yu. P. Rayzer, *Physics of Shock Waves and High Temperature Processes*, Moscow: Nauka, 1966, pp. 359-382 (in Russian).

54.   M. Y. Jaffrin, *Phys. Fluids.*, vol. 8, no. 4, 1965, pp. 606-625.

55.   B. Ahlborn and M. Salvat, *Z. Naturforsch.*, vol. 22A, no. 2, 1967, pp. 260-263.

56.   R. F. Avramenko, A. A. Rukhadze, and S. F. Teselkin, *Pisma v Zh. Exper. i Teor. Fiz.*, vol. 34, no. 9, 1981, pp. 485-488 (in Russian).

57.   G. I. Mishin, A. P. Bedin, and I. P. Yavor, *Pisma v Zh. Tekhn. Fiz.*, vol. 8, no. 3, 1982, pp. 182-185 (in Russian).

58.   R. F. Avramenko and G. A. Askar'yan, *Ibid.*, vol. 8, no. 20, 1982, pp. 1254-1256 (in Russian).

59.   F. G. Baksht and G. I. Mishin, *Zh. Tekhn. Fiz.*, vol. 53, no. 5, 1983, pp. 854-857 (in Russian).

60.   A. A. Rukhadze, V. P. Silakov, and A. V. Chebotarev, *Kratk. soobch. po fiz.*, no. 6, 1983, pp. 18-23 (in Russian).

61.   G. V. Vstovskii and G. I. Kozlov, *Zh. Tekhn. Fiz.*, vol. 56, no. 8, 1986, pp. 1536-1542 (in Russian).

62.   V. I. Dejiev, S. A. Maiorov, and S. I. Yakovlenko, *Fiz. plazmy*, vol. 13, no. 8, 1987, pp. 939-945 (in Russian).

63.   V. I. Dejiev, S. A. Maiorov, and S. I. Yakovlenko, *Ibid.*, vol. 13, no. 9, 1987, pp. 1050-1057 (in Russian).

64.   V. L. Bychkov and G. K. Gurev, *Chim. Fiz.*, vol. 7, no. 2, 1988, pp. 282-283.

65.   A. A. Filinkov, *Astron. Zh.*, vol. 65, no. 5, 1988, pp. 1094-1097 (in Russian).

66.   E. M. Barkhudarov, et al., *Pisma v Zh. Tekhn Fiz.*, vol. 10, no. 19, 1984, pp. 1178-1181 (in Russian).

67.   A. M. Galkin, N. N. Sysoev, and F. V. Shugaev, *Vestn. Mosk. Univ.*, ser. 3, vol. 26, no. 2, 1985, pp. 77-79. (in Russian).

68.   A. M. Galkin, N. N. Sysoev, and F. V. Shugaev, *Zh. Tekhn. Fiz.*, vol. 56, no. 3, 1986, pp. 596-598 (in Russian).

OPTICAL TOMOGRAPHY OF AERODYNAMIC OBJECTS. DENSITY RECONSTRUCTION

V. A. Komissaruk, N. P. Mende, L. N. Popov

Abstract: This paper examines optical tomography of aerodynamic objects–density reconstruction under the following headings; Introduction, Problem Formulation, Reconstruction of Selected Cross-Section Boundary and Flow Region Boundaries, Determination of True Path Difference, Current Cross-Section Division into Ring Regions and Zones, Interpolation of Path Difference, Estimating Density Expansion Parameters in Ring Zone, Computation of Outer Ring Zone Contribution to Path Difference Function, Results of Technique Application to Density Determination in Flow about Cone in Supersonic Flight, Conclusions.

1. Introduction

Tomography as a method for studying the spatial inner structure of an object by penetrative radiation probing has found wide applications in various fields of science and technology, following its remarkable success in medical diagnostics. First studies on the density field reconstruction for three-dimensional flows around objects were performed in the USSR in the early 1960s [1] simultaneously with research carried out by Cormack [2]. The approaches used by these authors were alike. However, the research begun in aerodynamics was not advanced. One of the reasons is the flow nonuniformity in wind tunnels used for the experiments and, as a consequence, low accuracy of the initial data. Gas density is a parameter of less practical interest than pressure or temperature. It can be used as a benchmark for numerical simulation of the gas flows and with this end in view high accuracy is naturally required. In this sense, a ballistic experiment provides the best opportunity, since ambient gas is quiescent and free of perturbations. However, in contrast with the wind tunnel where the orientation of the object can be varied, in a ballistic experiment we need a tomograph — an apparatus that provides several simultaneous projections of the object. Such a facility as made at the Physical Technical Institute of the Academy of Sciences of the USSR in late 1981 [3].

The optical tomograph for ballistic experiments consists of several (5 in our case) optical devices for gas nonuniformity visualization. The PTI system uses Schlieren devices (IAB-451) re-equipped as grating interferometers that allow us to obtain five interferograms of the flying object with nonsimultaneity of about 10^{-8} sec.

The initial processing of the interferograms, i.e., reconstruction of the wave surface shapes of the probing radiation, is carried out essentially as for axisymmetric objects [4]. Note that numeration methods for fringes in the vicinity of gasdynamic discontinuities were developed which should be described.

Several criteria were applied to the mutual congruence of coordinate systems in all photographs and to the accuracy of the scale determination and the view angles.

After the shapes of wave surfaces for all view angles are determined and arranged as the sets of discrete optical path differences and coordinates, further processing is the subject of computerized tomography.

Of the numerous methods for the reconstruction of the object structure from the set of its projections [5-10] the authors prefer to use an early approach [1, 2].

According to the reviews by Lewitt and Censor [6] methods based on integral transforms (ITM) are more suitable for medical problems. Methods based on the series expansions (SEM) seem to be more efficient for gasdynamic applications. This choice is dictated by the presence of inner singularities. When integral transform methods are used the reconstruction problem is solved in the continuous form up to the calculation stage when the discrete presentation is dictated by calculation algorithm. In the case of a small number of views the presence of boundaries such as gasdynamic discontinuities in medium density results in the boundaries smearing due to the difficulty (and sometimes the absurdity) of the unknown image distribution representations by continuous functions. Thus, the discretization of the function sought which is the most significant distinction between SEM and ITM, as was mentioned in [7] is prescribed by the structure of the physical object.

Another advantage of SEM, as we see it, is the wider possibility to take into account *a priori* information about the object. This is especially important in the case of limited-view tomography. As was already stated, the ballistic experiment as well as any high speed process requires several simultaneous object projections. When large-scale devices are employed it is technically impossible to obtain more than 7-10 projections. Hence it is imperative to use all the available information regarding the nature and geometry of the object — to take into account the discontinuities of the unknown function, the presence of symmetry planes, etc.

The approach by Tatarenchik [1] generalizes the Shardin zone method to the spatial problems and admits a decomposition of the two-dimensional cross-section problem by means of its reduction to the sequence of problems for several single-argument functions with narrow ring discretization. Use is made of the fact explicitly noted by Cormack that only the "outer" projection fragments are needed in order to reconstruct the outer layers of the object. Thus the size of the projection matrix is $M \times L$ where L is the number of views and M is the number of terms in the unknown function expansion in azimuthal coordinate (the number of basis functions).

The determination of the density distribution function in the cross-section reduces to the sequence of the regression problems for each ring zone

$$\bar{\bar{Q}} \cdot \rho = \mathbf{S} \cdot b + \mathbf{l},$$

where in our case ρ is the M-dimensional vector of the density expansion parameters, $\bar{\bar{Q}}$ is the projection matrix, $\mathbf{S}$ is the vector with the measured optical path differences as the components, $\mathbf{l}$ is the vector of geometrical path lengths in the ring for different projections, b is the scalar parameter determined by specific experimental conditions (here and throughout the paper a bold symbols denote vectors, two bars above the symbol a matrix).

In the present paper as well as in [1] the zone method (or alternatively, the method of layer splitting) is used, but the statistical criteria of quadratic optimization apply for the choice of adequate density distribution models and the estimation of an importance of the unknown parameters. Somewhat different density expansions and flow boundary representations adopted in our work as compared to [1] simplify the calculation of the projection matrix elements and improve iteration convergence when reconstructing the boundaries.

The algorithm developed copes with the inner gasdynamic discontinuities and includes radial and azimuthal smoothing of the path difference during ring-to-ring transition.

2. Problem Formulation

The density reconstruction problem for the interior of the three-dimensional object reduces to the two-dimensional problem by dividing the object into plane slabs. The directions of the probing light rays are chosen to be parallel to the current slab, i.e., to the cross-section of the object for which the density distribution is to be determined. Figure 1 displays the cross-section bounded by the contour $R(\gamma)$. The probing ray $L(y, \varphi)$ in the cross-section is specified by its distance, y, from the coordinate system pole and its orientation φ. The x-axis is normal to the figure plane and is parallel to the object velocity vector.

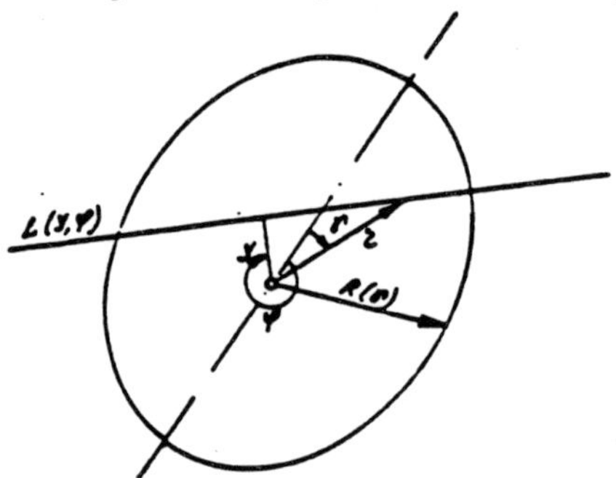

Fig. 1. Coordinates and notation.

The computerized diagnostics fundamental equation is $Az = u$ where A is a certain operator, z is the unknown function, and u is the measured function. In our case the equation relates the variation of the probing ray's optical path — path difference and the refraction index along the trajectory. The ray's path in the gaseous non-uniformities is usually assumed to be straight. The above equation reads

$$S(y, \varphi) = -\frac{1}{\lambda} \int_{L(y,\varphi)} \left[n(r, \gamma) - n_\infty \right] dL. \qquad (1)$$

here $S(y, \varphi)$ is the measured path difference; λ is the light wavelength; $n(r, \gamma)$ is the refraction index for the object interior which depends on both the radial r and azimuthal γ coordinates; n_∞ is the refraction index for the environment; the integral is performed along the light ray $L(y, \varphi)$. The Radon transform of the refraction index function $n(r, \gamma)$ is the path difference.

Equation (1) for the segment $l_j^{(k)}$ of the ray $L(y, \varphi)$ that has an index j in the kth projection and lies inside the contour $R(\gamma)$ can be written as

$$S(l_j^{(k)}) = S_j^{(k)} = \frac{1}{\lambda} \int_{l_j^{(k)}} (n(l) - n_\infty)dl. \tag{2}$$

By using the known gas relation

$$\frac{n-1}{\rho} = \text{const},$$

we rewrite Equation (2) as follows

$$\int_{l_j^{(k)}} \rho^*(l) \cdot dl = S_j^{(k)} \cdot b + l_j^{(k)}, \quad k = 1, 2, ..., L, \tag{3}$$

where $b = \dfrac{\rho_0 \cdot \lambda}{\rho_\infty(n_0 - 1)}$; ρ_0, ρ_∞ are the undisturbed base densities under normal and experimental conditions, respectively; n_0 is the refraction index for the light wavelength λ at normal conditions; ρ^* is the gas density scaled by ρ_∞; $l_j^{(k)}$ is the geometric path length of the ray in the nonuniformity.

The cross-section in question is divided into narrow rings in such a way that we can assume the density within each ring to depend on the azimuthal coordinate only. The reconstruction algorithm we shall use is the successive one: when the density distribution in the outer ring is determined its contribution can be subtracted from the path difference (the "removal" of the outer zone) and the density reconstruction can be carried out for the next ring. Thus at the ith stage we are interested in the path differences for the rays intersecting only the outer i^{th} zone (subscript j is reserved for other rays and path differences). Of all such rays only those tangent to the inner zone contour are considered because they contain the maximum information regarding the density distribution and have the maximum path length inside the ring.

The object we study (the body of revolution at angle of attack) has the plane of symmetry. Taking this fact into account we represent the relative density ρ^* in the outer ring at the ith stage of reconstructing procedure by truncated even Fourier series

$$\rho_i^* = \sum_{m=1}^{M_i} \rho_m^{(i)} \cdot \cos[(m-1) \cdot \gamma] \tag{4}$$

where γ is the polar angle. Then Equation (3) for the current i^{th} outer zone is written as follows

$$\int_{l_i^{(k)}} \sum_{m=1}^{M_i} \rho_m^{(i)} \cdot \cos[(m-1)\cdot\gamma]dl = S_i^{(k)} \cdot b + l_i^{(k)} \tag{5}$$

Integration and summation operations can be rearranged because the integrand is a finite sum of the integrable functions
Introducing new notation we write

$$\sum_{m=1}^{M_i} Q_{km}^{(i)} \cdot \rho_m^{(i)} = S_i^{(k)} \cdot b + l_i^{(k)} \tag{6}$$

where

$$Q_{km}^{(i)} = \int_{l_i^{(k)}} \cos[(m-1)\cdot\gamma] \cdot dl, \quad k = 1, 2, ..., L \tag{7}$$

By using matrix notation we arrive at the equation mentioned in the introduction

$$\bar{\bar{Q}}_i \cdot \rho_i = S_i \cdot b + l_i. \tag{8}$$

In order to calculate the elements of projection matrix $\bar{\bar{Q}}_i$ and components of the geometric path length vector l_i it is necessary to reconstruct the cross-section geometry of the object from its projections.

3. Reconstruction of Selected Cross-Section Boundary and Flow Region Boundaries

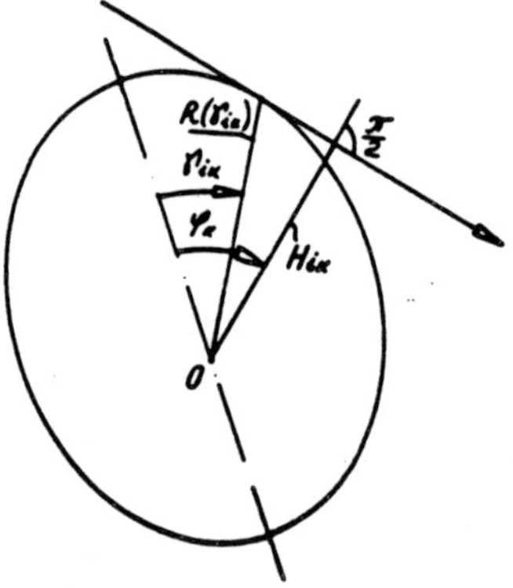

Fig. 2. On the derivation of Equation (10).

Initial data needed to reconstruct the region boundaries in the cross-section (Figure 2) are unknown contour projections H_{ik} and angles φ_k which specify projection planes orientation (two angle values φ_k and $\varphi_k + \pi$ correspond to each probing direction). Subscript k is the projection ordinal number, subscript i is the region boundary number (there may be several boundaries in supersonic discontinuous flows) starting from the outer boundary of the cross-section.

Following Ref. [1] and also assuming the contour to be convex we seek the contour equation for the ith discontinuity in the form

$$R_i(\gamma) = \sum_{m=1}^{M_i} R_{im} \cdot \cos[(m - 1)\gamma), \quad i = 1, 2, ..., J^* \tag{9}$$

where according to Ref. [1] the cosine powers are replaced by the cosines of multiple angles. Here J^* is the total number of the discontinuous boundaries. If the opaque object is present in the current cross-section, its contour is numbered as the last discontinuity. Otherwise a circle of small radius with the center in the cross-section pole is introduced for this purpose. The radius of this circle is chosen to be small as compared to calculation grid spacing in radial coordinate. It follows from the geometry shown in Figure 2 that

$$\sum_{m=1}^{M_i} R_{im} \cdot \cos[(m - 1)\gamma_{ik}] = \frac{H_{ik}}{\cos(\varphi_k - \gamma_{ik})}, \quad k = 1, 2, ..., N \tag{10}$$

where γ_{ik} is the polar angle of the point of the ray contact with the contour.

By using the expression for the angle between the tangent to the contour $R_i(\gamma)$ and the radius-vector of the contact point we obtain a relation

$$\gamma_{ik} = \varphi_k - \text{arctg} \frac{R'(\gamma_{ik})}{R(\gamma_{ik})}, \quad k = 1, 2, ..., N, \tag{11}$$

where trait denotes the derivative with respect to the polar angle. Equations (10) and (11) determine expansion coefficients (9).

It is clear that the choice $M_i = N$ in Equations (9), (10), (11) means an interpolation. A more correct approach is to consider the system (10), (11) as the least mean square method equations and starting with low frequency harmonics, to choose M_i by minimizing the mean square residual up to the standard level of H_{ik} measurement uncertainty. This approach reduces a danger of contaminating the contour equation with high frequency harmonics which do not correspond to its real shape.

The normal equations of the least square method derived from (10) read as follows:

$$\sum_{m=1}^{M_i} R_{im} \sum_{k=1}^{N} \cos[(m - 1)\gamma_{ik}] \cdot \cos[(l - 1) \cdot \gamma_{ik}] =$$

$$\sum_{k=1}^{N} \frac{H_{ik} \cdot \cos[(l - 1)\gamma_{ik}]}{\cos(\varphi_k - \gamma_{ik})}, \quad l = 1, 2, ..., M_i \tag{12}$$

Equations (11) do not change. Coefficients R_{im} for a fixed M_i are determined by means of iterations. First we assume $\gamma_k = \varphi_k$ and obtain R_{im} from Equations (10). These values of R_{im} are used to update the polar angles of contact points γ_{ik} in accordance with Equations (11). The convergence of the iteration process is checked by the difference of γ_{ik} values at two successive iterations.

Equations (10) and (11) contain 2N measured quantities H_{ik} and φ_k for the ith contour. Unknown values R_{im} and γ_{ik} are related by $M_i + N$ equations: M_i normal Equations (12) and N Relations (11). Thus the number of degrees of freedom when determining the unknown parameters is

$$\nu = 2N - M_i - N = N - M_i.$$

Hence the approximation variance can be estimated as

$$\widehat{\sigma}_i^2 = \frac{\sum_{k=1}^{N} \left\{ \left[\sum_{m=1}^{Mi} R_{im} \cdot \cos(m-1)\gamma_{ik} \right] \cdot \cos(\varphi_k - \gamma_{ik}) - H_{ik} \right\}^2}{N - M_i} \tag{13}$$

Comparing Estimate (13) to the hypothetical variance σ^2 of the H_{ik} measurements on the basis of χ^2-criterion

$$\chi^2 = (N - M_i)\widehat{\sigma}_i^2/\sigma^2 \tag{14}$$

($\widehat{\sigma}_i^2$ is the variance estimate according to (13), σ^2 is its hypothetical value) we can judge with given confidence probability on the adequacy of the determined approximation $R_i(\gamma)$ to the measured H_{ik} values (the latter are assumed to be normally distributed). If the observed criterion value (14) exceeds its critical table, we proceed to improve Approximation (9) by adding one additional harmonic. However, the importance of the coefficients already found in Equation (9) must be checked. This check (with a given confidence probability) is carried out with the use of Student's Random quantity distribution

$$t = \frac{\widehat{R}_{im} - \widetilde{R}_{im}}{\widehat{\sigma}_{R_{im}}}, \tag{15}$$

where $\widetilde{R}_{im}$ is the unknown true coefficient value and $\widehat{R}_{im}$ is its estimate. To estimate the mean-square deviation $\widehat{\sigma}_{R_{im}}$ of the coefficient R_{im} the linear regression analysis is applied in the following way. Equation (10) is rewritten in vector form

$$\overline{\overline{A}} \cdot \mathbf{R} = \mathbf{B}, \tag{10'}$$

where the notation is clear from comparison with Equation (10). Then Equation (12) becomes

$$\bar{\bar{A}}^T \cdot \bar{\bar{A}} \cdot \mathbf{R} = \bar{\bar{A}}^T \cdot \mathbf{B}, \tag{12'}$$

and its solution is

$$\hat{\mathbf{R}} = (\bar{\bar{A}}^T \cdot \bar{\bar{A}})^{-1} \cdot \bar{\bar{A}}^T \cdot \bar{\bar{B}}.$$

Variance estimate of the vector $\mathbf{R}$ i^{th} component R_{im} is expressed as

$$\hat{\sigma}^2_{R_{im}} = [(\bar{\bar{A}}^T \cdot \bar{\bar{A}})^{-1}]_{i,\,i} \cdot \hat{\sigma}^2_i,$$

where the term in the square brackets is the ith diagonal element of the variance matrix and σ^2_i is given by Equation (13).

Equation (15) allows estimation of the confidence interval of the parameter sought by using the number of degrees of freedom $N - M_i$ and the critical value t taken from Student's distribution table for a given confidence probability P:

$$|\hat{R}_{im} - \tilde{R}_{im}| < \sqrt{[(\bar{\bar{A}}^T \cdot \bar{\bar{A}})^{-1}]_{ii} \cdot \hat{\sigma}^2_i} \cdot t_{\text{crit}}.$$

If the confidence interval contains zero value, the correspondent coefficient must be regarded as unimportant. (An error is greater than 100%). It is possible, however, that there is still no adequacy between the model and experimental data. In that case we must improve the approximation by another method (to skip the next harmonic or to try another approximation). If these remedies do not help it is concluded that either the hypothetical measurement variance is found incorrectly or there are systematic errors. In the first case we must check the measurement variance using a real test object so that projections may be calculated before the measurements. The variance value thus obtained is to be accepted as its estimate. It is characterized by definite number of degrees of freedom, so the adequacy control according to Criterion (14) is no longer valid. It should be replaced by the variances comparison with the use of a Fisher distribution whose observed critical value is expressed as

$$F_{\text{observ}} = \frac{\hat{\sigma}^2_i}{\hat{\sigma}^2_{\text{test}}} \tag{16}$$

The adequate contour approximation procedure just described, along with the importance of the control of parameters will also be applied to choose an adequate mathematical model for an azimuthal density distribution in each zone. We stress once more the necessity of a simultaneous application of the adequacy and importance criteria since the adequate models (several such models can be formulated for the same initial data) may contain high-frequency harmonics with unimportant coefficients. If unknown density and measured path difference are related by the integral operator, the density field reconstruction results in significant error growth when high-frequency harmonics are present.

4. Determination of True Path Difference

It is well known that quantitative processing of the interferograms of supersonic flows around aerodynamic objects encounters difficulty in interference fringes numbering through a boundary between the disturbed and undisturbed domains and through the inner refraction index discontinuity images. In the vicinity of gasdynamic discontinuities such as shock waves and contact discontinuities, fringes either disappear in the optical device due to strong beam deflection or concentrate excessively and cannot be resolved in photography.

In such cases the interferogram measurements allow us to determine the shape of the wave surface between the discontinuity boundaries with an accuracy up to a constant that ought to be equal to zero, but really is an arbitrary integer. As a result, fictitious path difference discontinuities arise which can be removed by means of calculations [11]. We use a modification of this method [12] that has been generalized to the three-dimensional case. An approach described in [4] is used to treat the inner discontinuities.

The essence of the method is an extrapolation of the conditional path difference dependence on the geometrical path length of the probing beams to the vicinity of zero path length value, i.e., to the vicinity of the discontinuity. The use of the cross-section chord (i.e., geometrical path length) as an independent variable removes the singular path difference behavior (see [12]) and makes the above-mentioned extrapolation feasible. The determined conditional path difference value at zero chord length is rounded off to an integer and is accepted as the difference between the conditional and true path differences.

True path difference at the outer discontinuity being determined, the refraction index in the region bounded by the outer discontinuity and the next inner one is calculated following an approach in Ref. [4]. The contribution of this region is subtracted from the initial path difference. Then the procedure of the true path difference determination is applied to the next discontinuity. Thus, the removal of the conditional path difference discontinuities in the vicinities of the inner density discontinuities reduces to the solution of this problem for the outer discontinuity by means of successive subtracting from the path difference function of the contributions due to the nonuniformity layers between two adjacent discontinuities one of which is considered the outer.

In practice, the analysis becomes complicated by the fact that, if the contour is approximated by equation (9) with $M_i < N$, the contour projections on the rays φ_k will differ from the measured values H_{ik}. These discrepancies are small, but their effect on chord length is significant since, as was already noted, the length increases rapidly in the contour vicinity with the distance from the contour. This fact can invalidate the path difference function extrapolation.

These difficulties can be overcome with the help of the method which proved useful in practice. Its main point is the deformation (stretching or shrinking) of the H_{ik} interval in such a way that its new length would be exactly equal to the contour projection given by Equation (9) with the determined coefficients. The interference fringe coordinates are changed accordingly. If deformation is small (the length variation is comparable with the measurement error, i.e., is not greater than 0.1-0.2 mm in our experiments while the length H_{ik} amounts to tens of millimeters), there is no effect on refraction index calculation. The extrapolation remains valid since the slight variation of the slope of the path difference dependence on the chord length does not result in noticeable error in the conditional path difference jump. It is possible to diminish the deformation effect on the refraction index with the use of nonlinear stretching that decreases rapidly with distance from the jump.

Since the extrapolation is a correct technique it is applied repeatedly with various number of nodes and power of the approximating polynomial. The mean value of all these results is accepted as final.

The complete algorithm for the true path difference determination is presented in Figure 3.

5. Current Cross-Section Division into Ring Regions and Zones

If there are inner refraction index discontinuities in the current cross-section it is natural to divide the cross-section into regions bounded by the discontinuity contours. The necessity of this operation is predetermined by the accepted reconstruction method: the uniform approximation of the density function in the zone crossing by a discontinuity is senseless; the discontinuities must be fitted rather than smeared when the stepwise approximation function in the radial direction is used.

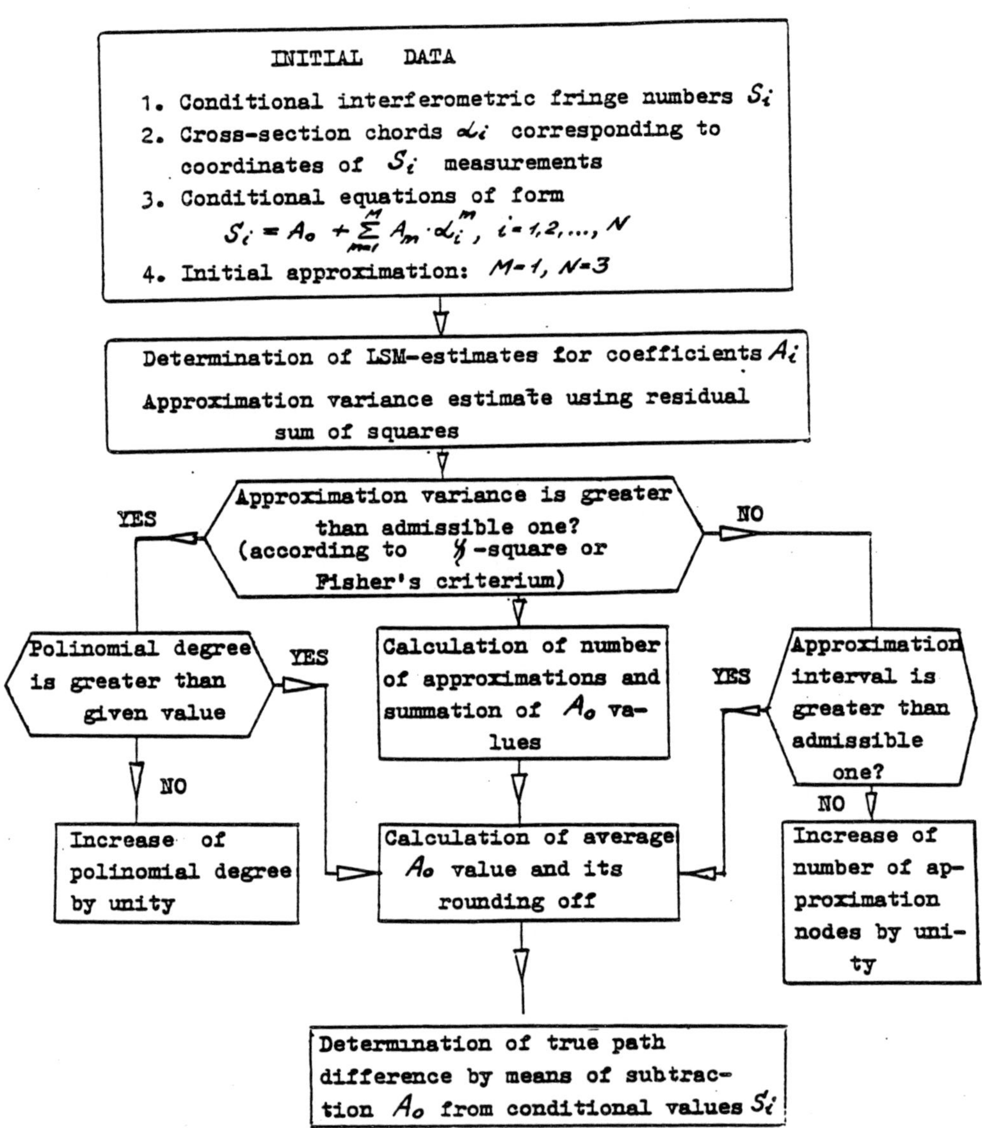

Fig. 3 Algorithm for true path difference determination.

In addition to the discontinuities the regions of specific density variation are to be distinguished. Experience of the two argument density function expansion in two sets of basis functions [13, 14] suggests that the cross-section division into the regions of smooth density variation is also necessary in the cases of steep density gradients since the number of terms in the expansion is limited by a small number of views.

In order to accomplish the division we introduce the dimensionless radial coordinate

$$\xi = \frac{J^* - i}{J^* - 1} - \frac{1}{J^* - 1} \cdot \frac{R_i(\gamma) - r(\gamma)}{R_i(\gamma) - R_{i+1}(\gamma)}, \; i = 1, 2, ..., J^* - 1, \qquad (17)$$

where $r(\gamma)$ is the current radius, i is the current discontinuity number, and J^* is the total number of the discontinuity contours including the opaque object contour.

In accordance with the assumed radial density distribution approximation by stepwise constant function we introduce the discrete grid by dividing the domains between the singularity contours into ring zones. For J fixed values of ξ corresponding to the zone boundaries the zone contour equations read as

$$r_j(\gamma) = \left(\xi_j - \frac{J^* - i}{J^* - 1} \right) \cdot (J^* - 1) \cdot \left[R_i(\gamma) - R_{i+1}(\gamma) \right] + R_i(\gamma) \qquad (18)$$

$$i = 1, 2, ..., J^* - 1, \quad j = 1, 2, ..., J$$

6. Interpolation of Path Difference

In the refraction index calculation path difference values corresponding zone boundary projections are used. Naturally these values do not coincide with interference fringe coordinates obtained by photography processing. Hence an interpolation becomes necessary. Additional difficulties are associated with the already mentioned singularity of the path difference near the discontinuous cross-section boundary which can result in large interpolation errors. Here we again employ the chord length at the correspondent points instead of the coordinate in the projection plane. The efficiency of such substitution can be analytically proved for an axisymmetric case [12]. The only distinction of other cases is the following: the dependence of path difference versus chord length no longer needs to be an odd function, nevertheless the derivative singularity is removed. We present the following example for illustration. Providing the refraction index is constant throughout the cross-section, the path difference function versus chord length is linear with the slope depending on the refraction index value for arbitrary cross-section shape.

Thus, for the path difference interpolation we use the following simple expression

$$S = a_1\alpha + a_2 \cdot \alpha^2, \tag{19}$$

which provides a good description of the function behavior near the discontinuity, i.e., in the inner zone adjacent to the discontinuity; S is the path difference, α is the cross-section chord length. Coefficients a_1, a_2 are found by the least square method using 3-5 nodes. The value calculated by Equation (19) is accepted as the path difference for the inner boundary of the current zone. When the next zone is considered we repeat the above procedure after subtracting the current exterior zone contribution from the path difference. Efficiency of the method is provided by the multistage interpolation using a relatively small part of the function $S(\alpha)$ which allows us to employ the simple approximation expression.

7. Estimating Density Expansion Parameters in Ring Zone

Consider again Equation (8). We have not yet calculated projection matrix elements (7). We transform this equation in accordance with

$$dl = y_i^{(k)} \frac{d\gamma}{\cos^2(\varphi_k - \gamma)} \tag{20}$$

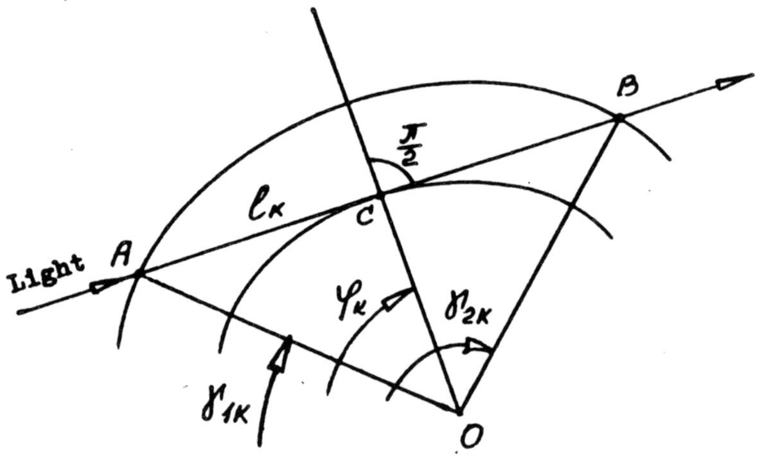

Fig. 4. Fragment of a ring zone.

where $y_i^{(k)}$ is a distance from current chord to the pole at the ith stage for the kth projection to obtain

$$Q_{km}^{(i)} = \int_{\gamma_{1,k}}^{\gamma_{2,k}} \cos[(m-1) \cdot \gamma] \frac{y_i^{(k)} d\gamma}{\cos^2(\varphi_k - \gamma)}, \quad k = 1, 2, ..., L \qquad (21)$$

The integration limits $\gamma_{1,k}$, $\gamma_{2,k}$ correspond to the angle coordinates of the chord and outer contour intersection points (see Figure 4). In this figure $l_i^{(k)} = AB$, $y_i^{(k)} = DC$.

The quantities $Q_{km}^{(i)}$ can be now reduced by means of the replacement to the form suitable for the integral computation coding. We have $\gamma = \varphi_k = \psi$

$$Q_{km}^{(i)} = y_i^{(k)} \left[\cos(m-1)\varphi_k \cdot \int_{\psi_{1,k}}^{\psi_{2,k}} \cos(m-1)\psi \cdot \frac{d\psi}{\cos^2\psi} \right.$$

$$\left. - \sin(m-1)\varphi_k \int_{\psi_{1,k}}^{\psi_{2,k}} \sin(m-1)\psi \frac{d\psi}{\cos^2\psi} \right], \qquad (22)$$

where $\psi_{1k,2k} = \gamma_{1k,2k} - \varphi_k$.

Integrals in the right-hand side of Equation (22) are computed with the help of recurrent relations

$$A_n = \int_{\psi_1}^{\psi_2} \frac{\cos n\psi}{\cos^2\psi} d\psi = \frac{4\sin(n-2)\chi}{n-2} \Big|_{\psi_1}^{\psi_2} - 2A_{n-2} - A_{n-4},$$

$$B_n = \int_{\psi_1}^{\psi_2} \frac{\sin n\psi}{\cos^2\psi} d\psi = -\frac{4\cos(n-2)\chi}{n-2} \Big|_{\psi_1}^{\psi_2} - 2B_{n-2} - B_{n-4}, \qquad (23)$$

where n is an integer.

For initial values $n = \overline{0, 4}$ we obtain immediately

$$\int \frac{d\psi}{\cos^2\psi} = \operatorname{tg}\psi, \quad \int \frac{\cos\psi}{\cos^2\psi} d\psi = \operatorname{lntg}\left(\frac{\psi}{2} + \frac{\pi}{4}\right),$$

$$\int \frac{\sin\psi}{\cos^2\psi} d\psi = \frac{1}{\cos\psi}, \quad \int \frac{\cos^2\psi}{\cos^2\psi} d\psi = 2\psi - \operatorname{tg}\psi,$$

$$\int \frac{\sin^2\psi}{\cos^2\psi} d\psi = -2\ln \cos\psi, \quad \int \frac{\cos^3\psi}{\cos^2\psi} d\psi = 4\sin\psi - 3\operatorname{lntg}\left(\frac{\psi}{2} + \frac{\pi}{2}\right),$$

$$\int \frac{\sin^3\psi}{\cos^2\psi}d\psi = -4\cos\psi + \frac{1}{\cos\psi}, \quad \int \frac{\cos^4\psi}{\cos^2\psi}d\psi = 4\cos2\psi - 4\frac{1}{\cos^2\psi},$$

$$\int \frac{\sin^4\psi}{\cos^2\psi}d\psi = 4\sin2\psi - 4\frac{\sin\psi}{\cos\psi}.$$

Vector l_i components in Equation (8) are calculated using the coordinates of the points of the beam intersection with the current zone outer boundary.

Thus the equation

$$\bar{\bar{Q}}_i \cdot \rho_i = S_i \cdot b + l_i \tag{24}$$

the components of vector ρ_i are to be estimated using the measured and computed right-hand side values and elements of matrix $\bar{\bar{Q}}_i$.

Generally, all terms in Equation (24) are subjected to random error. S_i is the result of direct measurements and interpolations. Elements of $\bar{\bar{Q}}_i$ and l_i are computed using the geometric measurements and hence also contain random errors. However, in order to simplify the problem we assume that $\bar{\bar{Q}}_i$ and l_i are fixed for each cross-section. The geometric errors will then cause the increase of the path difference variance, the latter being estimated with the use of the mean-square residuals of the density approximation. This technique is known as "noise reduction to system exit", in our case the density values, the "entry", and the path difference values are the "exit." The accepted assumption reduces our problem to a regression model [15]. The only question left to be discussed is the variance matrix of the right-hand side of Equation (24). Then the unknown vector ρ_i can be estimated by the standard methods.

Since the components of vector S_i are obtained in the course of direct independent measurements with equal accuracy (slight differences in the image scales in photographs of different projections may be neglected) variance and covariance matrices of vector S_i components are diagonal for the external zone of the object cross-section. Due to the variance uniformity the variance matrix is expressed as follows

$$\bar{\bar{D}}(S_1) = \sigma_0^2 \cdot \bar{\bar{I}} \tag{25}$$

where σ_0^2 is the variance of the path difference measurements, $\bar{\bar{I}}$ is the unit matrix. Other terms in the right-hand side are assumed to be fixed. According to the known property of a linear transform [15], the variance matrix of the right-hand side of Equation (24)

$$\overline{\overline{D}}(S_1 \cdot B + l_1) = b^2 \cdot \sigma_0^2 \cdot \overline{\overline{I}} \tag{26}$$

is also diagonal with the uniform variances of components. Thus the outer zone problem reduces to a classical regression model [15]. However, we shall not write here an estimate for the unknown vector ρ_i since a generalized regression model will be considered for all subsequent zones and the first (outer) zone solution will follow as a particular case. When we take the second zone the contribution due to outer zone represented by correlated random quantities obtained from the first zone density distribution must be subtracted from the independent measured path differences. As a consequence, covariances of components of the remaining path difference vectors will differ from zero after the removal of the precedent zone. It means that a generalized regression model must be employed. Following [15] by analogy with (25) we represent the variance matrices of the remainder path difference vectors as

$$\overline{\overline{D}}(S_i) = \widetilde{\sigma}_{i-1}^2 \cdot \overline{\overline{W}}_i \tag{27}$$

where parameter σ_{i-1}^2 is estimated by the deviation of the remainder path differences in the preceding zone from the values calculated using the chosen density distribution model (see below); $\overline{\overline{W}}_i$ is a positive matrix taking into account the change of variances and covariances of the remainder path difference components as the zones are removed. An analog of equation (26) for all zones starting from the second is written as

$$\overline{\overline{D}}(S_i \cdot b + l_i) = b^2 \widetilde{\sigma}_{i-1}^2 \cdot \overline{\overline{W}}_i \tag{28}$$

The generalized least mean square estimate (LMS-estimate) of the vector ρ_i for the model described by Equations (24) and (28) is given by expression (see Ref. [15])

$$\widehat{\rho}_i = (\overline{\overline{Q}}_i^T \cdot \overline{\overline{W}}_i^{-1} \cdot \overline{\overline{Q}}_i)^{-1} \cdot \overline{\overline{Q}}_i^T \cdot \overline{\overline{W}}_i^{-1}(S_i \cdot b + l_i), \tag{29}$$

and the variance matrix is written as

$$\overline{\overline{D}}(\widehat{\rho}_i) = b^2 \cdot \widetilde{\sigma}_{i-1}^2 \cdot (\overline{\overline{Q}}_i^T \cdot \overline{\overline{W}}_i^{-1} \cdot \overline{\overline{Q}}_i)^{-1}, \tag{30}$$

where T denotes transposition.

Estimate (29) minimizes the expression

$$\Phi(\widehat{\rho}_i) = (S_i \cdot b + l_i - \overline{\overline{Q}}_i \cdot \rho_i)^T \cdot \overline{\overline{W}}_i^{-1} (S_i \cdot b + l_i - \overline{\overline{Q}}_i \cdot \widehat{\rho}_i), \tag{31}$$

which allows us to obtain the estimate $\tilde{\sigma}_i^2$ for the next step $i + 1$

$$\tilde{\sigma}_i^2 = \frac{\Phi(\hat{\rho}_i)}{L - M_i} \tag{32}$$

where $L - M_i$ is the number of degrees of freedom, L is the number of projections, M_i is the number of vector ρ_i components.

As for the computation of the matrix $\overline{\overline{W}}_i$ elements at each step, the question is discussed in detail in [16]. We shall present here only the final result of rather cumbersome derivation, but in order to do this we need the notation which will be introduced in the next section.

The choice of an adequate mathematical model for the density distribution in each zone along with the estimate of vector ρ components consistency is carried out by the technique described in Section 2 for the cross-section boundaries reconstruction from the projections. Naturally, the path difference variance is subjected to a statistical estimate instead of a variance of geometrical quantities.

If the density distribution parameters (29) as well as their variances and covariances (30) are known, we can obtain the dependence of the density variance in the current zone versus polar angle γ. The vector form of Equation (4) for the density in the current zone is

$$\rho_i^*(\gamma) = \mathbf{F}^T(\gamma) \cdot \hat{\rho}_i,$$

where $\mathbf{F}(\gamma)$ is vector with the components $\cos[(m - 1)\gamma]$ for different $m = 1, 2, ..., M_i.$

Employing the well-known property of the linear transform of a random vector ρ_i we find

$$\tilde{\sigma}_{\hat{\rho}(\gamma)}^2 = \mathbf{F}^T(\gamma) \cdot \overline{\overline{D}}(\hat{\rho}_i) \cdot \mathbf{F}(\gamma) = b^2 \cdot \tilde{\sigma}_{i-1}^2 \cdot \mathbf{F}^T(\overline{\overline{Q}}_i^T \cdot \overline{\overline{W}}_i^{-1} \cdot \overline{\overline{Q}}_i)^{-1}$$

The scalar form of the last expression reads as

$$\tilde{\sigma}_{\hat{\rho}(\gamma)}^2 = b^2 \cdot \tilde{\sigma}_{i-1}^2 \sum_{l=1}^{M_i} \sum_{k=1}^{M_i} d_{lk} \cdot \cos[(l - 1)\gamma] \cdot \cos[(k - 1)\gamma],$$

where d_{lk} are the elements of matrix $(\overline{\overline{Q}}_i^T \cdot \overline{\overline{W}}_i^{-1} \cdot \overline{\overline{Q}}_i)^{-1}$.

The necessary estimates being obtained, the adequacy of the determined density model as well as the importance of the density expansion parameters must be examined with the help of Fisher's and Student's distributions in a way described in Section 2 for the cross-section contour equation.

8. Computation of Outer Ring Zone Contribution to Path Difference Function

To calculate the contribution of the removable i^{th} ring zone along the jth ray we solve Equation (24) for S_i:

$$S_i = \frac{1}{b}\bar{\bar{Q}}_i \cdot \rho_i - \frac{l_i}{b} \tag{33}$$

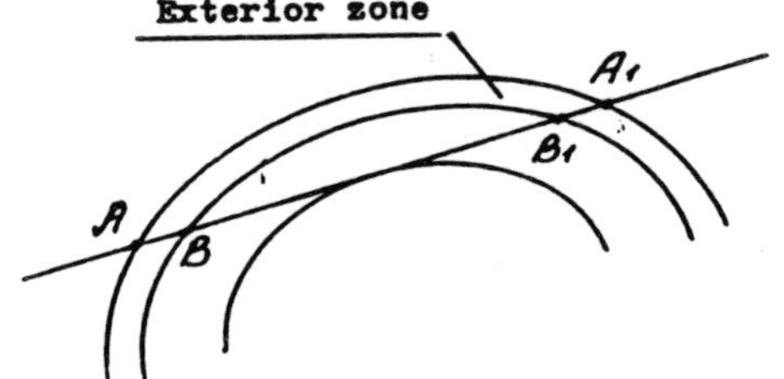

Fig. 5. On the calculation of the outer ring zone contribution to the path difference function.

Relation (33) gives the contribution to the path difference for the components of the vector l_i which are chords of the outer zone contour and are tangent to its inner contour. We seek the outer zone contribution corresponding to the ray segments AB and A_1B_1 between the outer zone contours depicted in Figure 5. To obtain these values it is sufficient to replace vector l_i in (33) by the sum of vectors uniting the above-mentioned segments for different views and to calculate integrals (7) along these segments to replace matrix $\bar{\bar{Q}}_i$ by their matrix sum. In obvious notation we obtain

$$\Delta S_i^{(j)} = \frac{1}{b}(\bar{\bar{Q}}_{1,i}^{(j)} + \bar{\bar{Q}}_{2,i}^{(j)}) \cdot \hat{\rho}_i - \frac{l_{1,i}^{(j)} - l_{2,i}^{(j)}}{b} \tag{34}$$

The superscript j alone is to be explained. Subscript i as above refers to the current zone to be removed, superscript j is used to number the rays of the remaining inner zones. Evidently $j \geqslant i$ when numbering starts from the outer boundary at $i = j$ Equation (34) reduces to Equation (3) and $\Delta S_i^{(i)} = S_i$.

Denoting by $S_i^{(j)}$ the path difference vector on the jth ray for i^{th} current outer zone, we write the remainder path difference after the removal of the ith zone as follows

$$S_{i+1}^{(j)} = S_i^{(j)} - \Delta S_i^{(j)}, \tag{35}$$

where $\Delta S_i^{(j)}$ is defined by Equation (34). Thus the remainder path difference after the removal of the ith zone is found.

Now we consider again the change of the variance matrices of the path difference vectors due to Operation (35) assuming the variance matrices of vectors $S_i^{(j)}$ to be known.

Applying the generalized covariance operator [17] to the vectors given by Equation (35) we obtain

$$\overline{\overline{D}}(S_{i+1}^{(j)}) = \overline{\overline{D}}(S_i^{(j)}) + \overline{\overline{D}}(\Delta S_i^{(j)}) - \overline{\overline{cov}}(S_i^{(j)}, \Delta S_i^{(j)}) - \overline{\overline{cov}}(\Delta S_i^{(j)}, S_i^{(j)}), \tag{36}$$

where $\overline{\overline{cov}}$ (**A**, **B**) denotes the matrix of the covariance operator applied to vectors **A** and **B** (the operator is noncommutative as regards to its vector arguments).

The first term in the right-hand side of Equation (36) is known. The second can be computed according to Equation (34) since the variance matrix of vector $\hat{\rho}_i$ is already calculated in (30) and the remaining terms in the right-hand side of Equation (34) are fixed:

$$\overline{\overline{D}}(\Delta S_i^{(j)}) = \frac{1}{b^2}(\overline{\overline{Q}}_{1,i}^{(j)} + \overline{\overline{Q}}_{2,i}^{(j)}) \bullet \overline{\overline{D}}(\hat{\rho}_i) \bullet (\overline{\overline{Q}}_{1,i}^{(j)} + \overline{\overline{Q}}_{2,i}^{(j)})^T. \tag{37}$$

However, we need to express $D(\Delta S_i^{(j)})$ as a function of the path difference vectors instead of the variance matrix of the density parameters vector. To this effect we substitute expression (29) into Equation (34) and, introducing new notation, obtain

$$\Delta S_i^{(j)} = \overline{\overline{H}}^{(j)} \bullet S_i^{(j)} + L_i^{(j)} \tag{38}$$

where

$$\overline{\overline{H}}_i^{(j)} = (\overline{\overline{Q}}_{1,i}^{(j)} + \overline{\overline{Q}}_{2,i}^{(j)}) \bullet (\overline{\overline{Q}}_i^T \bullet \overline{\overline{W}}_i^{-1} \bullet \overline{\overline{Q}}_i)^{-1} \bullet \overline{\overline{Q}}_i^T \bullet \overline{\overline{W}}_i^{-1} \tag{39}$$

and $L_i^{(j)}$ is a fixed vector defined by the cross-section geometry.

In our notation Equation (37) is rewritten as

$$\overline{\overline{D}}(\Delta S_i^{(j)}) = \overline{\overline{H}}_i^{(j)} \bullet \overline{\overline{D}}(S_i^{(i)} \bullet \overline{\overline{H}}_i^{(j)})^T. \tag{40}$$

The second term in the right-hand side of Equation (36) is determined.

The last two terms in (36) are also to be expressed using the vectors $S_i^{(i)}$ and $S_i^{(j)}$. Omitting the derivatives we write the final result

$$\bar{\bar{D}}(S_{i+1}^{(j)}) = \bar{\bar{D}}(S_i^{(j)}) + \bar{\bar{H}}_i^{(j)} \cdot \bar{\bar{D}}(S_i^{(j)}) \cdot \bar{\bar{H}}_i^{(j)T} - $$

$$\bar{\bar{H}}_i^{(j)} \cdot \overline{\overline{\text{cov}}}(S_i^{(j)}, S_i^{(j)}) - \overline{\overline{\text{cov}}}(S_i^{(j)}, S_i^{(j)}) \cdot \bar{\bar{H}}_i^{(j)T} \tag{41}$$

and at last we present without a derivation the formula for the calculation of the covariance operator matrices at the $i + 1$ stage using its values in the preceding i^{th} stage.

$$\overline{\overline{\text{cov}}}(S_{i+1}^{(2)}, S_{i+1}^{(p)}) = \overline{\overline{\text{cov}}}(S_i^{(2)}, S_i^{(p)}) + \bar{\bar{H}}_i^{(r)} \cdot \bar{\bar{D}}(S_i^{(i)} \cdot \bar{\bar{H}}_i^{(p)T} - $$

$$\bar{\bar{H}}_i^{(2)} \cdot \overline{\overline{\text{cov}}}(S_i^{(i)}, S_i^{(p)}) - \overline{\overline{\text{cov}}}(S_i^{(2)}, S_i^{(i)}) \cdot \bar{\bar{H}}_i^{(p)T}. \tag{42}$$

Note that Equation (42) reduces to (41) if $r = p = j$. When calculating, operations (41) and (42) are performed with the use of the elements of matrix $\bar{\bar{D}}$ instead of the variances and covariances of matrices $\bar{\bar{W}}$. These quantities differ by a scalar factor (see (27)) which is estimated for each step in accordance with formulas (31) and (32).

As the practical calculations presented in the next section reveal, there is a significant increase of the variances of the path difference vector components as the central cross-section zone is approached (in the presented example an increase is about 20% of the measurement variance for 18 zones). Such a small growth of the path difference variance is explained by the influence of the negative correlation terms in Equation (41) and also by the "superfluous" information (the number of projections exceeds the number of terms in the density expansion (4)). Each projection produces two values of the path difference per zone (for φ_k and $\varphi_k + \pi$), hence we have ten measurements with our five-view equipment while the number of density expansion terms never exceeds three.

9. Results of Technique Application to Density Determination in Flow about Cone in Supersonic Flight

The results of five interferometric measurements correspondent to five views were used as initial data. A typical set of interferograms is shown in Figure 9 at the end of the paper. The quality of the tomograph performance can also be judged by the illustrations in [20] where the interferometer equipment has been described in detail.

The cone was flying in air under atmospheric pressure, its velocity exceeds twice the speed of sound (M = 2). Reconstruction of the probing beam wave surfaces was carried out for two cross-sections shown in Figure 6 by dashed lines. The cross-sections were normal to the cone velocity vector.

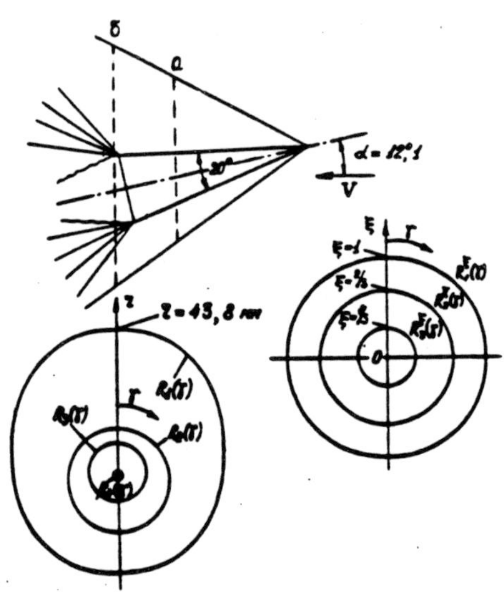

Fig. 6. The cross-section positions and division of the cross-section behind the cone base into the regions of smooth density variation $R_1(\gamma)$ - shock wave contour; $R_2(\gamma)$ - outer boundary of the rarefaction fan region; $R_3(\gamma)$ - boundary of the stagnation flow region; $R_4(\gamma)$ - fictitious contour of a small radius.

The results of the density reconstruction in these cross-sections without correction for the growth of the path difference variance as the zones are removed and also without statistical estimates are presented in Ref. [14] (the cross-section contained in the cone) and [18] (the cross-section behind the cone base). In the first the reconstruction results are compared to those numerically computed obtained by the others of Ref. [19]. The maximum discrepancy was at the leeward side of the cone, but it was less than 5% at angle of attack equal to 11.-2°. The density distributions in the cross-section behind the cone base at an angle of attack equal to 12.1° are presented in the second paper. The latter results are re-examined in the present paper: the growth of the path difference variance when approaching the cross-section center is taken into account, the density confidence intervals are calculated at the importance level 5%.

The mean square deviation of the calculated shock wave contour projections from the measured ones does not exceed 0.2 mm, the interferometer receiver resolution being about 10 lines per millimeter. The cross-section behind the cone base was divided into three regions: the conical flow region bounded by a bow shock wave and an outer boundary of a rarefaction fan centered at the cone base edge; the rarefaction region adjacent to the base edge and bounded from the inner side by the base flow region; and, finally, the base flow region itself. The error of region contours reconstruction does not exceed 0.2 mm. The outer region was divided into 10 zones, each of the remaining ones into 4 zones. Density distribution models contain no more than three terms in angular coordinate expansions: the coefficients of higher frequency harmonics turned out to be unimportant within the confidence probability $P = 0.95$. The inconsistency of the parameters limited the number of harmonics.

The adequacy examination for the density distribution model in the outer zone using χ^2-criteria yielded the positive result: for the hypothetical variance $\sigma_0^2 = 0.005$ the obtained approximation variance estimate is $\widetilde{\sigma}_i^2 = 0.006$ with the number of degrees of freedom $v = 7$. The zero-hypothesis H_0: $\widetilde{\sigma}_1^2 > \sigma_0^2$. The critical value $\chi^2_{\text{crit}} = 14.1$ for the number of degrees of freedom $v = 7$ and the confidence probability $P = 0.95$. Since

$$\chi^2_{\text{observ}} = (L - M_i) \cdot \widetilde{\sigma}_i^2 / \sigma_0^2,$$

the ratio of the variance estimate to its hypothetical value must be not greater than 2 for $v = L - M_i = 7$.

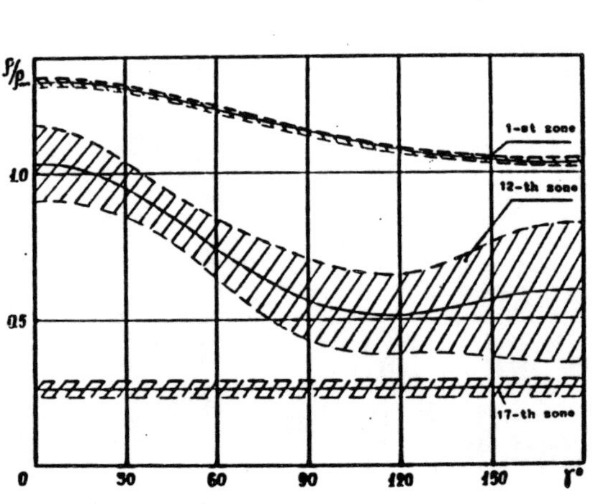

Fig. 7. Density distribution versus polar angle for the cross-section behind the cone base.

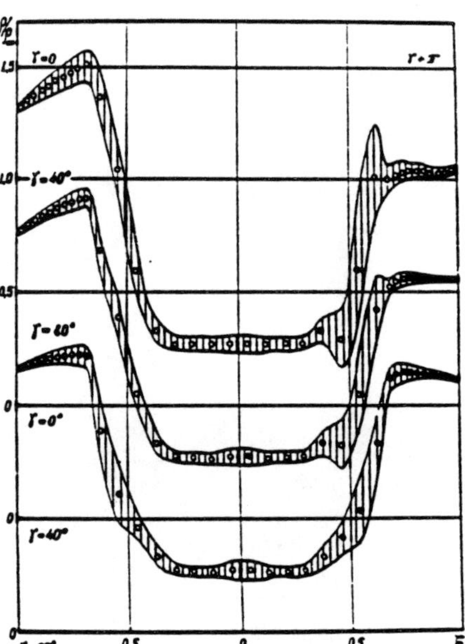

Fig. 8. Density distribution versus dimensionless radial coordinate.

Starting the second zone, we compared the estimate of the density approximation variance in the current zone with its value in the preceding one. In that case Fisher's distribution is to be used because both quantities are finite samplings and are characterized by known numbers of degree of freedom v_{i-1} and v_i. If $v_i = v_{i-1} = 7$ and confidence probability $P = 0.95$, Fischer's distribution critical value is $F_{\text{crit}} = 3.8$. The condition $F_{\text{observ}} < F_{\text{crit}}$ is satisfied by the variances in all adjacent zones in spite of gradual but significant growth of the parameter $\widetilde{\sigma}_i^2$ that becomes 22 times greater than σ_0^2

in the 12th zone (0.111 versus 0.005). This, naturally, influences the density confidence intervals in the expansion wave. The maximum half-width of that interval reached the value of 41% in the 12th zone against 1.25% in the first zone. However, the increase of $\widetilde{\sigma}^2_{12}$ alone cannot account for this error growth. Another reason is nearly degeneration of the leeward part of the expansion wave (as compared to the windward one), and as a consequence, a considerable decrease of the optical path in correspondent zones. Nevertheless, large density errors in this region did not result in a further error growth in the base flow region due to the small contribution of these zones to the path difference. The estimates $\widetilde{\sigma}^2_i$ began to decrease in the last zones ($i > 12$) and in the base region ($i \geqslant 15$) approached the hypothetical value of the measurement variance (at constant density within the zone). The substantial increase of the parameter $\widetilde{\sigma}^2_i$ in the intermediate zones ($i = 7$–15) is not a result of the error accumulation due to zone removal (this effect is

taken into account with the use of matrix $\overline{\overline{W}}$ and, as was already mentioned, does not exceed 20%), but seems to be associated with the coordinate systems disagreement in different projections, which tends to be more fatal in the regions of drastic path difference function variations ($i = 10$–15). At tempts to accomplish a correction by means of small displacements of region boundaries were not successful.

The density reconstruction results, along with the confidence inter vals are presented in Figures 7 and 8.

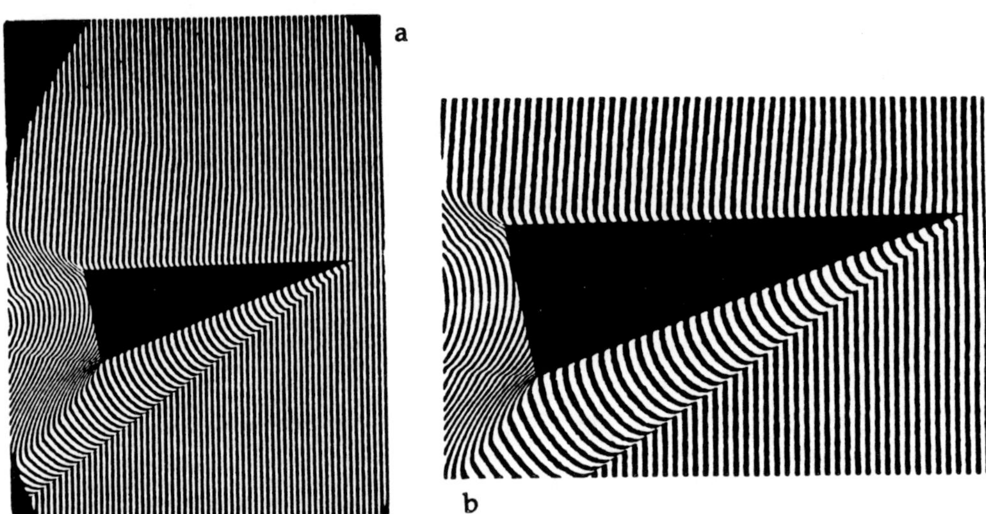

Fig. 9. The set of interferograms. Vertical plane projections.

a - complete pattern; b - fragment in the same scale to all projections. Diameter of the cone base - 20 mm;

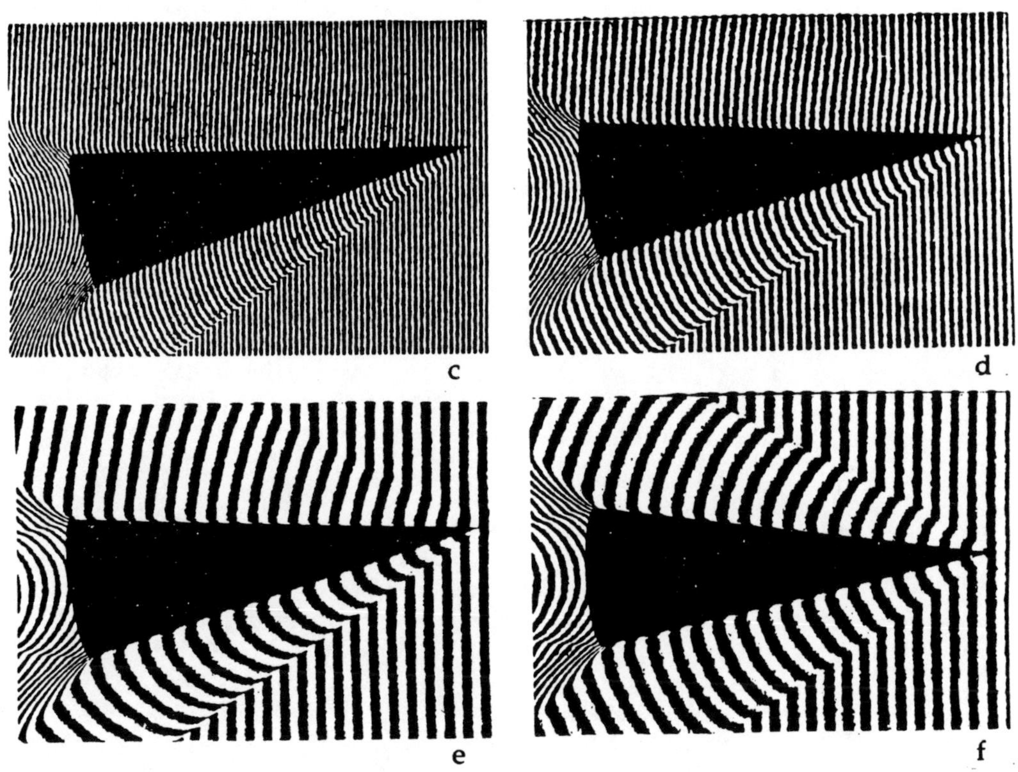

Fig. 9 (continued)c - 30° slope to horizon projection; d - 45°; e - 60°; f - horizontal plane projection.

10. Conclusions

The comparison of the density reconstruction results for the flow about the cone at an angle of attack from interferometer projections to the numerical computation results [14, 19] reveals that five projections may be sufficient to investigate the flows around bodies of revolution at an angle of attack. The applied method of the density representation by continuous functions in the ring zones requires cross-section division into the regions of smooth density variation. In addition, we assumed the boundaries of regions and ring zones to be convex. In practice, more complex cases may be encountered; for example, the group objects in motion, for which our technique will either need essential modification or be invalid. It is plausible that an approach based on polygon discretization [21] will be more flexible for complex discontinuous flows, although less accurate.

Essential increase of the relative density error in the case of zone contraction, as in the leeward rarefaction regions in the above example, indicates that strategy of the region division into ring zones must take into account the function variation in the radial coordinate.

References

1. V. S. Taterenchik, In: *Three-Dimensional Gas Flow Research by Optical Methods*, Ed. by S. M. Belotserkovsky, Proc. N. E. Zhukovskii Air Force Engineering Academy, no. 1059, 1964, pp. 21-34 (in Russian).

2. A. M. Cormack, *J. Appl. Phys.*, vol. 34, 1963, pp. 2722-2727.

3. P. I. Kovalev, V. A. Komissaruk, and N. P. Mende, *Optics and Laser Tech.*, vol. 15, no. 3, 1983, pp. 141-144.

4. V. A. Komissaruk and N. P. Mende, In: *Optical Methods in Ballistic Experiments*, Ed. by G. I. Mishin, Leningrad, 1979, pp. 178-194.

5. G. T. Herman, *Image Reconstruction from Projections: The Fundamentals of Computerized Tomography*, New York: Academic Press, 1980.

6. *The Special Issue on Computerized Tomography*, Proc. IEEE, vol. 71, no. 3, 1983.

7. Y. Censor, In: *Methods of Reconstructing Images. Ibid.* pp. 409–419.

8. T. S. Melnikova and V. V. Pikalov, *Radon Inversion in Emission Tomography of Nonstationary Plasma*, Preprint no. 80-82, Inst. Theor. and Appl. Mech., Novosibirsk, 1982, 51 pp. (in Russian).

9. T. S. Melnikova and V. V. Pikalov, *Tomography Study of Electric Arc*, Preprint no. 99-83, Inst. Theor. Appl. Mech., Novosibirsk, 1983, 47 pp. (in Russian).

10. A. I. Sedelnikov and E. I. Chernov, *Computerized Tomography Measurement of Gasdynamic Density Fields*, Preprint no. 20-88, Inst. Theor. Appl. Mech., Novosibirsk, 1988, 42 pp. (in Russian).
11. E. P. Kazandzhan and B. C. Sukhorukikh, In: *Optical MEthods in Ballistic Experiments*, Moscow: Nauka, 1979, pp. 158-170.
12. A. G. Belyaev and S. V. Milovidova, *Ibid.*, pp. 171-177.
13. A. G. Belyaev, S. v. Milovidova, G. B. Maksimov, et al., In: *Abstr. III All-Union Holographic Conf.*, Leningrad, 1978, pp. 161-162 (in Russian).
14. A. G. Belyaev, V. A. Komissaruk, and N. P. Mende, In: *Problems of Reconstructive Tomography*, Novosibirsk Acad. Sci. USSR, 1985, pp. 3-7 (in Russian).
15. S. M. Ermakov and A. A. Zhiglyavskii, *Mathematical Theory of Optimal Experiment*, Moscow: Nauka, 1987 (in Russian).
16. N. P. Mende, *Computational Tomography: On Accumulation of the Optical Path Difference Error in Zone Methods Application*, Preprint Phys. Tech., Inst., Leningrad, 1989.
17. G. Seber, *Linear Regression Analysis*, New York: Wiley, 1977.
18. V. A. Komissaruk, N. P. Mende, and L. N. Popov, In: *Reconstructive Tomography*, Kuibyshev: Kuibyshev Aviation Inst., 1987, pp. 34-39 (in Russian).
19. N. S. Bachmanova, V. I. Lapygin, and Yu. M. Lipnitskii, *Izv. AN SSSR, Mekh. Zh. i Gaza*, no. 6, 1973, pp. 79-84 (in Russian).
20. V. A. Komissaruk and N. P. Mende, In: *Optical Methods in Ballistic Experiments*, Ed. by G. I. Mishin, Leningrad: Nauka, 1979, pp. 91-113 (in Russian).
21. A. A. Vereninov, *Limited-View Computed Physical Tomography*, Preprint no 1237, Phys. Tech. Inst., Leningrad, 1988, 29 pp. (in Russian).

A STUDY OF BLUNT BODY SUPERSONIC MOTION IN A GAS CONTAINING SOLID OR LIQUID PARTICLES

I. A. Dukhovskii, P. I. Kovalev

Abstract: This paper examines the study of blunt body supersonic motion in a gas containing solid or liquid particles under the following headings; Some Aspects of Gas-Solid Particle Flows about Bodies in the Presence of Particle Disintegration, A Single Solid Particle Impact on the Flat Surface of Moving Object, Droplet Impact on Supersonic Sphere.

1. Some Aspects of Gas-Solid Particle Flows About Bodies in the Presence of Particle Disintegration

In the majority of theoretical investigations developing models of two-phase flows about bodies do not account for dispersed particle disintegration during particle-body surface impact. The efficiency and validity of such an approach are certain and are illustrated by, for example, Refs. [1–3]. However, at high numerical concentration of the particles $\varepsilon \sim 10^3$ $1/sm^3$ the phenomenon mentioned above leads to a significant change of particle dispersion and, therefore, to noticeable reorganization of the flow structure.

To understand the flow peculiarities appearing under these conditions, ballistic range experiments have been carried out. The motion of a tempered and polished steel sphere 10 mm in diameter in suspensions containing chrome, bronze, and electrocorundum particles has been investigated. The velocity of the sphere launched by a cannon was $\simeq 800$ m/s. Gas dusting and particle concentration measurement were conducted by a technique described in Ref. [4].

Some experimental results are presented in photographs (Figures 1–3, 7, 9, 10).

Fig. 1. The flow of the gas containing chrome particles about the sphere. M $\simeq 2.37$.

At low free stream particle concentration (less than 30%) a local variation of the gas phase parameters in the shock layer, displayed by kinking of interferometric fringes, is observed (Figure 1). Nevertheless, the interferometric measurements show that the gas density in the domain mentioned above is nearly the same as in one-phase flow.

The particle loading increasing the structure of the supersonic flow is changing. For instance, at numerical concentration $n \sim 10^3$ 1/sm^3 (electrocorundum powder) a dust train is formed which extends to the near wake (Figure 2). The body is shielded by the dust cloud and its location can be determined by a visible bow shock wavefront.

Fig. 2. The sphere motion outside the dust domain containing electrocorundum particles $n \simeq 10^{-3}$ $1/m^3$, M $\simeq$ 2.41.

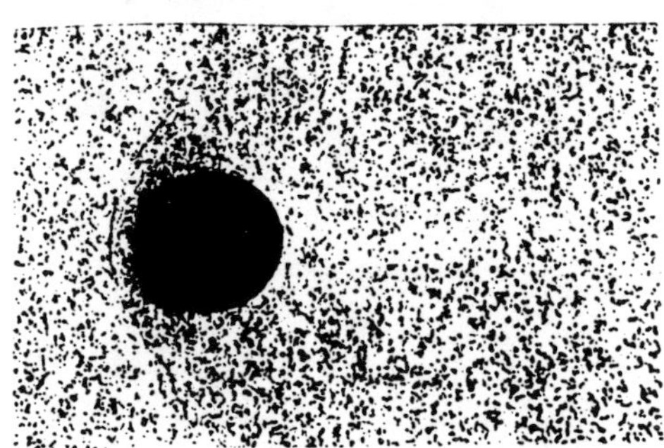

Fig. 3. The sphere motion outside the dust domain containing spherical bronze particles $n \simeq 10^{-3}$ $1/m^3$, M $\simeq$ 2.45.

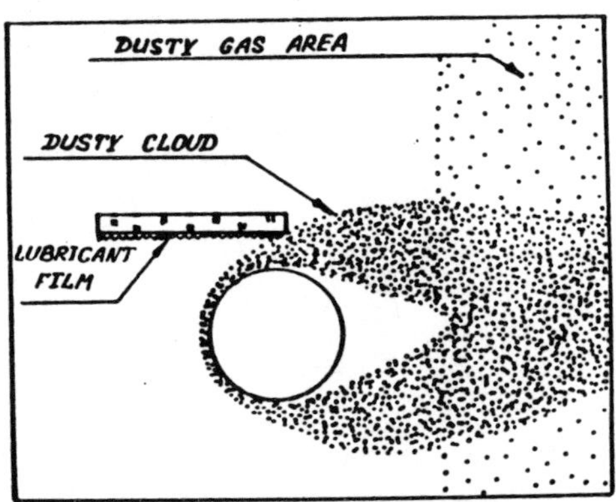

Fig. 4. The scheme of experiment for determination of dust cloud dispersion.

The dust train was observed in the case of an electrocorundum suspension and was absent in a spherical bronze particle suspension (Figure 3).

Concerning this effect it was supposed that the phenomenon was the result of particle disintegration during the impact on the body surface [5].

In further experiments a direct selection of the dispersed particles from the dust cloud was carried out. Figure 4 represents the experimental scheme. At a distance of 8.5 mm over the sphere trajectory a glass plate was located in a pure gas region not far from the dusty gas region. Therefore, the dispersed particles of the dust cloud carried out from the dusty gas region by the sphere, sedimented on the plate. Photographs corresponding to the moment of the sphere passing by the plate are presented in Figure 5a and b. Microscopic measurements of the obtained samples allow reconstruction of the size distribution function of the particles (Figure 6). A comparison of the measurement data with the initial characteristics of the powder (particle size of the main fraction was 100–125 μm) indicated the dramatic increase of the fine particle fraction.

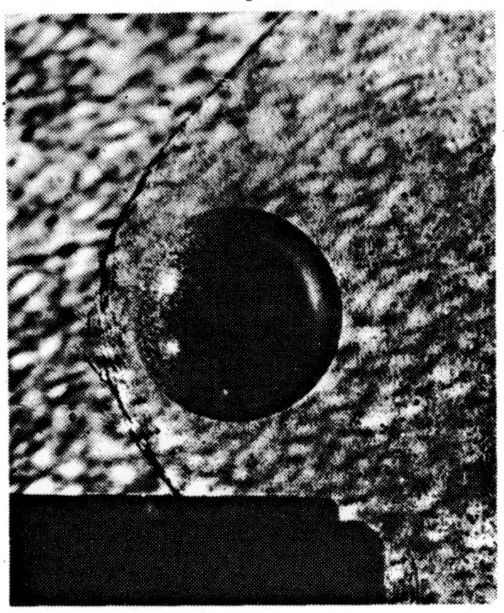
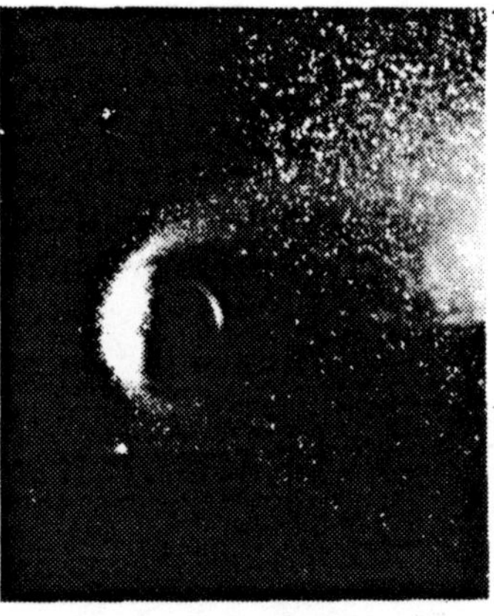

a   b

Fig. 5. a – Photograph taken at the moment of the sphere motion below the plate. The "laser knife" and shadow techniques are used simultaneously, electrocorundum, $n \simeq 10^{-3}$ 1/m^3, M $\simeq$ 2.49. b – Photograph taken simultaneously with photograph 5a, the "laser knife" technique only.

Such a result confirms the assumption about the disintegration of the particles and provides a crude estimate of the particle numerical concentration. After the impact it increases 10^3 times as compared with the initial value. This is an upper estimate of the concentration; in fact, the impact is accompanied by formation of both fine and sufficiently large particles, as

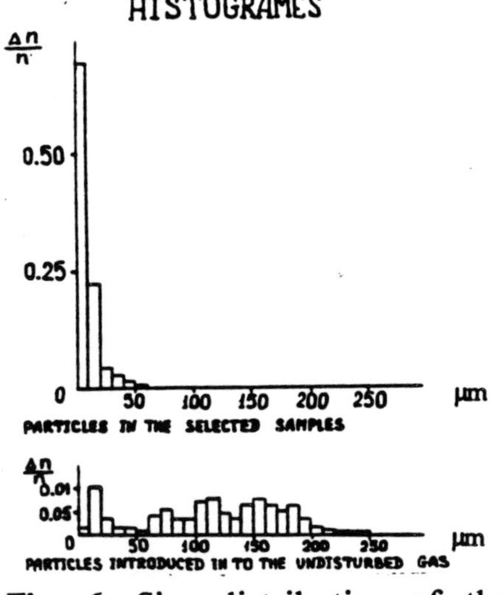

Fig. 6. Size distribution of the electrocorundum particles.

well as body material fragments flight out of the shock layer. A particle trajectory can be presented as follows: after collision and disintegration the large particle, whose Stokes number Sk ≥ 1, moves through the shock layer with a rebound velocity. At the moment of flight outside the shock layer the particle oncoming stream parameters jump and gas phase deceleration effect decreases. After a while the decelerated particle enters the shock layer and the free stream parameters jump again, but now deceleration decreases.

The shock wave crossing by the particles rebounding near the stagnation point leads to bow shock wave destruction and formation of the Mach cone shock wave structure. This effect has been noticed and explained in Refs. [2, 6]. In the series of experiments a number of rebounded particles having flown outside the shock layer far from the flow axis were observed. The flows about these particles were accompanied by Mach cone formation, however shock destruction, mentioned in Refs. [2, 6], was not observed.

Turning to the problem of dust cloud formation it can be seen that the formation of numerous fine particles within the shock layer causes a qualitative change in the two-phase flow structure.

The characteristic feature of the described experiments is the flow regime change due to particle disintegration. For the initial dispersion the Stokes number is approximately equal to 10^3, i.e., the "frozen" flow regime takes place. Estimates show that Stokes numbers approximately equal to 1 and 5 correspond to the disintegrated particles of 1 and 3 μm, respectively. Thus, the efficiency of interphase momentum transport increases sharply after disintegration. In contrast to the incident particles, reaching the sphere surface practically without the interphase momentum exchange, the rebounded, disintegrated fine particles are involved in gas flow.

Figure 7 shows the meridional cross-section of the dust cloud using the "laser-knife" technique. (In this case the above technique permits us, except in layer-by-layer visualization, to exclude optical effects of the dispersed phase density gradients.)

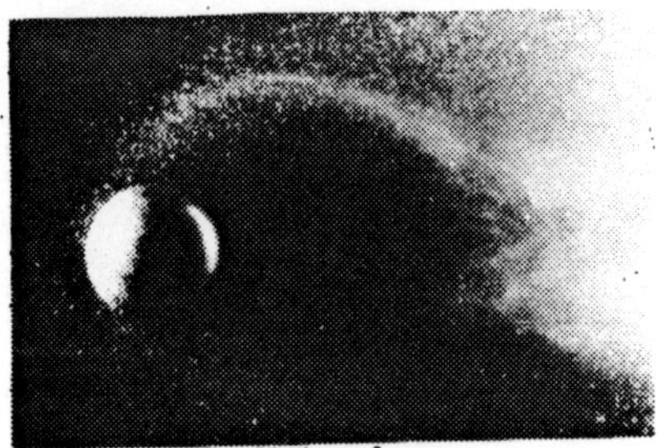

Fig. 7. The sphere flight outside the dust domain, the laser knife technical, electrocorundum $n \simeq 10^{-3}$ $1/m^3$, M $\simeq$ 2.31.

Two characteristic domains can be seen in the photograph, namely the fine particles, moving along the curveline trajectory, come to the nearest wake, reaching the flow axis, the larger particles occupy a domain near the shock wave.

The existence of a curveline surface of the fine particle location can be explained in the following manner.

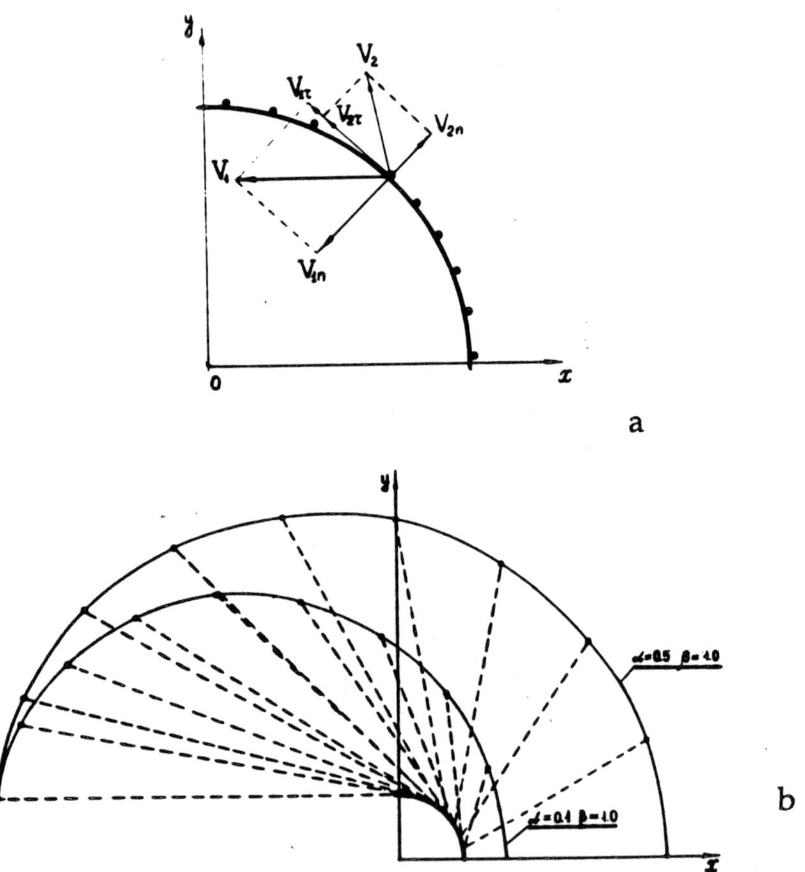

Fig. 8. a – The scheme of the particle-body surface impact. b – The trajectories of the particles after spherical dust layer-body surface impact.

Let the spherical layer of the particles incident the sphere surface at moment $t = 0$. Let us consider the meridional plane containing the sphere velocity vector (the frame of reference originates in the sphere center). Figure 8 represents the incident and the rebounded particle velocity vectors V_1 and V_2, respectively). Accommodation coefficients α and β for normal and tangential velocity components of the rebounded particles are assumed to be independent of the incident angle φ. Then the following relations are valid:

$$V_{2\tau} = -\beta V_1 \sin\varphi$$

$$V_{2n} = \alpha V_1 \cos\varphi$$

Projecting on the Cartesian coordinates (x, y) we can obtain:

$$V_x = \alpha V_1 \cos^2\varphi - \beta V_1 \sin^2\varphi$$

$$V_y = (\alpha + \beta)V_1 \sin\varphi\cos\varphi.$$

Assuming that the particles move without deceleration these relations can be integrated:

$$x = (\alpha\cos^2\varphi - \beta\sin^2\varphi)V_1 t + R\cos\varphi$$

$$y = (\alpha + \beta)\sin\varphi\cos\varphi V_1 t + R\sin\varphi$$

Figure 8 represents the curves connecting the rebounded particles at moment $t = 40$ μs for $\alpha = 0.5$, $\beta = 1.0$, and $\alpha = 0.1$, $\beta = 1.0$, respectively. Dashed lines show the particle trajectories.

Naturally, the curves presented in Figure 12 only illustrate the phenomenon observed experimentally, because the particle deceleration was not taken into account and the accommodation coefficient α was chosen on the basis of crude estimates obtained using the ballistic experimental data for suspension at the gas phase pressure equal to 5 mbar (coefficient β was not estimated).

The above discussion allows us to assume that the photograph (Figure 7) shows the result of the impact of the sphere and "the last" dust particle layer located on the dusted domain boundary. The particle decelerated by the gas phase lag behind the sphere.

To verify the suggested model for formation of a dispersed phase high concentration layer due to particle disintegration, experiments have been carried out at low pressure (~5 mbar). In this case particle deceleration can be ignored. The experiments not only justified the model, but provided new information on the incident and rebounded particle interaction as well. It was found that this interaction changed the shape of the dust cloud. Incident particles collide with the rebounded particles within the dust cloud and carry the latter to the body surface.

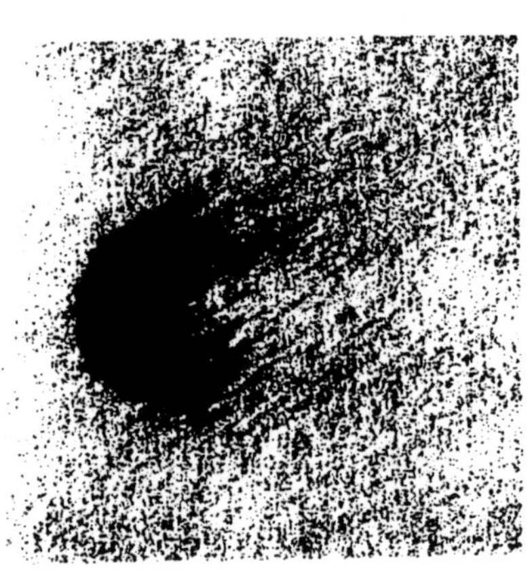

Fig. 9. The sphere motion within the dust domain, electrocorundum, $n \simeq 10^{-3}$ 1/m^3, the gas pressure is equal to 5 mbar, M $\simeq$ 2.5.

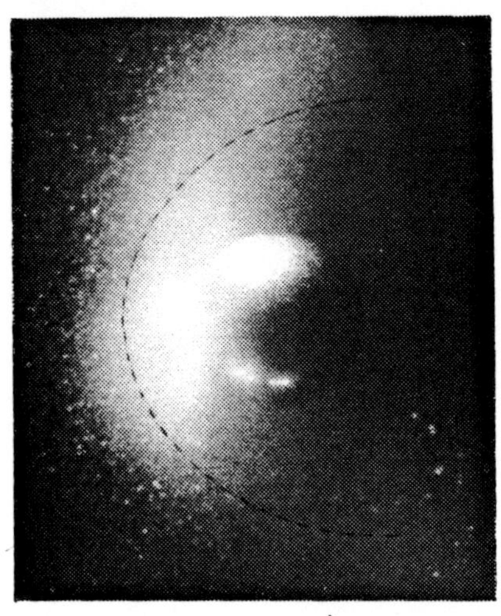

Fig. 10. The sphere motion outside the dust domain, conditions correspond to those of Figure 9.

Photographs (Figures 9 and 10) represent the "laser knife" and shadowgraph patterns of the flow. The photograph in Figure 10 was taken when the sphere had flown outside the dust domain. The left edge of the photograph is the boundary of the dust domain. A comparison of the dust cloud shape observed at this moment to that in the dust domain (dashed line in Figure 10) illustrates the above statement.

The experiments performed at low pressure prove that the main reason for the highly dispersed phase concentration in the dust cloud is not the incident two-phase flow stratification in the shock layer, but for the most part, particle disintegration during impact with the body surface.

Formation of the rebounded particles moving in "opposite directions" to the gas phase flow and possessing an extended interphase surface in contrast to the incident particles, produces a reverse effect on the flow. A grating of the rebounded particles causes gas deceleration in the shock layer, resulting in an increase of the layer thickness. In the series of experiments it was determined that its value increases by 20%. An opportunity for the shock wave distance variation occurring in the dusty gas flow about a plate follows from computations [7]. The monograph mentioned above treats the problem in which the shock layer thickness is decreased by the incident particles and increased by the rebounded particles. In our experiments the shock wave distance increases due to fine particle fraction flight on the one hand and the "frozen" incident particle motion on the other.

Thus, the joint effect of a number of phenomena such as particle disintegration, carrying of the fine particles by the gas flow, and the incident-rebounded particle collisions, gives rise to formation of the dispersed phase high concentration cloud. The result of reverse influence of the rebounded particles is the variation of the shock wave distance.

2. A Single Solid Particle Impact on the Flat Surface of a Moving Object

In order to make the model of dust cloud formation more accurate, experiments on a single-particle impact on the flat surface at various incident angles were carried out [3].

Estimates obtained by the authors have shown that the majority of the disintegrated particle fragments had a size of less than 10 μm. Therefore it is necessary to remove particle deceleration by the gas to measure the fragment velocities. In studies [9, 10] the solid particle-surface interaction was investigated. The particles were accelerated by means of a gas-jet. The maximum impact velocity was 100 m/s. In this study the ballistic range facility which permitted us to exclude the gas phase influence and to obtain an impact velocity approximately equal to 850 m/s was used. The study was made of a freely-falling particle colliding with the projectile launched by a laboratory accelerator. Projectiles were steel tempered cylinders with a slanted front base. Electrocorundum particles ~180 μm were used. The colliding process was investigated in the hermetically sealed range at low pressure (~0.5 mbar). Photo-registration of the flying particles was made by the "laser knife" technique.

Some experimental results are presented in Figures 11 and 12.

In the range of incident angles from 0 to 80° (an incident angle is the angle between the incident velocity vector and the lo-

Fig. 11. The fragments of particle flight after normal impact on the plane face of a projectile

cal normal to the surface of a projectile) disintegration of the particles took place.

At normal impact axisymmetrical flight of the fragments mainly in radial direction (Figure 11) is observed. It shows that solid particle disintegration as well as droplets [15] is related to the shock wave originated at the particle-projectile contact region coming out to the particle surface. The following unloading process is accompanied by the development of strain stresses exceeding the rupture limit. This leads to disintegration and flight of the fine particles. The tangential component of the fragment velocity to the impact velocity ratio is equal to 0.4.

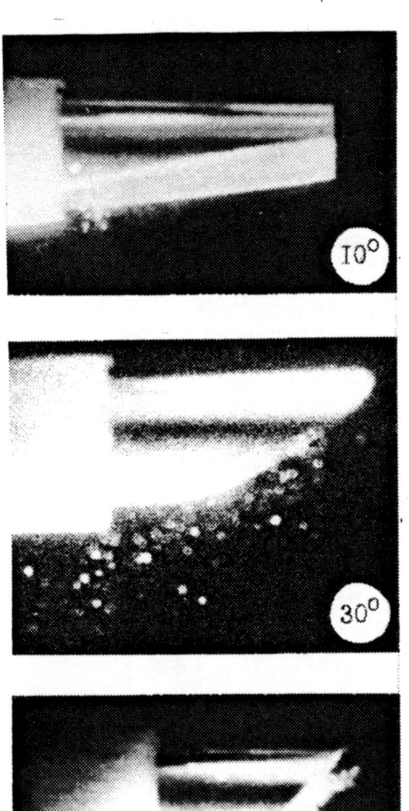

Fig. 12. The fragments of particle flight after oblique impact.

Increasing the incidence angle γ the cloud of fine dust extends along the tangential impact velocity vector V_τ. Then the incidence angle is equal to 40° (Figure 12) in the interaction domain there were fragments moving along and opposite the τ-axis (the reference frame is shown in Figure 13a). The asymmetry of the dust cloud is characterized by the ratio $V_{max+}/V_{max\tau-}$. Here $V_{max\tau+}$, $V_{max\tau-}$ are the maximum tangential velocities of the fragments along and opposite the τ-axis, respectively. At $\gamma = 40°$ $V_{max\tau+}/V_{max\tau-} \sim 2$.

Further increasing the incidence angle up to $\gamma = 60°$ (Figure 12) a steep increase in the above-mentioned ratio takes place, up to ~20. At $\gamma = 80°$ all the fragments move along the τ-axis (Figure 12).

The fine particle velocity estimates for γ belonging to the 20-80° range show that the value of $V_{max\tau+}$ in all cases does not exceed V_τ. This effect is explained by the scheme presented in Figure 13b. The fragment velocity at the dust cloud edge (Figure 13b) is the sum of vectors $k \cdot \bar{V}_\tau$ and $\bar{V}_{pi}$, here k_τ is the accommodation coefficient for the tangential velocity of impact, $\bar{V}_{pi}$ is the flight velocity correspondent to the normal velocity of impact $\bar{V}_n$.

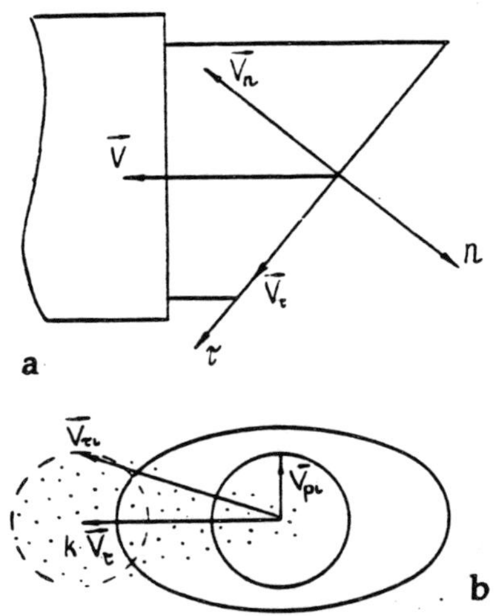

Fig. 13. The scheme of particle-plane surface impact.

a - (τ, n) reference frame; b - scheme of the fragment motion in the plane parallel to the surface of the projectile.

In addition, measurements indicate that the particles having velocities less than the sum of $k \cdot \bar{V}_\tau$ and $\bar{V}_{pi}$ are present in the dust cloud. This is related to secondary impact of the fragments on the projectiles. Evidence in favor of this suggestion comes from the secondary craters observed along the tangential impact velocity direction. Another fragment deceleration mechanism is possible. Within a particle penetrating the projectile material, shear stress develops because of the oblique impact. This stress causes particle disintegration but fragment tangential velocities in this case decrease.

It was seen from investigations that the maximum fragment velocity was determined by the sum of the radial component of flight velocity and tangential impact velocity. Experiments carried out in a wide range of incidence angles showed that the disintegrated dispersion could significantly contribute to gas phase flow field change in a dusty gas flow about a body.

3. Droplet Impact on a Supersonic Sphere

Proceeding to investigations concerning the liquid particle impact on the projectile surface it is necessary to pay attention to some common features of the phenomenon discussed above and those under study in spite of a number of essential differences. In particular, flow about body disintegrated particle flight did not cause internal shock waves to rise in the interaction zone even if rebounded particle velocities exceeded the local sound velocity. Apparently, the width of the shear was so small that the visualization method used in experiments failed to register these waves. We suggest that the interphase exchange mechanisms for finely dispersed particles originating after disintegration of solid or liquid particles have a number of common features.

The problem of droplet impact on the solid body moving in a gas can be conditionally divided into three stages. At each stage different factors play dominant roles, effecting the process as a whole as well as individual stages.

At the first stage a droplet enters the shock layer passing through the bow shock. At certain conditions the flow about the droplet could lead to its disintegration [12]. Proceeding from disintegration type classification and taking into account the flow conditions (standard conditions: sphere diameter - 30 mm, flight velocity - 650 m/s, diameter of water droplet - 3 mm) it was established that the droplet disintegration was of the strip type [13].

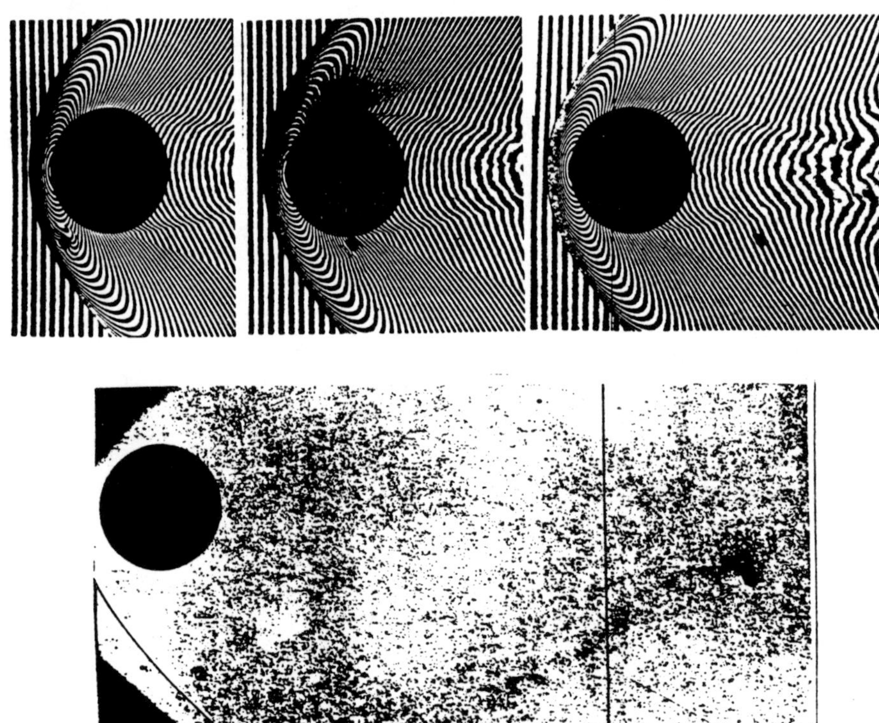

Fig. 14. Droplet disintegration in the flow behind a flying sphere. The droplet diameter is 3 mm. Standard conditions, the sphere velocity is 650 m/s.

Study [13] showed that the droplet disintegration time in strip regime could be defined by:

$$t_c = \frac{d}{v}\left(\frac{\rho_l}{\rho}\right)^{1/2}$$

here v is the droplet-gas relative velocity, d is the droplet diameter, ρ_l is the liquid density, ρ is the gas density. This time is equal to $123 \cdot 10^{-5}$ s, which is much greater than the time of droplet motion along stagnation streamline before impact. The estimate presented above is confirmed by experimental data shown in photographs (Figure 14).

Thus, the droplet remains spherical and behaves as a solid body in the gas flow up to impact on the projectile surface (Figure 15). The diffraction of the bow shock on the droplet surface results in shock wave formation near the droplet. This is similar qualitatively to results presented in [14]. Within the shock layer the flow about the droplet causes local decrease of the bow shock intensity. The photograph (Figure 16) shows that in a droplet wake zone the intensity of the bow shock decreases.

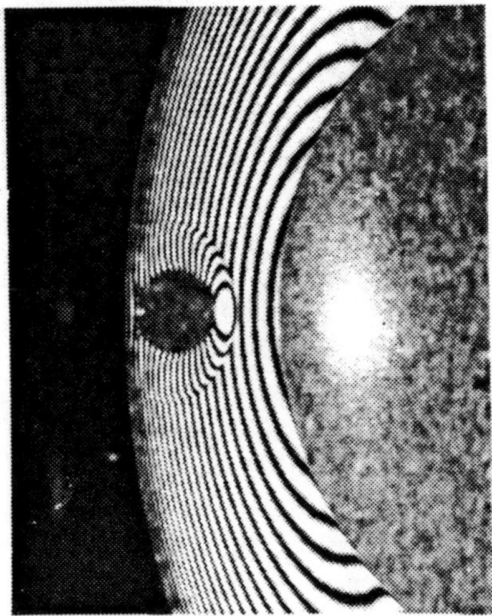

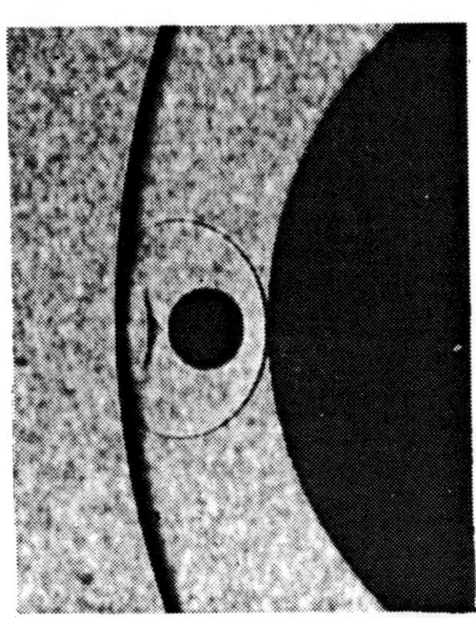

Fig. 15. Droplet entering the shock layer, 5 μs later the bow shock crossing.

Fig. 16. Droplet entering the shock layer, 7 μs later the bow shock crossing.

The second stage begins at the moment of impact. The impact is accompanied by compression and unloading processes in the droplet [13, 15].

First the shock wave attached to the perimeter of the contact zone is formed. The rate of contact zone expansion is greater than the internal shock wave velocity and there are no perturbations within the droplet outside the internal shock. At a certain moment (for the plane surface $t = 0.005$

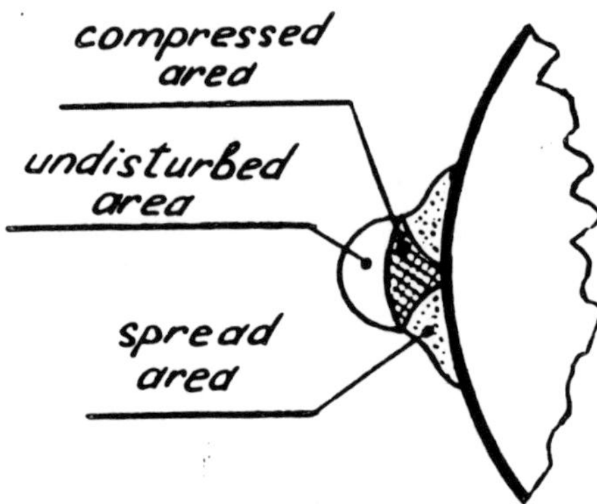

Fig. 17. Droplet effusion on the sphere surface, frontal photograph taken 2.8 µs after impact.

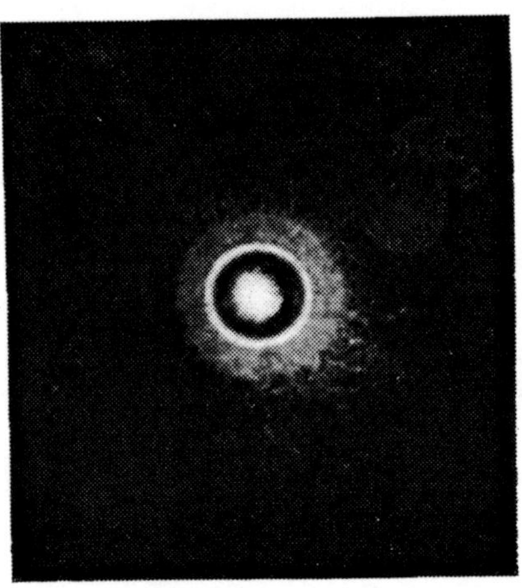

Fig. 18. Droplet effusion on the sphere surface, frontal photograph taken 2.8 µs after impact.

v/d) the shock wave velocity becomes equal to the rate of the contact zone expanding and this shock wave reaches the droplet surface. A depression-type wave moves from the surface into the droplet. Liquid is accelerated along the direction normal to the droplet surface. This situation is presented in Figure 17. As follows from [13] the stretching can exceed the breaking stress within the depression domain, this produces the cavity and as a result the break of the droplet.

In one of the ballistic experiments the disintegration stage illustrated in Figure 17 was photographed in reflected light. The photograph was taken in a direction opposite to the sphere flight at an angle approximately equal to 10° to the trajectory of flight. In the photograph (Figure 18) taken 8 µs after impact we can see three concrete zones differed by the light reflection conditions. A comparison of the photo with an interferogram made at this moment in a direction normal to the sphere trajectory permits us to characterize each zone.

Internal zone (domain A) corresponds to the droplet surface free of perturbations related to the shock wave within the droplet.

Circular zone (domain B) situated around the A-domain is a result of shock wave-droplet surface interaction.

Periphery, diffusively reflecting zone (domain C) consists of dispersed liquid. Photographs (Figures 19, 20) show that motion in the radial direction is predominant.

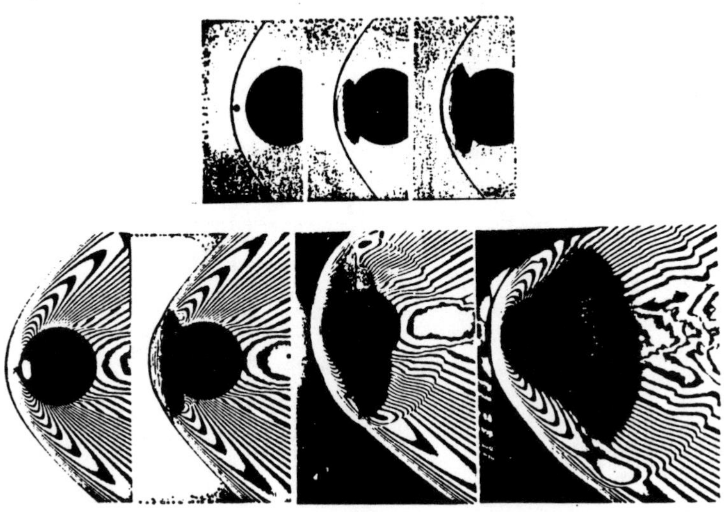

Fig. 19. Motion of the dispersed fraction in the flow about the sphere, axisymmetrical interaction.

Fig. 20. Motion of the dispersed fraction in the flow about the sphere, 15 μs after impact.

In Ref. [13] mentioned above, a calculated value of the initial velocity of the liquid flow in the radial direction is presented for the moment $Vt/d = 0.045$ this velocity has a range of 10.5 to 12.8 of the impact velocity.

In our experiments the fine fraction velocity was measured for later moments. Figure 21 represents the dependence of nondimensional radial location of fine dispersion (R/d) on nondimensional time at conditions mentioned above. The results indicate that for moment $Vt/d = 0.93$ the ratio of flight velocity to the impact velocity is 1.5; this value is far less than that for moment $Vt/d = 0.045$. The difference can be explained by droplet disintegration and fine dispersion deceleration.

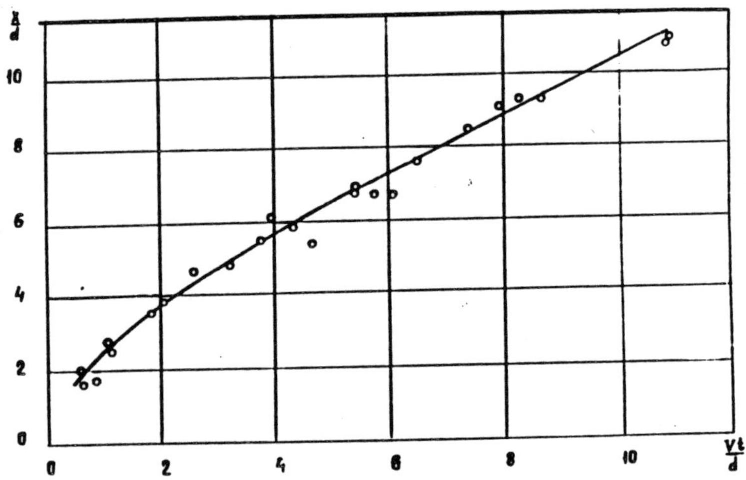

Fig. 21. Dependence describing the dispersed fraction motion in the gas flow field.

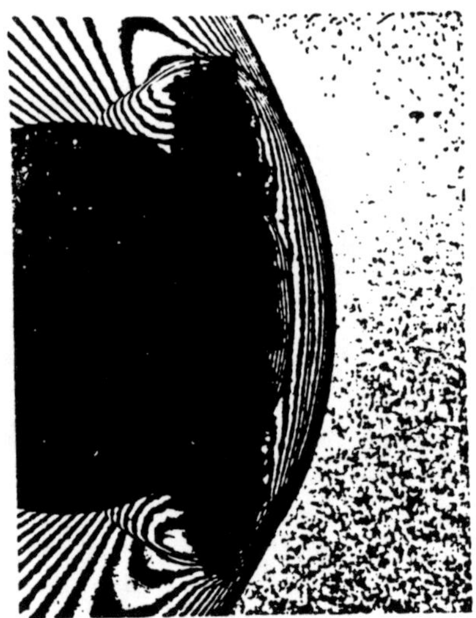

Fig. 22. Interferogram of the dispersed fraction motion in the flow about the sphere, 28 μ later than impact.

Estimates performed for initial stage of disintegration (Figure 20) using the dependence shown in Figure 21 show that the dispersed phase motion is supersonic. A nearly spherical shock wave originates in dispersed phase-gas-phase interaction. This wave attached to the other edge of the dispersed phase cloud moves towards the bow shock as can be seen in Figure 20.

The causes for a second wave shown in Figure 20 can be found after processing the interferogram (Figure 23).

Figure 23 shows the density distribution along the flow axis for undisturbed flow about the sphere (squares) and for disturbed flow (in 33 μs after impact (circles)). In Figure 22 we can see that by the moment 27.9 μs after impact the first shock wave has caught the bow shock and has interacted with it, the photo shows a distortion of the bow shock. The resultant shock wave is of greater intensity, which can be seen from the tendency of the curve to vary at interval X_c = 0.8-1 (lower figure).

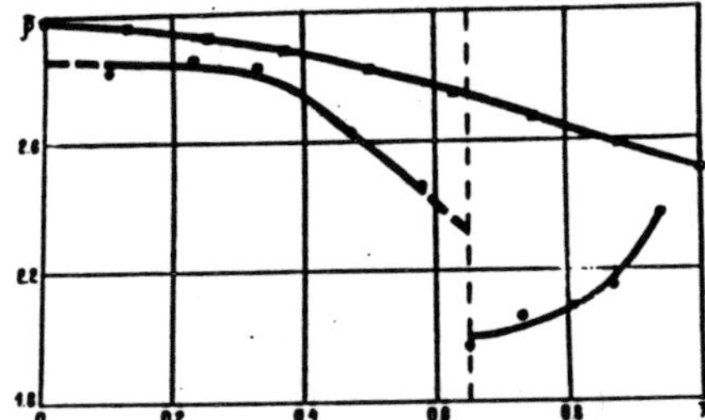

Fig. 23. Density distribution.

The sharp decrease of the gas density between the bow shock and the second shock (dashed line) shows a considerable rarefaction of the gas in the domain. AFter the second shock wave (X_c = 0.65), the density profile smoothly approaches the profile correspondent to the undisturbed flow.

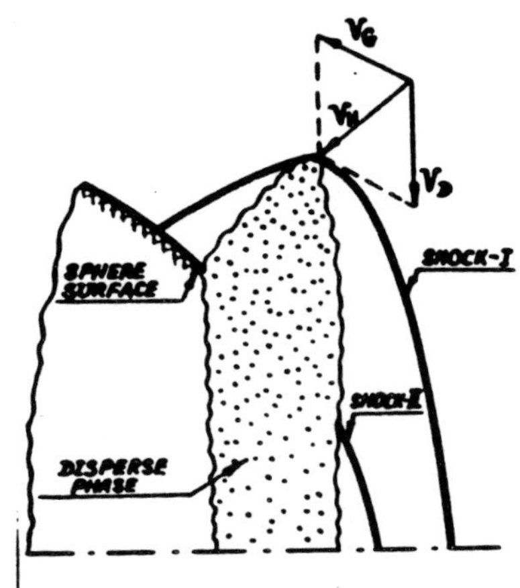

Fig. 24. Motion of the dispersed fraction in the flow about the sphere.

Results presented above show that in the domain between the bow shock formed after catching the interaction X_c = 1 and the second shock shown by the dashed line, gas rarefaction takes place. Therefore, conditions arise behind the front of the shock wave moving in a concurrent flow which promote acceleration of the gas inflowing through the first shock wave. In this domain the gas velocity relative to the sphere is greater than the local sound velocity. An interaction of the accelerated gas flow with the sphere and the dispersed cloud leads to a second shock wave formation (Figure 24).

Processing of the interferograms for later times display that in ~90 μs after the impact at the stagnation point all the parameters come to their initial values. By this time the liquid particles which have lost their kinetic energy obtained during droplet disintegration are carried by the gas flow to the nearest wake. Initial flow conditions are established without additional shocks.

The study of asymmetric process continues the undertaken investigations. The goal of the study is to display the effects related to the tangential velocity component.

In the presence of tangential velocity and essential wettability of the projectile surface it seems to be reasonable to assume an asymmetric cloud of the dispersed phase. The cloud will extend along the tangential component of impact velocity.

The photographs shown in Figure 25 represent successive stages of interaction taken in pairs from various experiments. The incidence angle was in a range of 37–45° with the other parameters being the same as in the case of axisymmetric impact.

From the pattern of droplet disintegration it can be seen that at initial stage up to moment $Vt/d = 1.15$ the dispersed phase flow is axisymmetric relatively normal to the projectile surface at the impact point. The wettability effects are not displayed. The succeeding stages show that the gas flow field is the predominant factor.

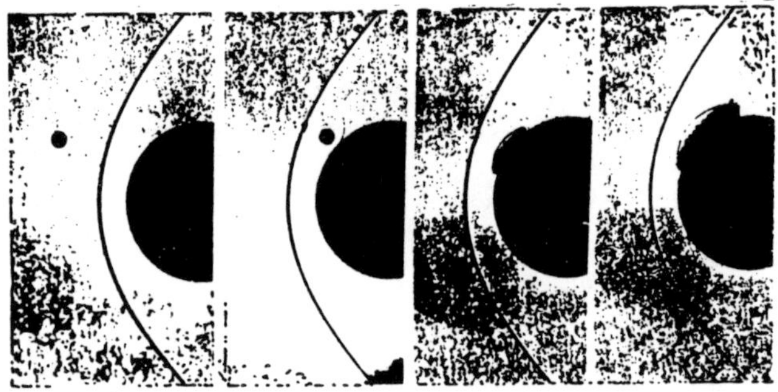

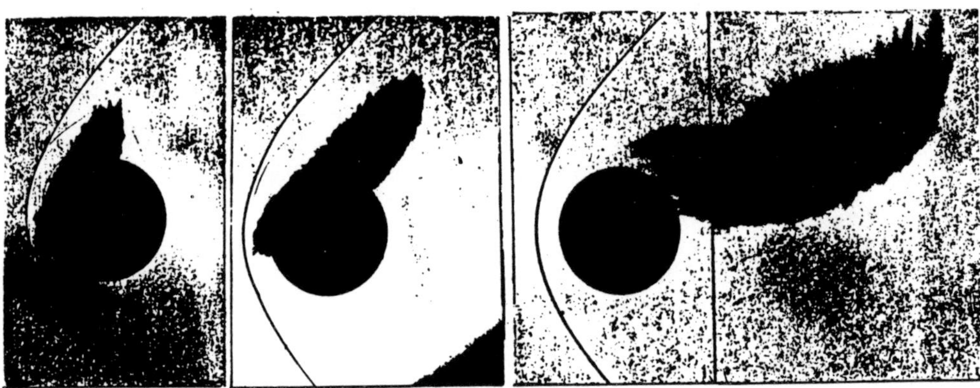

Fig. 25. The dispersed fraction motion in the flow about the sphere, oblique impact.

Similarly in the axisymmetric case, the high velocity particle flow acting as a piston forms the shock wave propagating towards the bow shock.

Photographs (Figure 25) corresponding to later moments show that the interphase momentum exchange is more intensive for particles moving to the initial symmetry axis. The shock wave separates from the dispersed phase cloud near the axis remaining to be attached at the cloud periphery. The other factor defying the difference in the shock distance is nonuniformity of the temperature distribution and, therefore, the sound velocity distribution as well.

There is a second shock wave in the flow about the sphere and the dispersed phase cloud. The reasons for this shock were considered above. Further process development occurs without additional discontinuities.

The investigations carried out show the essential role of particle disintegration in interphase transport. Quantitative results obtained are used in mathematical model formulation which will be discussed later.

References

1. S. M. Gilinskii and V. N. Tolsov, In: *Jet and Separation Flows*, Moscow: MGU, 1985, pp. 78-94 (in Russian).
2. D. T. Howe and A. A. Smith, *AIAA J.*, vol. 13, no. 7, 1975, pp. 947-948.
3. N. N. Yanenko, et al., *Supersonic Two-Phase Flows in the Presence of Particle Velocity Nonequilibrium*, Novosibirsk: Nauka, 1980, 160 pp (in Russian).
4. I. A. Dukhovskii, et al., In: *Turbulent Two-Phase Flows and Experimental Technique*, Tallinn, 1985, pp. 23-27 (in Russian).
5. I. A. Dukhovskii, et al., *Experimental and Numerical Study of Particle Disintegration Effect on Supersonic Dusty Gas Flow About Blunted Body*, Preprint FTI AN SSSR, no. 1300, 1988, 21 pp. (in Russian).
6. A. P. Alkhimov, N. I. Nesterovich, and A. N. Papyrin, *Zh. Prikl. Mech. i Tekhn. Fiz.*, no. 2, 1982, pp. 66-74 (in Russian).
7. R. I. Nigmatulin, *The Foundations of Heterogeneous Media Mechanics*, Moscow: Nauka, 1978, 336 pp. (in Russian).
8. I. A. Dukhovskii and P. I. Kovalev, *Pisma v Zh. Tekhn. Fiz.*, vol. 14, no. 17, 1988, pp. 1594-1596 (in Russian).
9. W. Tabakovv, M. F. Malak, and A. Hammed, *AIAA J.*, vol. 25, no. 5, 1987, pp. 721-726.
10. J. P. Armstrong, N. Collings, and P. J. Shayler, *AIAA J.*, vol. 22, no. 2, 1984, pp. 214-218.
11. I. A. Dukhovskii, P. I. Kovalev, and A. A. Schmidt, *Pisma v Zh. Tekhn. Fiz.*, vol. 10, no. 11, 1984, pp. 649-652 (in Russian).
12. A. A. Borisov, et al., *Inzh Fiz. Zh.*, vol. 40, no. 1, 1981, pp. 64-70 (in Russian).

13. M. Rosenblatt, J. M. Ho, G. E. Eggum, In: *Erosion Prevention and Useful Applications*, Philadelphia: Penn. Amer. Soc. Testing and Mat., 1979, pp. 15-26.

14. A. E. Bryson and K. W. F. Gross, *J. of Fluid Mech.*, vol. 10, pt. 1, 1960, pp. 1-16.

15. A. L. Gonor and V. Ya. Rivkind, In: *Itogi nauki i Tekhn. Mech. Zhidk. i gaza*, Moscow, vol. 17, 1982, pp. 86-159 (in Russian).

DETERMINATION OF GAS FLOW PARAMETERS AROUND BODIES IN FREE FLIGHT

P. I. Kovalev, S. G. Tomson

Abstract: Optical interferometry is one of the most informative methods available for the study of gas flows. It should be noted, however, that interferometric measurements do not provide complete information on the gas thermodynamic state.

This paper presents a new method [1] for determining flow parameters based on gas sounding by shock waves. Practical implementations are proposed [2, 3]. Verification in a ballistic experiment is presented.

1. Determination of Gas Flow Parameters by Means of Probing Perturbations

To clarify the principal idea of the method we consider a scheme presented in Figure 1 which depicts a solid surface in a gas flow. Two sounding shock waves are propagating along the surface. This flow can be described by a set of equations relating shock velocities v_1, v_2; their slopes α_1, α_2; Mach numbers M_1, M_2; flow velocity u; and sound velocity a

$$M_1 a - u \sin \alpha_1 = v_1,$$
$$M_2 a + u \sin \alpha_2 = v_2. \tag{1}$$

It is clear that we must measure the values of M_1, M_2, v_1, v_2, α_1, α_2 in order to determine gas velocity u and sound velocity a.

Recording of shock wave propagation in a gas by interferometric operation in a two-flow regime allows these measurements to be performed.

Indeed, processing of the interferograms gives the gas density distribution in the whole flow area under study. If the density ratios of both shocks are known, the values of M_1 and M_2 can be calculated for each shock for fixed moments. Simultaneously, shock coordinates are determined thus allowing us to obtain mean velocities v_1, v_2 and slopes α_1, α_2. Substituting

145

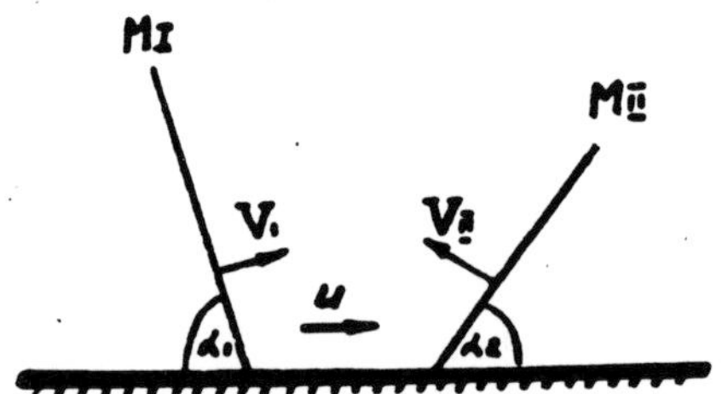

Fig. 1. Determination of gas flow parameters using sounding shock waves.

these values in Equations (1) we obtain gas velocity u and sound velocity a. Using the density distribution provided by the interferogram the gas pressure in the flow under study can be calculated.

The range of application of the proposed method may be enlarged by introducing additional perturbations in the flow.

2. Practical Implementation of the Method

A basic problem encountered in practical implementation of the method is to devise a way to introduce shock waves into the gas flow in question. This problem was solved for the objects shown in the interferogram (Figure 2). The main feature of the flow around such bodies is the presence of a front separation zone where dissipative processes takes place. This excludes the possibility of calculating other parameters using the density field without additional measurements.

Fig. 2. Interferograms of gas flow around a body with a front separation zone, $M_\Pi = 2.35$.

Fig. 3. Interferogram of gas flow around a blunted cone. $M_\Pi = 2.4$. The surrounding perturbation on the cone base is produced by laser breakdown.

To illustrate this point we note that such a possibility exists, for example, when studying stationary flows around bodies with a simple shape (Figure 3). If the Mach number of the bow shock is known, we can determine a constant in Poisson adiabat for a stagnation streamline and then calculate the pressure, temperature, and gas velocity along the lateral body surface using the well-known gasdynamic relations. Note that these calculations become impossible if either the streamline positions are unknown or the above-mentioned constant cannot be determined. Just such flow occurs around the object shown in Figure 2.

In order to devise the most convenient means for introducing sounding shock waves which do not break the axial symmetry, we have proposed several methods for initiating shock waves near the objects in free flight.

The first is to focus a pulsed laser beam at a given point located on the aiming line of the projectile system. When this point coincides with the surface of the projectile, laser radiation is generated and the propagation of the shock wave initiated by laser breakdown in gas is recorded. Recording of the wave position and intensity is performed at least twice in fixed time moments. Figure 2 displays a shock wave initiated in such a way in the base region of the blunted cone.

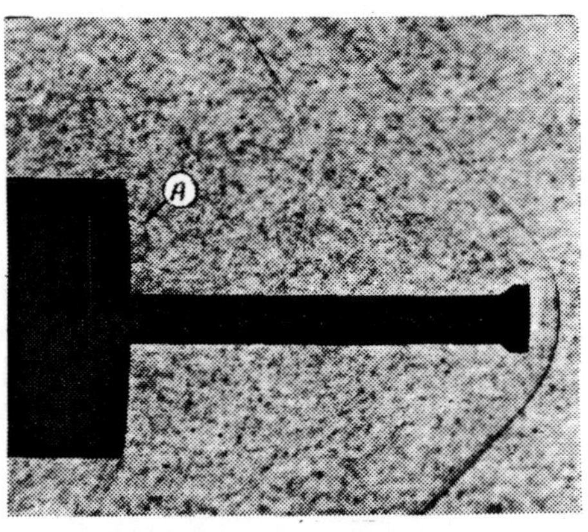

Fig. 4. Initiation of the sounding shock wave in the stagnation zone by means of a detonator. M_η = 2.1. "A" denotes the probing shock wave.

When the trajectory of the object cannot be determined beforehand with sufficient accuracy, for example, when its flight is disturbed by initial oscillations of a random nature, another method is more acceptable. It consists of mounting a detonator to be blown up at a distance in the inner cavity of the projectile with the drain at its surface. This method is convenient because the energy releasing "point" does not depend on the object's trajectory and can be placed at its symmetry axis. This method of implementation is illustrated in Figure 4. The sounding shock wave initiated by the jet from the drain orifice is seen near the front body surface.

However, neither of the methods just described is acceptable for introducing shock waves at the lateral surface of the body since they inevi-

Fig. 5. Initiation of the sounding shock waves on the surface of the body with a front separation zone. $M_\Pi = 2.31$.

tably break the axial symmetry. As a result processing of the interferograms becomes complicated and sometimes impossible without the application of a tomograph. We use the following technique to initiate sounding perturbations on a lateral surface. Placing a destructible screen normal to the flight trajectory and making a shelf on the projectile surface, we recorded the impact of the object to the screen. The measurements were taken in the undisturbed flow region between the reflected bow shock wave and the shock attached to the shelf. The photograph obtained in one such experiment is shown in Figure 5.

3. Verification Results Using Test Objects

The graphs in Figure 6 display the results of interferogram processing obtained with an exposure interval of 5.9 μs. The first one corresponds to the phase of the probing shock propagation shown in Figure 3. The cross-section which was close to the cone base and normal to the flow axis was processed. The ratio $\rho = \rho/\rho_\infty$ of the measured density to its free stream value and the ratio of the current radius value to the distance from the cone axis to the shock wave are plotted on the ordinate and abscissa axes, respectively. The dashed line marks the probing perturbation coordinate. The results presented of interferogram processing were used to determine the sound velocity in a gas adjacent to the base of the projectile, assuming it to be quiescent relative to the surface. Using the results of measurements both the gas temperature $T/T_\infty = 1.77$ and its density $\rho/\rho_\infty = 0.26$ were determined and employed to calculate the pressure $p/p_\infty = 0.46$. The parameters obtained are in good agreement with the results of Ref. [5].

As a test object for the lateral surface measurements a sharp cone with a divergence angle equal to 30° flying in air under normal conditions at Mach number M = 2.11 was taken (Figure 7). Simple supersonic flow allowed us to control all stages of the experimental data processing.

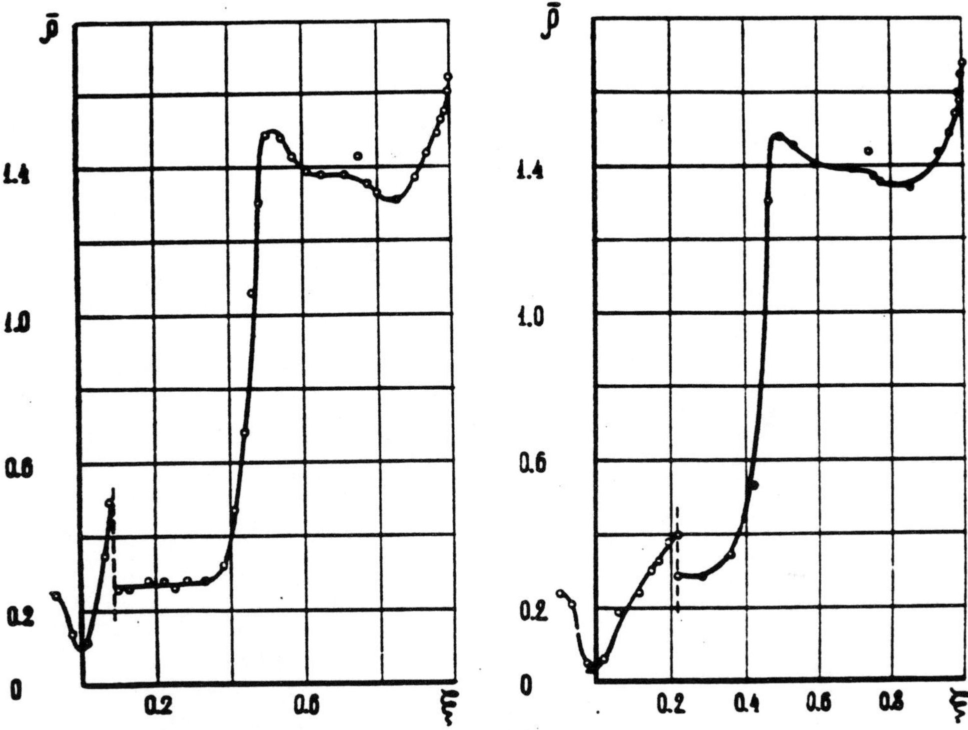

Fig. 6. Density distribution in the section containing the cone base.

Fig. 7. Initiation of the sounding shock waves on the surface of the sharp cone. $M_\Pi = 2.13$.

Fig. 8. Density distribution in the section crossing the reflected shock wave. The distance from the cone nose is 1.45 times the cone base diameter.

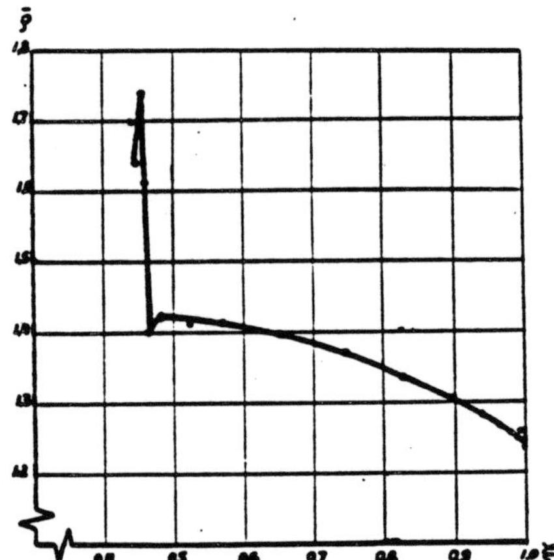

Fig. 9. Density distribution in the section crossing the shock wave attached to the shelf. The distance from the cone nose is 1.83 times the cone base diameter.

Indeed, because the gas parameters on the cone surface are constant their averaging during the measurements is excluded. As seen in Figure 8 the reflection of the shock attached to the cone nose from the screen produces a sonic wave. Its velocity near the surface must be equal to the sum $u + a$ where u and a are the velocities of gas and sound in the vicinity of the surface. The similarity of the reflected wave propagation allows control of the accuracy of its shape determination. The dimensionless density distribution in the section normal to the cone axis and crossing the shelf attached shock is shown in Figure 9. The Mach number of this shock as determined from the graph is equal to M = 1.1, its slope to the surface at the edge of the boundary layer $\alpha_2 = 37°$. We substitute these values along with velocity of the sonic wave $v_1 = 1028$ m/s measured with the use of two photographs made with an interval of 10.94 μs into Equation (1) which can be written as:

$$a + u = v_1$$

$$M_1 a - u \sin \alpha_2 = 0 \tag{2}$$

temperature $T/T_\Pi = 1.125$, and gas velocity $u/u_\Pi = 0.918$ on the cone surface were determined which coincide with the calculation results [6] to the accuracy of less than 5%.

The studies reported above show the efficiency and practical implementation of the method.

References

1. P. I. Kovalev and N. P. Mende, "A Method of Determination of Gas Flow Parameters", no. 1033954, In: *Bul. Izobretenii*, 1983, no. 29.
2. I. A. Dukhovski, P. I. Kovalev, N. P. Mende, and A. I. Razumovskaya, *Opt. Laser Tech.*, vol. 20, no. 5, 1988, pp. 259–262.

3. I. M. Dementiev, I. A. Kamalov, P. I. Kovalev, V. A. Komissaruk, A. N. Mikhalev, and S. G. Tomson, *Opt. Laser Tech.*, vol. 19, no. 6, 1987, p. 316.

4. I. A. Dukhovski, V. A. Komissaruk, P. I, Kovalev, and N. P. Mende, *Opt. Laser Tech.*, vol. 17, no. 3, 1985, pp. 148-150.

5. A. N. Mikhalev, *On the Parameters of Cone Near Wake in Supersonic Flow at Intermediate Reynolds Numbers*, Thesis: Phys. Tech. Inst., Leningrad 1979 (in Russian).

6. K. I. Babenko, G. P. Voskresenskii, A. N. Luibimov, V. V. Rusanov, *Three-Dimensional Ideal Gas Flows About Smooth Bodies*, Moscow: Nauka, 1964.

TURBULENT MIXING OF DRIVEN AND DRIVER GASES IN SHOCK TUBE

A. D. Zuev, R. V. Vasilieva, D. N. Mirshanov

Abstract: **The present work deals with the experimental study of a flow structure in a usual diaphragm shock tube. Primary attention was paid to the contact mixing region between driven and driver gases. The Mach number of the incident shock wave varied from 3.5 to 13.5. The initial pressure of a driven gas was in the range from 5 to 200 torr. The main method was X-ray diagnostics based on X-ray absorption in the studied medium. Density distributions of driven and driver gases in the contact region were obtained and the mass of driven gas in this region was determined. It was shown that test time diminution is caused to a greater extent by driven and driver gas mixing than by boundary layer leakage according to a widely used theory by Mirels. The evolution of the contact mixing region when moving along the shock tube channel was investigated. A mechanism for formation of a shock compressed uniform flow region is proposed that takes into account its shrinkage due to driven and driver gas mixing. A method is proposed for studying nonstationary shock tube boundary layers based on measuring the total driven gas mass flowing from different parts of the hot uniform flow region.**

1. Introduction

Wide use of shock tubes in gasdynamic and physical experiments implies obtaining more profound data on the flow structure in a shock tube channel. Until recently there were almost no data on the characteristics of the contact mixing region (CMR) which divides driven and driver gases in spite of the fact of its influence on the parameters of the shock compressed test flow region (TFR). In accordance with available data CMR flow is assumed to be turbulent due to the effects of turbulizing factors accompanying the diaphragm opening process. The main turbulizing factors are: flow separation near the fragments of the opening diaphragm, Kelvin-Helmholtz instability at the edge of the driver jet at its exhaust to the driven gas, series

153

of the oblique shock waves reflecting from the walls of the channel, intensification of gas mixing when the shock waves induced by the diaphragm rupture propagate through the turbulent area. In addition, substantial flow acceleration takes place during diaphragm opening and consequently Rayleigh-Taylor instability development is possible if the driver density near the contact surface is less than the driven gas density. As a result, turbulent flow is formed. The mean velocity of its boundaries is normal to the initial discontinuity surface, the external flow of driven and driver gases is shearless, the mean flow velocity in the turbulent area is equal to the external flow velocity. When the effect of the main turbulizing factors induced by the diaphragm rupture disappears, the turbulent area continues to smear. Such turbulent flow belongs to the class of problems which in turbulence develops on the normal surfaces. This class includes such problems as turbulent stage of Rayleigh-Taylor instability [1, 2], intensification of mixing at the two media boundary after the series of shock wave propagation [3-5], and smearing of the turbulent layer in a two-gas mixture [6]. Investigation of such nonstationary turbulent flows reveals large experimental and calculation difficulties. Evidently, that is why the development of the mixing area arising in the shock tube due to diaphragm rupture has been little studied.

The object of our investigation was the contact mixing region (CMR) of driven and driver gases at a distance from 15-90 diameters from the diaphragm, i.e., far enough from the diaphragm so the main turbulizing factors do not already function and the above region continues to expand when moving along the shock tube channel, and to capture the driven gas from the hot test flow region (TFR). Our purpose was to obtain the density distributions of driven and driver gases inside CMR, to determine the mass of driven gas in this region, to study the mixing effect on the hot TFR shrinkage, and to study the evolution of the contact area.

The study of gas flow in this region is important, on the one hand, since the shock tube is one of the main instruments of experimental gasdynamics and knowledge of the flow formation in it is necessary for correct organization of experiments and interpretation of experimental results. On the other hand, because of the wide use of shock tubes, the turbulent mixing area of driven and driver gases may be considered an appropriate model for nonstationary turbulent processes.

2. Experimental Facility

Experiments were carried out on a stainless steel diaphragm shock tube (see Figure 1). The length of the high-pressure chamber is 1.2 m; the low-pressure chamber is 5 m; the inner channel diameter is 5 cm. The diaphragm divides the high-pressure chamber and low-pressure channel. Aluminum diaphragms 0.1-1 mm thick were used. The diaphragms were opened under the pressure effect by two methods: with and without a special cross-like knife. Preliminary compression was performed by a hydraulic system. Gas pressure in the high-pressure chamber was in the range from 14-50 atm, gas temperature can be increased to 700 K. Hydrogen, helium, nitrogen, and a mixture (98.8% H_2 + 1.2% Xe) were used as driver gases. The low-pressure channel consists of separate sections. Measurements were performed at some cross-sections of the shock tube. Ar, Kr, Xe and various gas mixtures were used as the driven gas. Initial pressure was varied from 5 to 200 torr.

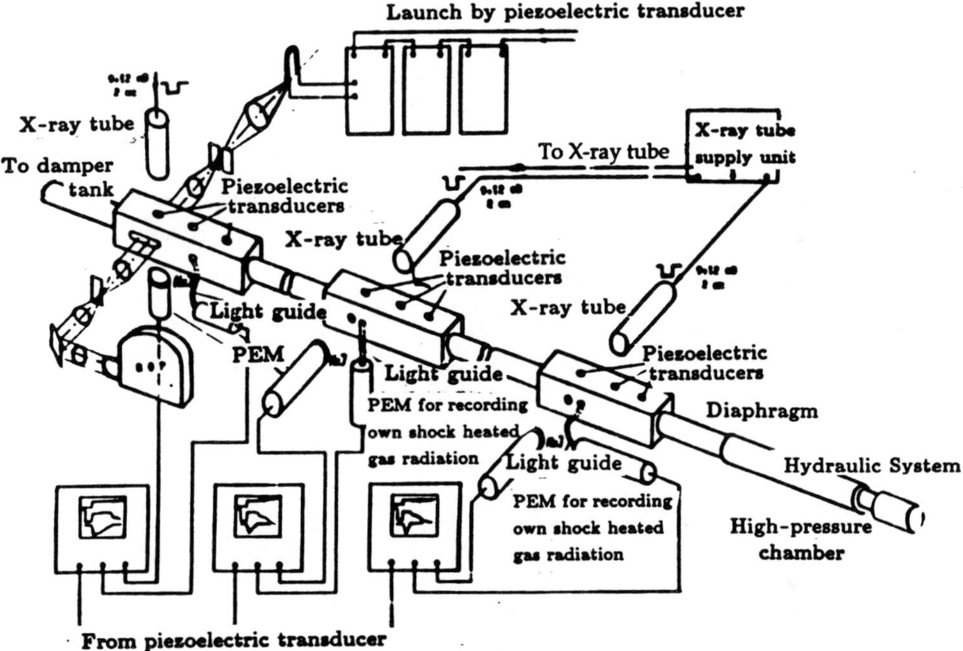

Fig. 1. Shock tube and instrumentation.

The velocity of the shock wavefront, pressure distribution, driven and driver gas densities behind the shock front, the radiation intensity of the shock heated gas in the optical wavelength range, X-t-diagrams of the flow shadowgraphs were recorded in the experiments. The shock front velocity was determined by means of the basis method, i.e., with the use of pressure and ionization gauges; and by means of the X-t-diagram method, i.e., from the shock front track inclination on X-t-diagrams of the shadow flow pictures. The pressure distribution behind the shock wavefront was obtained with the use of pressure gauges. Recording of plasma radiation allows us to determine the length of hot TFR of the shock compressed gas, to measure the length of the ionization relaxation zone, and to estimate the relative temperature variation in the regimes when the ionization relaxation length is far less than the hot TFR length and there is ionization equilibrium in the TFR.

Density measurement. Gas density is one of the main flow characteristics. Special attention was paid to the choice of the method used to determine the density in turbulized flows.

The method based on soft X-ray reabsorption in the studied gas seems to be most suitable. X-ray diagnostics are characterized by the following properties that make it the most convenient for studying the turbulent mixing layers: 1) due to small X-ray refraction it is possible to apply X-ray diagnostics in turbulent flows; 2) strong dependence of the X-ray absorption coefficient on the chemical element ordinal number allows us to obtain some definite component distribution in the turbulent mixing layer. On the whole, the practical implementation of the method is quite simple. Because of the above advantages, X-ray diagnostics essentially complements the optical methods of gas flow diagnostics and is worth further development. The main difficulty for its application is to obtain intensive soft X-ray radiation and to record this radiation effectively.

X-ray tubes with a tungsten anode providing a continuous spectrum were used in our experiments. The tubes operated in impulse regime, anode voltage was equal to 8-12 kV. In order to increase the intensity of X-ray radiation the tubes operated in over-incandescence regime. PEM were used as the receivers of X-ray radiation. The scintillators were made of polystyrene doped with $C_{18}H_{14}$ and of NaJ doped with thallium.

Special efforts were undertaken to minimize X-ray radiation losses in the inlet and outlet windows. The windows were made of beryllium plates glued or soldered to special brass insets. As a result of precautions the lowest measured xenon concentration behind the shock wavefront at diameter $d = 5$ cm is equal to $5 \cdot 10^{17}$ cm^{-3}. It can be seen that X-ray diagnostic sensibility for shock tubes has been increased by approximately an order of magnitude.

Figure 2 presents a scheme of the X-ray signal correspondent to the driven gas density variation behind the shock wave and a scheme of the pressure signal as well.

Interval 1-1' of the X-ray signal scheme conforms to the signal before shock wave arrival. This one is attenuated compared to the signal in vacuum due to its absorption in a gas filling the low-pressure channel before the experiment begins. When the shock wave passes through the measurement cross-section the density jumps from ρ_1 to ρ_2. A correspondent jump in the X-ray signal variation is denoted with 1'-2 in the scheme. Hot TFR is depicted by the interval 2-2'. Density distribution in this region depends on physical and chemical processes that occur in hot TFR, on the boundary layer growth on the shock tube walls, on the processes induced by diaphragm rupture. A further contact zone 2′-3-3′-4, i.e., driven and driver gases mixing region follows. X-ray absorption coefficient of the driver gases (H_2, He, N_2) is far less than that of xenon. So X-ray signal measurement in CMR corresponds to the driven gas density variation in this region. After the mixing region, the driver gas (H_2, He, or N_2) passes through the measuring cross-section. It nearly does not absorb X-rays, so the signal in domain 4-4′ may be considered as the signal in vacuum.

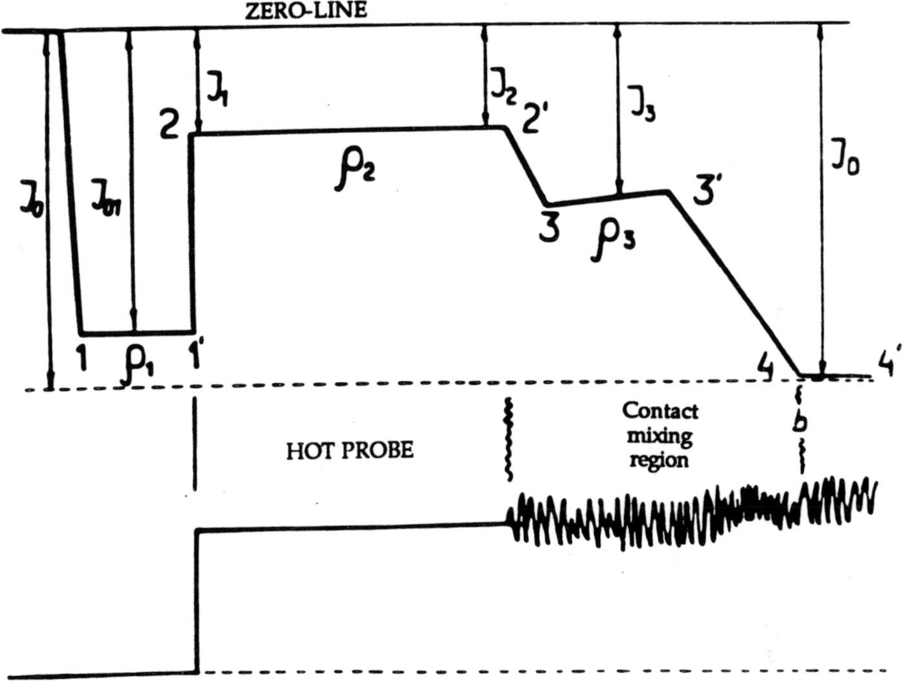

Fig. 2. X-ray signal (a) and pressure signal (b).

The pressure in hot TFR, CMR, and driver gas does not undergo large variations, but the pressure pulsation level increases significantly when the turbulized CMR arrives at the measuring cross-section.

A typical oscillogram of the X-ray signal is presented in Figure 3. Here the signal of shock heat gas radiation is also shown. Luminescence stops when the contact area arrives at the measuring cross-section. Note that hot TFR lengths determined by their own radiation signal and X-ray signal coincide.

Oscillograms of X-ray signal variation transforms to correspondent density variations with the use of the calibration curve. Calibration was carried out by means of static filling of the low-pressure channel with treated gas under various pressures. The X-ray tube worked in the pulsed regime. The measurements at the same pressure were fulfilled repeatedly in order to decrease the static error induced with the fluctuations of X-ray radiation.

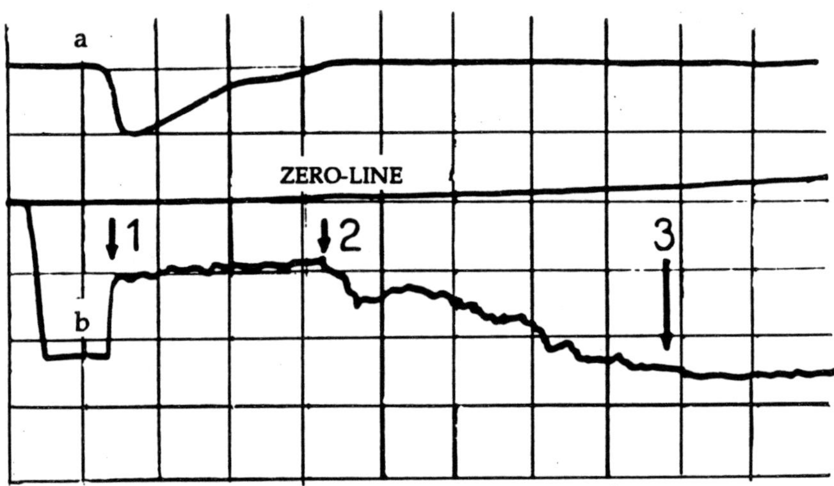

Fig. 3. Oscillograms of shock heated gas radiation (a) and X-ray signal (b). 1 - shock wavefront; 2 - end of hot TFR, leading edge of CMR; 3 - end of CMR, regime H_2-Xe, $p_1 = 25$ torr, M = 8.7.

Driven gas density values obtained after processing are the average density values on the probing X-ray:

$$\langle \rho \rangle = \frac{1}{d} \int\limits_0^d \rho(y)dy.$$

Here d is channel diameter.

As a result of density distribution processing the masses of driven gas in hot TFR and driver gas in the contact region were found. The mass of a gas in the tube volume between the control surfaces m, n (shock wave and front or back boundary of CMR) is determined as

$$M_{m,n} = \frac{\pi d^2}{4} \int\limits_m^n \rho(x)dx$$

where x is a distance from the shock wavefront. It was supposed that CMR end was located where X-ray signal recovered to 0.9 of its value in vacuum.

The error of the density determination by X-ray method is caused by statistic uncertainty and finite line width of the X-ray signal on the oscillogram, by inaccuracy of X-ray tube voltage determination during the experiment, and by the calibration error.

Consider these contributions to the total error.

1. If X-ray radiation is attenuated exponentially when passing through treated gas the uncertainty of the density determination is expressed by the formula

$$\Delta\rho/\rho = \frac{\Delta J/J_{01}(1 + J/J_{01})}{J/J_{01}\ln(J/J_{01})}.$$

The error is minimal at $J/J_{01} = 0.28$, then $\Delta\rho/\rho = 3.57\,(\Delta J/J_{01})$. If signal amplitude $J_{01} = 40$ mm and $\Delta J = 0.25$ mm we have $\Delta\rho/\rho = 2\%$.

2. X-ray tube voltage during the experiment was recorded with the oscillograph with an accuracy $\Delta U/U = \Delta I/I = 0.25/40 = 0.6\%$. Such voltage uncertainty results in the following error of density determination due to attenuation coefficient variation:

$$\frac{\Delta\rho}{\rho} = 3\cdot\frac{\Delta U}{U} = 2\%.$$

3. Calibration error is equal to 1%.

Hence, minimal total error is equal to 5%. This value is obtained for xenon at initial pressure $p_1 = 25$ torr. Initial pressure reduction to 5 torr results in an error increase up to 10%.

3. Shock Tube Flow Structure

When investigating the flow in a shock tube it is necessary to distinguish the TFR of shock compressed gas, CMR, and driver gas region. With the use of X-ray diagnostics the driven gas density distributions in the hot TFR were obtained and the driver gas density distributions in the CMR were obtained for the first time. The main result is that CMR contains a significant portion of the driven gas, its density in this region is comparable to the density in the hot TFR and the CMA length is the same or greater than the shock compressed TFR length.

Density distribution of driver gas was obtained by addition of a small amount (~1%) of heavy gas Xe to light driver gas H_2 [7].

Experimental results on the parameters of shock compressed TFR were compared with their values calculated in accordance with some theories. These theories, for the most part, concern the effects of the boundary layer, gas ionization, and energy loss due to radiation. The most important factors were taken into account according to the given conditions. Only the boundary layer influence on density distribution and shock compressed TFR length was taken into account at shock wave Mach numbers M < 8 in argon and xenon flows. Calculations were fulfilled according to Mirels theory [8] widely used for interpretation of the shock tube experimental result4s. At high shock wave Mach numbers ionization and energy losses by radiation mainly influence the parameter distribution in the TFR of shock compressed gas. So calculations for xenon at M > 8 were performed with the above factors taken into account.

Measurements were carried out at a distance of 50 diameters from the diaphragm. At a whole range of Mach numbers and initial pressures the experimental density values of driven gas near the shock wavefront are in satisfactory agreement with the values calculated with the use of measured shock wave Mach numbers.

Experimental density distribution of driven gas behind the shock wave at small Mach numbers is presented in Figure 4. Here the density distribution of driven gas in hot TFR according to Mirels theory is shown as well. Gas density variation in the hot TFR is not observed in experiment though calculations show its increase by 15%. Experimental hot TFR length is in satisfactory agreement with calculation results. However, this agreement is not sufficient to judge on Mirels theory because a substantial portion of driven gas is in the mixing region.

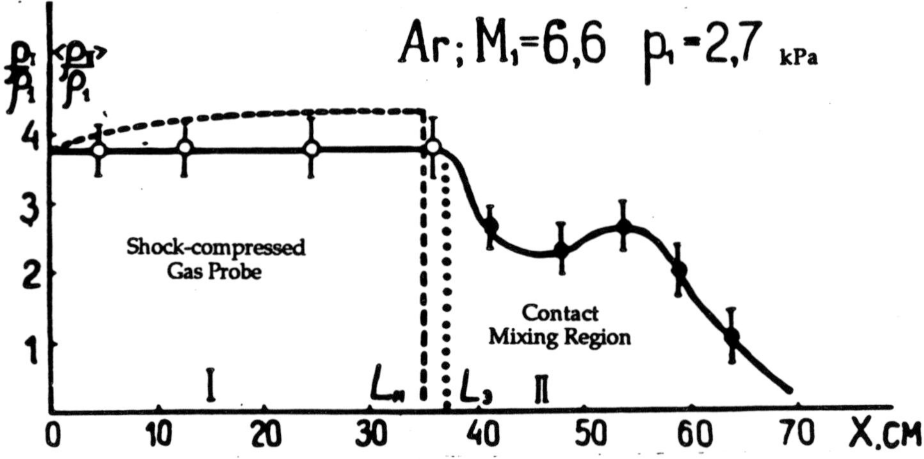

Fig. 4. Driven argon gas density distribution behind the shock wave.
Solid curve - experimental results; dashed curve - calculation result according to Mirels theory.

At high shock wave Mach numbers the temperature distribution was also obtained. Comparison of experimental and calculated values of gas density and temperature (Figure 5) shows that more intensive cooling is observed in experiment. Analogous to the case of small Mach number, the TFR lengths are two times less than calculated according to the elementary shock tube theory. CMR lengths are comparable with hot TFR length.

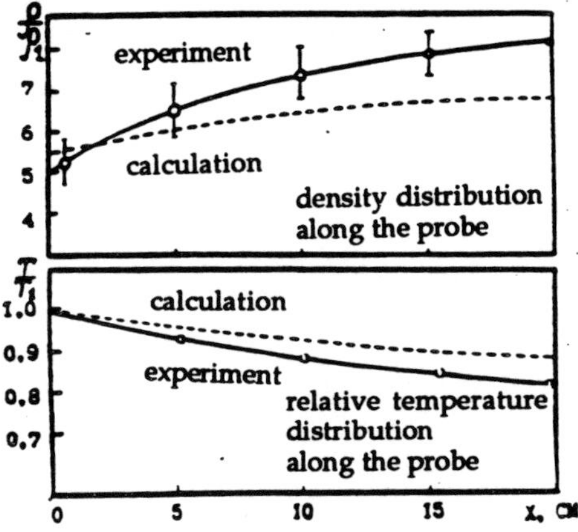

Fig. 5. Relative density and temperature distributions behind shock wavefront in xenon. M = 11.5, p_1 = 15 torr.

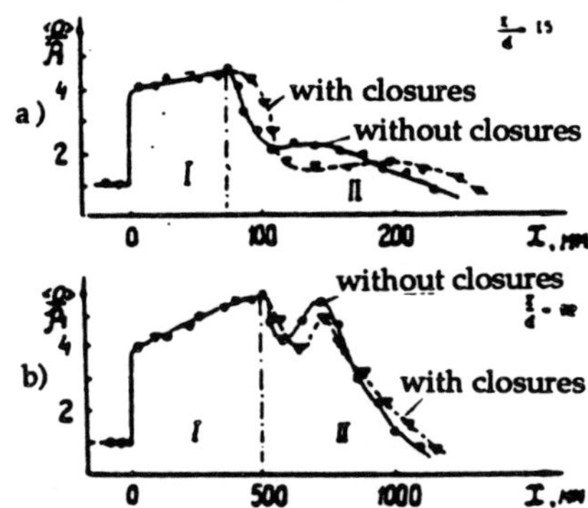

Fig. 6. Driven gas density distributions behind the shock wave in hot TFR 1 and CMR 2 in two cross-sections.

1 - x = 15 d; 2 - x = 92 d wit=hout and with closures.

Uniformity of the contact area was estimated by means of absorbing layer length variation, i.e., measurements were carried out at two values of absorbing layer length: at h = 50 mm with the windows made flush with the tube walls and at h = 30 mm with the windows protruded inside the tube. In the second case special insets were made that had a streamlined cross-section shape and jutted into the flow by 1 cm. The thickness of the inset mid-section is 5 mm, the diameter of the inner window is 2 mm. Beryllium is glued on the face. In this case the measurements were performed at distances of 15 and 92 diameters from the diaphragm. The results are presented in Figure 6. A notable reduction of driven gas density near the boundary between hot TFR and CMR at a distance of 15 d from the diaphragm is seen. The average density near CMR boundaries measured with the insets is higher than those without. This fact indicates that the average density of driven gas is somewhat higher in the mainstream than near the walls at CMR edges. At the cross-section x = 92 d the driven gas density near the front edge of CMR is comparable with the density at the TFR end. In addition, the difference between ⟨P⟩ values measured at two different h values is small.

Hence we must conclude that driven gas density distribution across the tube is rather uniform at the beginning of CMR and becomes less uniform at its end. Since *h* decreasing does not result in a notable density fluctuation increase it seems that the nonuniform scale in the cross-section is sometimes less than *d*, i.e., does not exceed 1 cm. Thus during shock wave propagation the density reduction near the front edge of CMR reduces and this boundary becomes more flat. The back boundary of CMR remains convex. The driven gas average density variation in the mixing region is nonmonotonous.

Density distributions of driven and driver gases at a distance of 50 diameters from the diaphragm are presented in Figure 7.

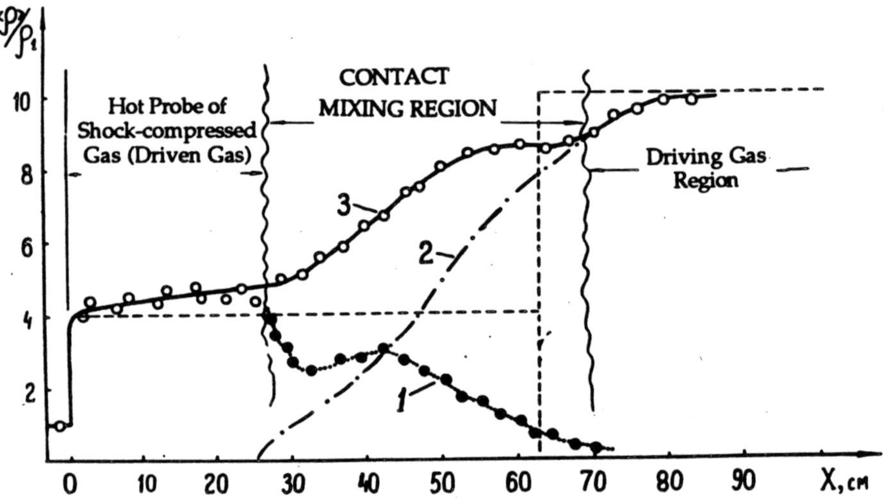

Fig. 7. Density distributions of driven and driver gases behind shock wave.

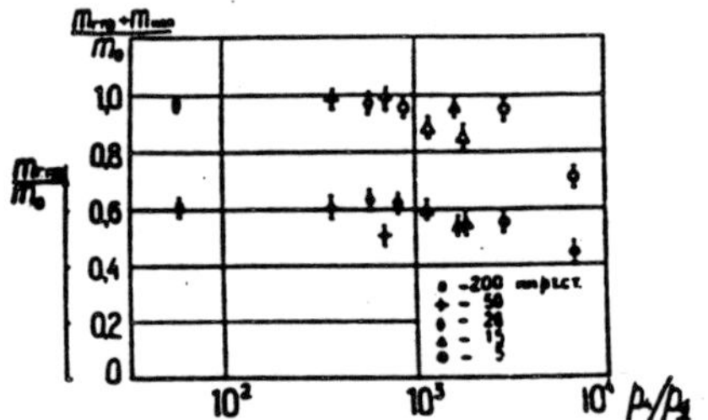

Fig. 8. Relative mass portion in hot TFR M_{TFR}/M_0 and relative mass portion in hot TFR along with CMR $(M_{TFR} + M_{CMR})/M_0$ versus diaphragm pressure ratio at different pressures.

The mass distributions of driven gas in TFR and CMR shown in Figure 8 were obtained by processing experimental data on driven gas density distributions.

Driven gas mass is determined as

$$M_{TFR} = \frac{\rho d^2}{4} \int_{(TPR)} \rho(x)dx$$

in TFR, and as

$$M_{\text{CMR}} = \frac{\rho d^2}{4} \int_{(\text{CMR})} \rho(x)dx$$

in CMR. Total mass of driven gas between the diaphragm and the measurement cross-section equal to the mass of ideal TFR is given by the formula:

$$M_0 = \frac{\rho d^2}{4} \rho_1 \cdot X = \frac{\rho d^2}{4} \rho_2 L_{id}$$

where L_{id} is an ideal TFR length. The difference between driven gas mass in ideal TFR and gas mass in TFR and CMR, $\Delta M_{bl} = M_0 - (M_{\text{TFR}} + M_{\text{CMR}})$ can be considered as a mass flowing out through the boundary layer and remaining outside CMR. The range of initial parameters was typical for the shock tube experiments; $5 \leqslant p_1 \leqslant 200$ torr, $50 \leqslant p_4/p_1 \leqslant 5000$. It was found that at pressure values $p_1 > 5$ torr the masses of driven gas in TFR and CMR amounted to 50-60 and 30-40% of the ideal TFR mass respectively, and this percentage did not depend on experimental conditions: initial pressure, pressure ratio at this diaphragm, choice of gas, Mach number, method of diaphragm rupture. As initial pressure decreases and pressure ratio at the diaphragm increases the boundary layer effect becomes more significant: for example, at $p_1 = 5$ torr, $p_4/p_1 = 7 \cdot 10^3$ mass loss due to a boundary layer leakage of 30%. TFR contains 45% and CMR 25% of the total driven gas mass.

4. Bulk and Wall Phenomena Effects on TFR Shrinkage

Of two principal phenomena resulting in shrinkage of the uniform flow in a shock tube (mixing in the contact region and boundary layer formation on the shock tube walls) primary attention was usually paid only to the second. The above considered results show that, firstly, CMR length is equal or even greater than the shock compressed TFR length; in the second place, driven gas density in CMR is comparable with gas density in TFR; and, in the third place, gas mass in CMR can be 50% of the ideal TFR mass. The high driven gas density in CMR cannot be explained by gas outflow through the boundary layer and gas accumulation in this region only. It is also necessary to take into account gas mixing at the leading edge of CMR and gas capturing from TFR. The methods applied today to study the shock tube boundary layers do not illustrate the boundary layer effects on uniform flow shrinkage and CMR formation. This is stipulated by the difficulty of studying a nonstationary boundary layer.

In order to clear up the effects of bulk and wall phenomena on the uniform flow shrinkage and CMA formation a special method was elaborated allowing us to control the development of various parts of TFR. The experiment was fulfilled in such a way that TFR consists of two parts: absorbing and non-absorbing X-rays. It allowed us to control the motion of the absorbing portion of TFR and, consequently, to study gas redistribution between this portion of TFR and CMR. The facility is shown in Figure 9. The principal element is that the low-pressure channel is divided by the moving closure in two sections that may be filled up with different gases. Closure is removed when both sections have been filled up.

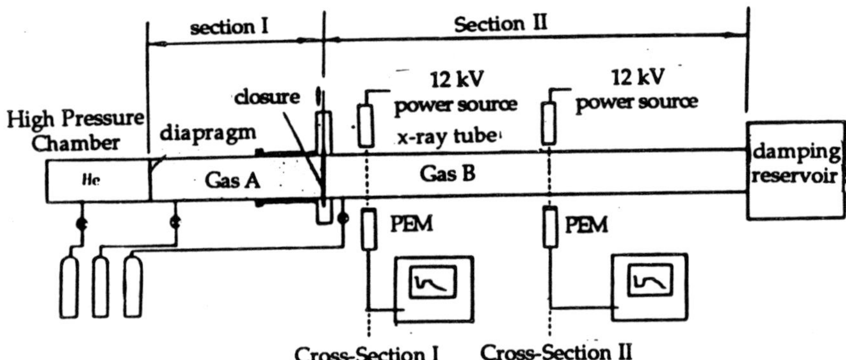

Fig. 9. Facility with two-section and low-pressure channel.

Three variants of low-pressure channel filling were used in experiments: 1 - absorbing gas fills the entire low-pressure channel, 2 - absorbing gas fills the second section (from closure to damping reservoir). Measurements were carried out at two cross-sections: just behind a closure at a distance of 1.12 m from the diaphragm (length of the first section is 1.07 m) and at a distance of 1.96 m from the diaphragm (the distance between closure and second measurement cross-section is 0.89 m).

When two sections of the low-pressure channel are filled with different gases the TFR consists of two parts: 1 - from the shock wave to the surface dividing working gases, and 2 - from this surface to the leading edge of CMR. Having arrived at the second measurement cross-section part 1 of the TFR is formed due to gas capturing by the shock wave and gas outflow through the boundary layer. Measuring gas mass in part 1 and comparing it with the gas mass between closure and working cross-section we can determine the gas mass that is flowed through the boundary layer. Part 2 of the TFR evolves due to gas outflow through the boundary layer and gas capturing to the turbulent zone when mixing at the leading edge of

CMR. Studying gas redistribution between part 2 of the TFR and CMR during their motion along the channel we can evaluate gas mass acquired by CMR due to turbulent mixing.

Thus, our method allows us to answer the following principal questions: what amount of gas is there in the CMR; what are the methods of gas transport to this region — due to boundary layer leakage or by the mixing process; what amount of gas is flowed out from hot TFR through the boundary layer.

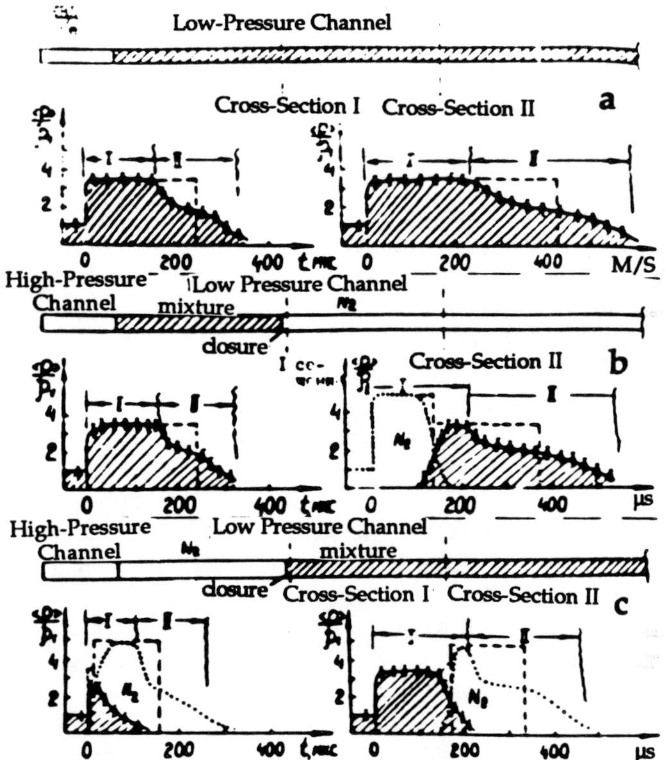

Fig. 10. Driven gas density distribution behind shock wave in two cross-sections at different low-pressure channel filling.

Dashed line — ideal TFR; broken line — nitrogen redistribution. Absorbing gas is shown by hatching; 1 — shock compressed TFR; 2 — CMR; $p_1 = 100$ torr, M = 5.

Special attention was paid to the choice of gas mixtures. Pairs of gases with equal molecular masses and equal or close heat capacities were used in order to reduce convective motion and to exclude a rise of additional shocks or expansion waves due to shock wave interaction at the gas boundaries. These phenomena are undesirable because they essentially complicate X-ray processing. Two pairs of gases were used in the experiments: 1) nitrogen + (2/3 Ar + 1/3 He) mixture with the same molecular mass value but different γ, absorption of nitrogen is negligible as compared to absorption of mixture caused by Ar presence; 2) Krypton + (Xe + He) mixture with the same molecular mass values and the same $\gamma = 1.67$, in this case absorption of krypton is less than the absorption of the (Xe + He) mixture by a factor of 2.5.

Analysis of mass redistribution in parts 1 and 2 of the TFR and analysis of the CMR shape variation due to effects of the above parts of the TRF allow us to compare the relative importance of bulk and wall effects on uniform flow shrinkage and CMR formation. Consider these questions for the regimes which in the first above-mentioned pair of gases was used as a working gas filling sections 1 and 2.

The variants of low-pressure channel filling and the correspondent density distributions at two cross-sections are presented in Figure 10. Figure 10a depicts density distributions at cross-sections 1 and 2 obtained without the closure. Figure 10b shows density distribution in the case when sections 1 and 2 are filled with a gas mixture and nitrogen respectively. In the latter case, the shock wave is formed in the mixture and then passes in nitrogen. Further growth of the hot TFR occurs due to nitrogen capturing by the shock wave. As seen in Figure 10b, density distribution in the contact region is not influenced by the leading edge of TFR. Density distributions at the second cross-section in the cases shown in Figures 10a and b coincide. When the absorbing mixture moves from the first cross-section to the second its redistribution occurs: in cross-section 1 the hot TFR contains 65% and CMR contains 35% of the total mixture mass $M_m = (\pi d^2/4)\rho_m \cdot X_1$, in cross-section 2 CMR contains 60%. In both cross-sections all mass M_m is concentrated TFR and contact region. This redistribution proves that gas capturing from the hot TFR into CMR due to turbulent mixing of driven and driver gases is the principal mechanism for the shock compressed TFR shrinkage.

In order to determine the boundary layer effect on the uniform flow shrinkage more exactly we used the following variant of the shock tube filling (see Figure 10c): the first section of the low-pressure channel was filled with nitrogen, the second with an (He + Ar) mixture. It is seen that the absorbing gas of the first part of TFR is not found in CMR up to the moment of the first part of TFR arrival at the second cross-section, i.e., gas flowing out through the boundary layer is not accumulated in CMR but rather is distributed along the channel as a shroud. Estimates show that all gas filling section 2 up to the measurement cross-section is found to be in the shock compressed TFR. Keeping in mind that the accuracy of mass determination in volume is equal to 10% we may assume that less than 10% of the total absorbing gas mass flowed out through the boundary layer.

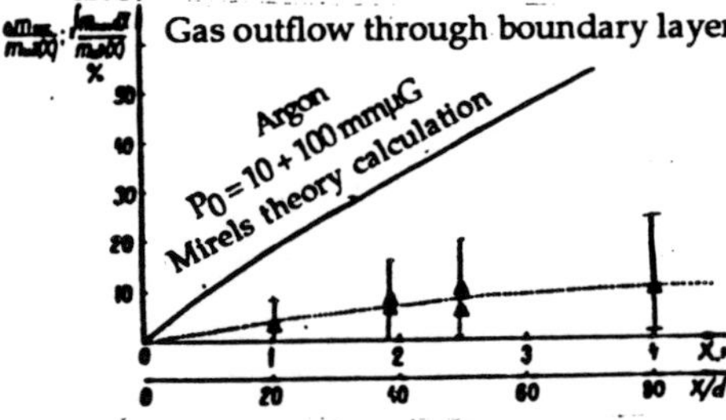

Fig. 11. Comparison of measured $\Delta M_{exp}/M_{id}(x)$ and calculated $\int M_{b.l.}(x)dx/M_{id}(x)$ values of driven gas mass flowing from hot TFR and CMR through the boundary layer, M = 9.

Figure 11 shows the total mass flowed through the boundary layer out of shock compressed TFR and CMR during the shock wave traveling from diaphragm to measurement cross-section. In this case the entire channel was filled with absorbing gas. Mass values shown for different shock tube cross-sections are referred to $M_{id}(x) = (\pi d^2/4)\rho_1 X$. The total mass of a gas flowing out of the shock' compressed TFR was also determined by means of calculations according to Mirels theory. These calculation results are shown in Figure 11 as well. It was found that measured gas outflow through the boundary layer is sometimes less than calculated. Thus, on the one hand, we observe quite satisfactory agreement of the experimental TFR lengths with those calculated according to Mirels theory; on the other hand, the experimental data show that this theory over-estimates gas outflow from the hot TFR. This contradiction is easily removed if we take into account that approximately one third of uniform flow shrinkage occurs due to gas outflow through the boundary layer but two thirds occur due to driven and driver gas mixing that results in mass transfer from the hot TFR and CMR.

5. Evolution of Turbulized Contact Mixing Region

Turbulent TFR formed during diaphragm opening and dividing driven and driver gases smears when moving along the channel [7]. To study this process measurements were carried out in several shock tube cross-sections. The results show that when moving along the shock tube CMR expands capturing new gas portions from the shock compressed in TFR (see Figure 12). An increase of driven gas mass in hot TFR and contact region is shown in Figure 13 along with variation of these region lengths. Both afore-mentioned masses increase at comparable rates. Relative shock TFR and CMR lengths are presented in the same figure. Hot TFR lengths agree well with the measurements by other investigators and are approximately two times less than calculated according to ideal theory.

Contact region smearing is accompanied with a motion of its leading edge in the shock compressed hot TFR. Flow patterns behind the shock wave were visualized with the use of the shadow method by Tepler in order to obtain quantitative data on the velocities of nonuniformities in CMR and TFR. The velocity of nonuniformities near the CMR leading boundary is found to be higher than flow velocity in the hot TFR. This effect was observed for all regimes considered.

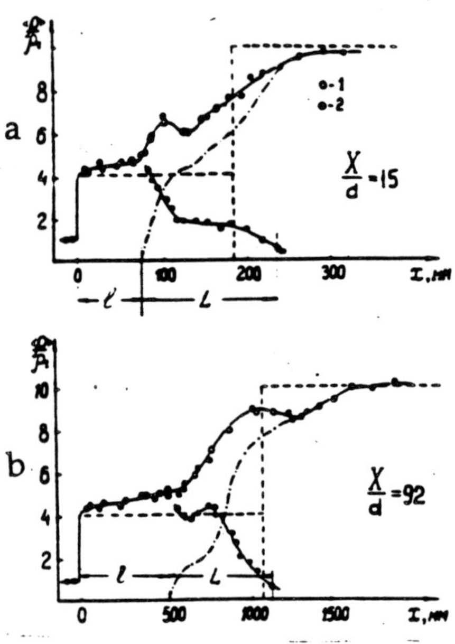

Fig. 12. Average density distribution of driven and driver gases behind shock wave at two distances from the diaphragm.

ρ_1 - density ahead of shock wavefront; l - the length of shock compressed TFR; L - CMR length. 1 - total gas density; 2 - driven gas density; broken line - driven gas density distribution in the mixing region; dashed line - density distribution of driven and driver gases in ideal shock tube.

$p_1 = 13$ torr, M = 9. p_1 is initial pressure in low-pressure channel. M is Mach number shock wavefront.

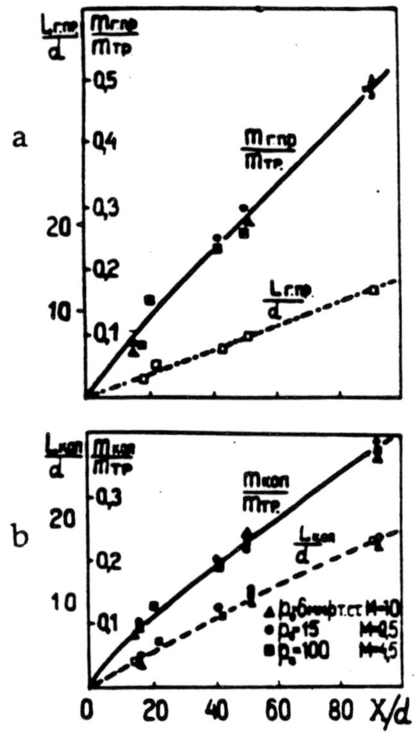

Fig. 13. a - Relative mass of driven gas in hot TFR and relative hot TFR length increase during TFR motion in shock tube channel. b - The same characteristics of CMR.

Quantitative data on CMR leading edge velocities can be obtained from Figure 14. The velocity of nonuniformities in hot TFR coincides with the calculated value of gas velocity behind the shock wave. It allows us to assume the velocity of nonuniformities in hot TFR to be equal to flow velocity. The velocity of nonuniformities near the CMR leading boundary exceeds the flow velocity in hot TFR by 10-15%.

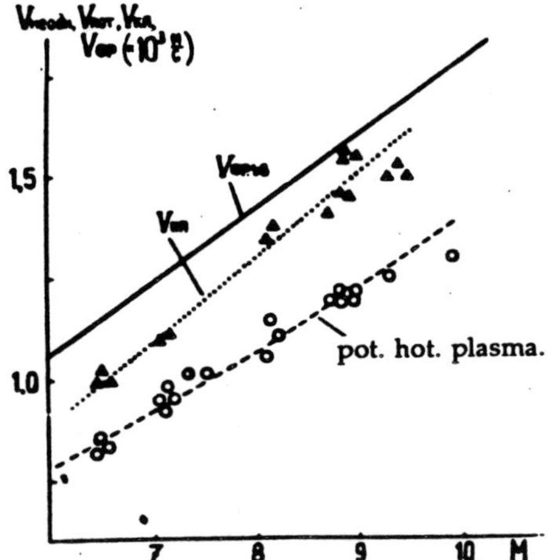

Fig. 14. Velocity of nonuniformities in hot TFR, CMR leading edge velocity versus shock wave Mach number.

o - velocity of nonuniformities in hot TFR; Δ - CMR leading edge velocity, ------ — calculated mean flow velocity; ····· — mean velocity of CMR leading edge.

The above results show driven gas transfer from the hot TFR to CMR due to turbulent mixing at the CMR leading edge. This process predominates over driven gas transfer to CMR through the boundary layer. This allows us to propose the following mechanism for shock compressed TFR formation; an increase of TFR length occurs due to gas capture by the shock wave while shrinkage is caused by gas transfer from hot TFR to CMR during its leading edge motion in the driven gas. Accordingly, contact region smearing is accompanied by gas reception from the hot TFR.

6. Rayleigh-Taylor Instability (RTI)

Rayleigh-Taylor instability is one of the best known phenomena that results in mixing at the two media boundary. For example, RTI may develop at the initial part of the shock tube where main flow acceleration occurs if gas density in hot TFR ρ_2 is higher than driver gas density ρ_3.

Using the approach of a small-scale turbulent diffusion, estimate the width of mixing area L when RTI develops at an initially undisturbed boundary dividing two gases with different densities. At the stage of diaphragm opening gas is accelerated from zero to 10^3 m/s per time t, which by an order of magnitude is equal to $2 \cdot 10^{-4}$ s. Then acceleration $a = 5 \cdot 10^6$ m/s^2. Acceleration being constant, the width of the turbulent mixing area is determined by formula

$$L = 270 \; \alpha^4 \ln\frac{\rho_3}{\rho_2} a t^2$$

where α is empirical constant. ρ_3/ρ_2 is driver and driven gas density ratio at the contact surface. Approximate values may be obtained according to formula $L \approx 0.1 \, at^2$. Substituting numerical values we obtain $L_0 \approx 2$ cm. At moment t_0 gas acceleration has stopped and further smearing of the turbulent layer occurs inertially in accordance with the formula

$$L = L_0 \left(1 + \frac{2}{n} \frac{t - t_0}{t_0} \right)^n$$

where n is a constant whose values are in the range from 0.25 to 0.29 in accordance with the estimates of Ref. [5]. Then assuming $n = 0.29$ we obtain the theoretical value of the turbulent layer width in the measurement cross-section $L \approx 6$ cm. In reality, CMR length exceeds the theoretical value calculated by nearly an order of magnitude with the use of a small scale diffusion approach.

In order to discuss RTI influence on CMR characteristics a comparative experimental investigation was performed for the cases $\rho_3/\rho_2 > 1$ and $\rho_3/\rho_2 < 1$. Different ratio ρ_3/ρ_2 values are determined by different shock wave Mach number M values that may be provided with the use of the same diaphragm due to variation in the shock tube. The function $\rho_3/\rho_2 = f(M)$ can be calculated for different regimes according to the ideal shock tube theory. When Mach numbers are small $\rho_3/\rho_2 < 1$ and flow acceleration provides the conditions for RTI development.

Density measurements in the cases $\rho_3/\rho_2 > 1$ and $\rho_3/\rho_2 < 1$ reveal that CMR integral characteristics (share of total driven gas mass in CMR and CMR length referred to length of the shock compressed TFR) are slightly dependent on the ρ_3 to ρ_2 ratio. The main RTI effects were found to be shrinkage of the observed non-uniformities and smearing of the CRM leading edge. Thus we conclude that RTI influence on CMR development is essentially less than the effects of the direct turbulizing factors at work during diaphragm opening.

RTI may be important as well far from the diaphragm when gas flows along the shock tube

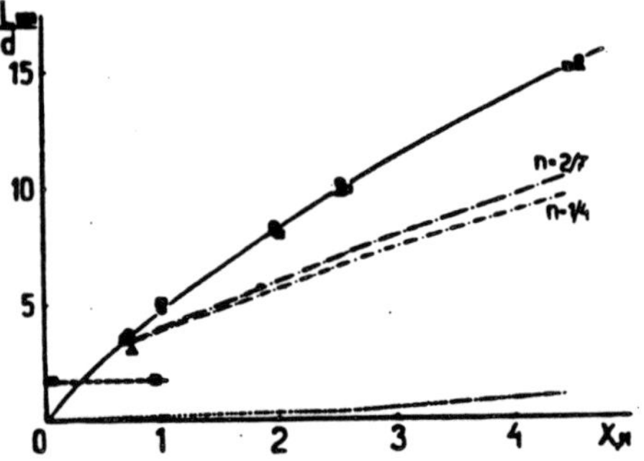

Fig. 15. CMR development in the shock tube.

Open circles — essentially turbulent region arising at diaphragm opening; solid circles — slightly turbulent region at closure removal. Calculations: ···· — RTI development; ········· — RTI attenuation after flow accelerating induced by diaphragm rupture has stopped; ——— according to formula $L = L_0(1 + (t - t_0)/t_0)^{0.7}$.

channel. In this case gas is accelerated due to the boundary layer effect and though this acceleration is significantly less than at the stage of diaphragm opening it works for a longer time.

Consider the following hypothetical case: mixing occurs at initial plane two media boundary, velocity variation is 5%/m of the low-pressure channel length, initial density ratio at the boundary is 2–3. Calculation results are shown in Figure 15 by a dashed line. Experimental data are also presented. We can see that the calculated CMR expansion due to RTI development for the hypothetical rapidly accelerated flow is essentially less than that measured.

7. Discussion of Results

On the basis of experimental data we can estimate some characteristics of turbulent flow in the mixing region. At $p_1 = 1.7$ kPa, $M_1 = 9$, flow velocity behind the shock wavefront $U = 1.2 \cdot 10^5$ cm/s. Estimate the velocity of turbulent region expansion. In cross-section $X = 4.6$ m mixing region length, $L_{CMR} = 90$ cm, then $v_{exp} = L_{CMR} \cdot U/X = 2.4 \cdot 10^4$ cm/s. Pulsation velocity v' is usually comparable with expansion velocity: $v' = (2-3) \cdot 10^4$ cm/s. In order to estimate the mixing path length we can use the formula: $l_{mix} = 0.1 \cdot L_{CMR} \approx 9$ cm. The turbulent diffusion coefficient is defined as $D_T = v' \cdot l_{mix} \approx 2 \cdot 10^5$ cm^2/s. It exceeds the molecular diffusion coefficient by an order of magnitude.

It is worth comparing inertial expansion of the essentially nonuniform CMR observed in experiment with theoretical results on smearing the turbulent mixing layer obtained within the framework of the small-scale turbulent diffusion approach. In our calculation we assume that CMR turbulization has finished the moment t_0 of its arrival on the first measuring cross-section and thereafter inertial smearing occurs according to the law of the small-scale turbulent diffusion approach [1, 6]. Calculation results are presented in Figure 15 by a broken line. It is seen that the calculated CMR increase in the shock tube end is considerably less than experimental. This demonstrates, first, an invalidity of the small-scale turbulent diffusion approach for the contact region in the shock tube, and, in the second place, the greater importance of energy accumulation in the turbulent region due to diaphragm rupture as çompared to possible RTI development. Time dependence of the turbulent region length presented in Figure 15 can be expressed by formula

$$L = L_0 \left(1 + \frac{t - t_0}{t_0} \right)^{0.7}$$

i.e., there is stronger dependence on time.

The degree of expansion of a slightly perturbed area during its motion in the shock tube can be obtained from experiments on a two-section shock tube with a closure. With the closure removed, the slightly turbulized translational region arises dividing the driven gases in the low-pressure channel. After the shock wave has passed this region it moves along the shock tube with a velocity equal to concurrent flow velocity. Variation of the length of this region during its motion from the first measurement cross-section to the second is shown in Figure 15 with solid circles. We can see a rather small variation of the slightly perturbed region length at a distance of approximately 20 diameters. Energy accumulation in CMR due to diaphragm rupture considerably exceeds the energy accumulated due to RTI development at the boundary of the gases with different densities. This conclusion is also confirmed by the analysis of x-t-diagrams of the flow shadowgraphs. The length of large-scale nonuniformities that are seen distinctly in the x-t-diagrams is about 1 cm. It is known (see [9]) that energy is mainly transferred by large eddies, so the length scale of nonuniformities may to some extent characterize energy accumulated in CMR. Flow in the 5 cm diameter channel with the length of nonuniformities equal to 1 cm cannot be considered small. In addition, a mixing layer develops in the channel with a diameter comparable or even less than the mixing layer width. Thus the flow considered differs from the known types of turbulent flows. To describe such flows we must repudiate the small-scale diffusion approach and elaborate new approaches that take into account the flow structure (length scales of large nonuniformities) and cross-size of the channel in which gas flows.

8. Conclusions

As a result of this investigation, data were obtained on some peculiarities of a shock tube operating as a device for scientific research and on driven and driver gas mixing evolution. The principal results and conclusions can be formulated as follows.

1. The range of X-ray diagnostic application in shock tube experiments is expanded to lower gas densities.

2. A method is developed to determine gas mass flowing from the different portions of TFR through the boundary layer.

3. Density distributions of driven and driver gases in CMR are first obtained. It is found that for a wide range of initial parameters the contact region contains about 30-40% of the ideal TFR mass.

4. It is shown that driven gas mass flowing through the boundary layer out of hot TFR and CMR is $\simeq 10\%$. The boundary layer influence on

hot TFR shrinkage becomes comparable to the mixing effect at an initial pressure of 5 torr.

5. The main process providing driven gas transfer to CMR is gas capturing from the hot TFR when the CMR leading edge moves relative to the driven gas. The same process results in hot TFR shrinkage. Shock compressed TFR is formed in the following way. When moving along the shock tube channel the shock wave involves new portions of driven gas into hot TFR, while some part of the driven gas leaves the hot TFR due to mixing at the leading edge of the contact region.

6. Quantitative laws for CMR evolution are stated. CMR length dependence on time is expressed by the approximation formula.

7. Flow character in CMR and its development describe the invalidity of a small-scale turbulent diffusion approach in this region. Because of the wide use of shock tubes CMR is considered to be an object for special treatment to obtain the necessary experimental data for working out complex models of turbulent flows.

References

1.　V. E. Neuvazhayev, *Zh. Prikl. Mekh. i Tekhn. Fiz.*, no. 5, 1983, pp. 81-88 (in Russian).

2.　B. Sturtevant, In: *Shock Tubes and Waves*, Proc. 16th Intern. Symp., Ed. by H. Grönig, Aachen, 1987, pp. 89-100.

3.　S. G. Zaitsev, E. V. Lazareva, V. V. Cherruha, and V. M. Belyaev, *Teplofiz. Vesok. Temp.*, vol. 23, no. 3, 1985, pp. 535-541 (in Russian).

4.　A. N. Aleshin, E. G. Gamalii, S. G. Zaitsev, et al., *Kvantovaya Elektronika*, vol. 14, no. 11, 1987, pp. 2299-2303 (in Russian).

5.　D. Besnard, M. Bonnet, S. Gauthier, et al. In: *Shock Tubes and Waves*, Proc. 16th Intern. Symp., Ed. by H. Grönig, Aachen, p. 709.

6.　V. E. Neuvazhayev, *Zh. Prikl. Mekh. i Tekhn. Fiz.*, no. 2, 1988, pp. 46-51 (in Russian).

7.　R. V. Vasilieva, A. D. Zuev, V. L. Moshkov, et al., *Zh. Prikl. Mekh. i Tekhn. Fiz.*, no. 7, 1985, pp. 128-133 (in Russian).

8.　H. Mirels, In: *Shock Tube Research*, Proc. 8th Intern. Symp., Ed. by . L. Strollery, A. G. Gaydon, and P. R. Owen, London, 1971, pp. 6.1-6.30.

9.　L. D. Landau and E. M. Lifshits, *Hydrodynamics*, Moscow: Nauka, 1986 (in Russian).

SPECIFIC FEATURES OF SHOCK WAVE DYNAMICS
IN SLAB EXPLOSION

E. V. Nazarov, M. P. Syschikova

Abstract: Up to the present numerous studies concerning strong explosions have been carried out. A point explosion theory describing processes due to an instant energy release in a small volume (as compared to an ambient one) appears to be the most advanced. The fundamental problem is a point explosion similarity problem formulated and solved by L. I. Sedov [1, 2]. In the following studies reviewed and analyzed in monographs [3, 4], the point explosion in a homogeneous atmosphere has been treated both analytically and numerically with a finite ambient pressure, gas viscosity, and thermal conductivity; radiative energy transfer, physical–chemical processes in the shock waves are taken into account.

However, a point explosion model is far from always acceptable. If the length scales of an energy release region and the problem character length are comparable, other models are preferred, for example, a model of a uniform instantly heated region. In that case new parameters appear which characterize the initial shape, size, and medium properties of the exploding volume.

In finite-volume explosion problems the flow pattern is much more complicated as compared to point explosion due to the presence of inner discontinuities, their interactions, and the appearance of new ones. Thus, the basic method for investigating the problem is numerical simulation followed by generalization of the computed results to provide their application to the expanded range of initial conditions. Special attention is paid to the asymptotic behavior of the solutions and their comparison with point-problem solutions [5-8].

This paper presents an analytical study of some peculiarities of shock wave dynamics in the finite width slab explosion depending on the new determinant parameters absent in the point explosion problem. Some results were published in [9].

The slab explosion problem is formulated as follows. At the moment $t = 0$, the flow starts from a slab of width $2 R_d$ with gas parameters P_1, ρ_1, k in the quiescent gas with parameters P_0, ρ_0, γ. Here P is gas pressure, ρ is gas density, k, γ are specific heat ratios. Shock wave (SW) arises in the medium, rarefaction wave (RW) propagates toward the slab center. The flow development in the medium is considered. The slab explosion is the flow that occurs in a normal shock tube. The initial stage of the wave propagation as well as the late acoustic stage of the slab explosion process can be treated analytically. The complete solution can be obtained numerically.

It is known that the determinant dimensionless parameters for the plane point explosion problem in a perfect gas are r, t, γ; the spatial and temporal scales usually relate to the initial energy of the explosion E_0:

$$r^0 = \frac{E_0}{P_0 \alpha^0}, \quad \tau^0 = \frac{r^0 \gamma^{1/2}}{a_0}.$$

Here a_0 is the sound velocity in the medium, α^0 is the similarity constant [3]. The dependence of the pressure at the shock wavefront from the point explosion on the distance from the explosion center is governed by the parameter γ.

In intrinsic spatial scale R_d arises in the slab explosion problem. To compare the slab explosion solution with that for plane point explosion we reserve r^0, τ^0 as the scales. Then we can take as the determinant parameters for the volume explosion the quantities r, t, r_d, a, γ, k, where r_d is the dimensionless half-width of the slab, a is the ratio of initial sound speed values in the slab and ambient medium. For the perfect gas

$$E_0 = \frac{2P_1 R_d}{k-1}, \quad r^0 = \frac{2p_i R_d}{(k-1)\alpha^0}, \quad r_d = \frac{k-1}{2p_i}\alpha^0.$$

Here $p_i = P_1/P_0$.

In the case of the slab explosion the shock wavefront parameters depend on not only the spatial coordinate r and the specific heat ratio γ. New, when compared to the point explosion case, parameters r_d, a, k influence the initial discontinuity evolution and the dynamics of the explosive zone development. Let us consider the effects of these parameters at the initial stage of the explosion.

The shock wave parameters remain constant until the rarefaction wave induced by initial discontinuity catches the shock wavefront after reflection from the slab center. The relative compression at the shock wavefront p_f on the interval where the pressure is constant as well as the length of this interval are easily calculable with the use of the initial parameters of the problem [10]. The results of calculations for $k = \gamma = 1.4$ presented in Figure 1 allow us to compare the shock wavefront pressure $p_{fd} = p_f$ at $r = r_d$ and the pressure $p_{fs} = p_f$ in the moment when the rarefaction wave catches the shock wave (at distance $r = r_s$) to the point explosion pressure p_f and to consider the effect of the determinant parameters r_d, a. Pressure variation at the shock wavefront versus a distance from the center according to the plane point explosion solution [3] is shown in Figure 1 for curve 1. Solid lines 2–7 correspond to $p_{fd}(r_d)$ at $a = \infty$, 10, 3.73, 2, 1, 0.5, dashed lines 3′–7′ to $p_{fs}(r_s)$ at $a = 10$, 3.73, 2.1, 0.5, respectively. In the case of the slab explosion in shock tube curves 2–7 provide the values of the initial shock wavefront pressure for a given ratio of sound velocities in a channel and a chamber as a function of the pressure difference at a diaphragm. When $r_d \to 0$ ($p_i \to \infty$) the front pressure is given by

$$p_{fd} \to \lim_{r_d \to 0} p_{fd} = 1 + \frac{2\gamma}{k-1}\,a\left\{\frac{\gamma+1}{2(k-1)}\,a + \sqrt{\left[\frac{\gamma+1}{2(k-1)}\,a\right]^2 + 1}\right\}$$

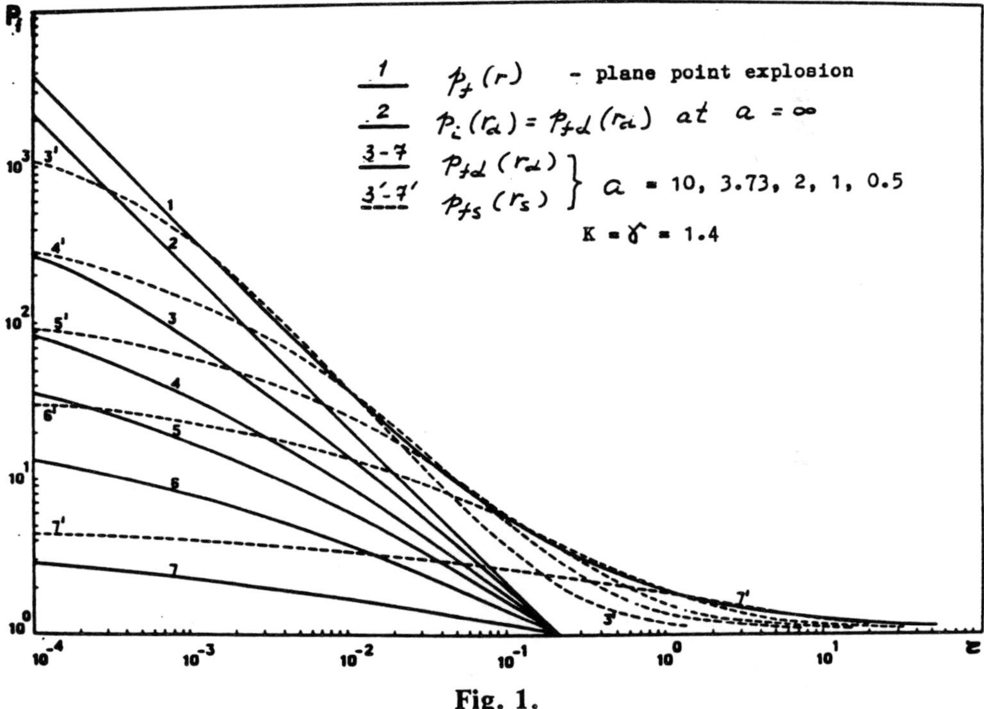

Fig. 1.

Since $p_{fd} \to p_i$ when $a \to \infty$ for any value of r_d a straight line 2 in Figure 1 relates p_i to r_d. The set of dependences presented in Figure 1 allows us to determine p_i, p_{fd} as well as the length $\Delta r_s = r_s - r_d$ of the interval of constant p_f for given values of r_d and a. If r_d is prescribed, the intersection of the ordinate r_d with curve 2 gives the value of p_i while the intersection with curves 3-7 gives the initial value of p_{fd} for a corresponding a. The length of Δr_s is determined as the difference of the coordinates of points belonging to curves 3-7 and 3'-7' corresponding to $p_{fs} = p_{fd}$ for a given a.

The comparison of the shock wavefront pressure p_{fs} with the plane point explosion pressure p_f reveals that for any given a there is a certain range of r_d values for which p_{fs} is close to the point-explosion value. The same property of the slab explosion solution is also found for other values of $k = \gamma$.

Calculation results for $k = \gamma = 1.1$ are presented in Figure 2. Here the function $p_f(r)$ for the plane point explosion is shown by a curve 1; the functions $p_{fd}(r_d)$ and $p_{fs}(r_s)$ for $a = 10, 4, 1$, and 0.1 are shown by curves 3-6 and 3'-6'; curve 2 corresponds to $p_{fd} = p_i(r_d)$ for $a = \infty$.

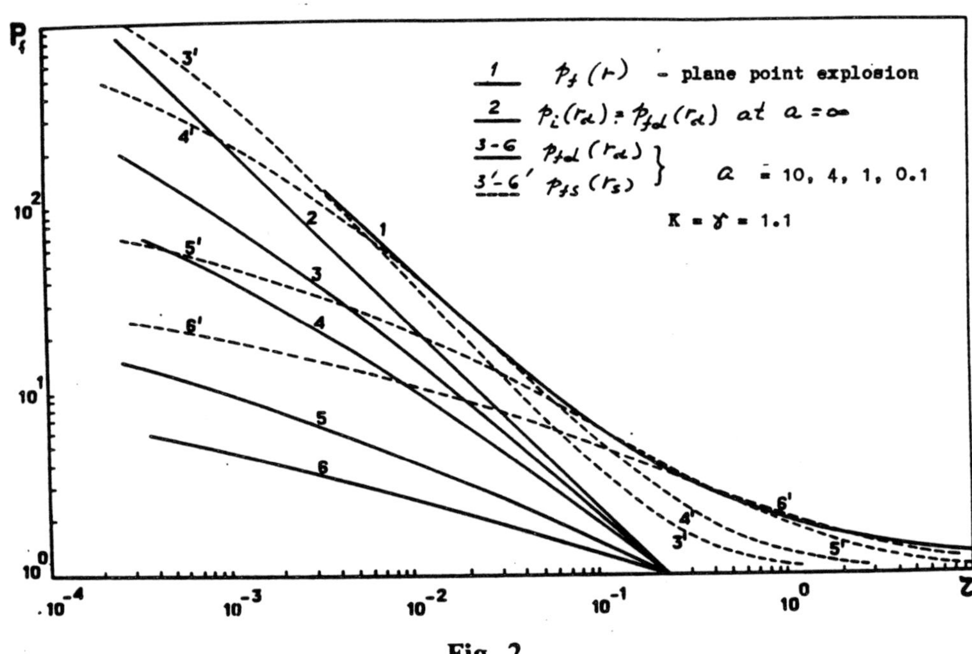

Fig. 2.

As seen in Figures 1 and 2, one-parameter family of curves $p_{fs} = f(r_s, a)$ has an envelope in the p_f, r plane and the envelope is close to the point explosion dependence $p_f(r)$ at $k = \gamma$.

The envelope equation cannot be obtained explicitly in the general case. However, several interesting features can be drawn from the analysis of the equations governing the envelope behavior.

The functional relation for the family $p_{fs}(r_s, a)$ can be easily derived from the equation for coordinate of the point where the rarefaction wave reflected from the slab center catches up with the shock wave

$$R_s = R_d + D \cdot \tau_s \tag{1}$$

where D is shock wave velocity, τ_s is the time of catching (dimensioned).

Using the time of intersection of the reflected head the rarefaction wave characteristic with that of the tail (see, for example [10])

$$\tau_c = \frac{R_d(1 + 2n)^{1+n}a_1^n}{[(1 + 2n)a_1 - U_{cc}]^{1+n}} \tag{2}$$

we obtain

$$\tau_s = 2\tau_c + \frac{2(D - U_{cc})\tau_c}{U_{cc} + a_{cc} - D} = \frac{2a_{cc}}{U_{cc} + a_{cc} - D} \frac{R_d}{a_1} \left[1 - \frac{U_{cc}}{(1 + 2n)a_1} \right]^{-(1+n)} \tag{3}$$

Here $2\tau_c$ is the time needed for the rarefaction wave to get the contact surface, subscript "cc" refers to the concurrent flow parameters.

Substituting τ_s into Equation (1) and using the dimensionless variables we obtain the function describing the family $p_{fs}(r_s, a)$:

$$F(p_{fs}, r_s, a) = r_s - \frac{a^0}{(1 + 2n)p_r p_{fs}} \left[1 + \frac{2M}{1 - M_{cc}} \frac{1}{a} p_r \right]^{\frac{1+n}{3+2n}} = 0 \tag{4}$$

Here $M = \dfrac{D}{a_0} = \sqrt{\dfrac{Np_{fs} + 1}{N + 1}}$ is the shock wave Mach number;

$M_{cc} = \dfrac{D - U_{cc}}{a_{cc}} = \sqrt{\dfrac{p_{fs} + N}{p_{fs}(N + 1)}}$ is the concurrent Mach number;

$p_r = p_i/p_{fs}$ is the initial pressure ratio in the rarefaction wave;

$N = \dfrac{\gamma + 1}{\gamma - 1}$ is the limit compression in the shock wave.

For the plane nonstationary rarefaction wave

$$p_r = \left[1 - \frac{1}{1+2n} \frac{U_{cc}}{a_1} \right]^{-(3+2n)} = \left[1 - \frac{1}{1+2n}\frac{N-1}{N}\frac{M^2-1}{M}\frac{1}{a} \right]^{-(3+2n)} \quad (5)$$

The envelope equation $p_{fen} = g(r_s)$ is obtained from Equation (4) by excluding a with the use of a condition $\partial F / \partial a = 0$ which results in the following equation for p_{ren} on the curve $g(r_s)$.

$$F(p_{ren}, n) = \frac{1+2n}{3+2n} \left[(3+n)p_{ren}^{\frac{1+n}{3+2n}} - (2+n)p_{ren}^{\frac{1+n}{3+2n}} \right] =$$

$$= \frac{N-1}{2N} \frac{(M^2-1)(1-M_{cc})}{M^2} = f(p_{fs}, N) \quad (6)$$

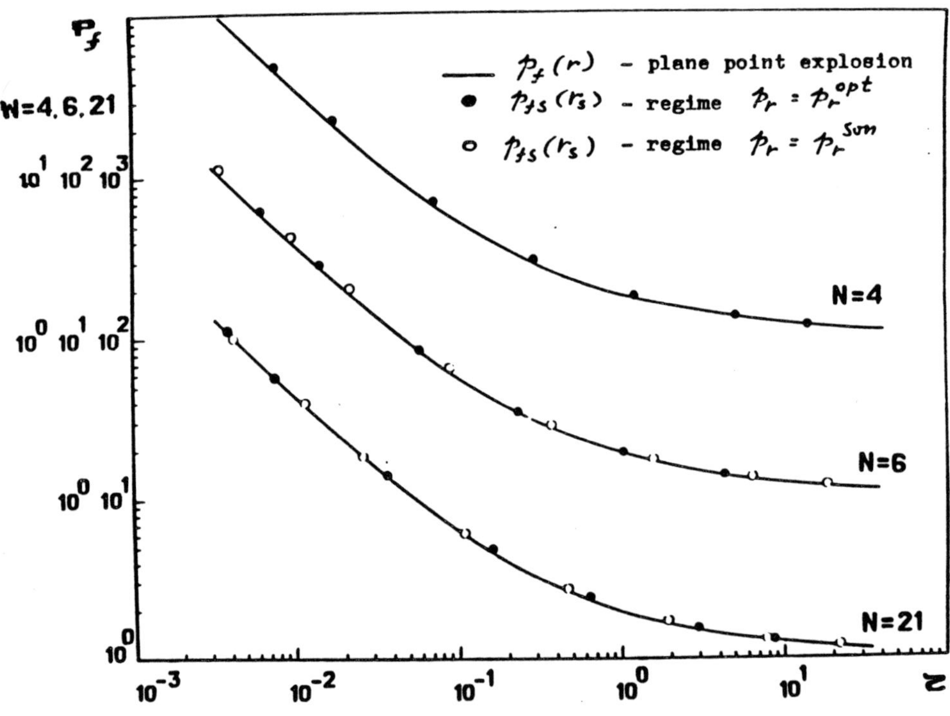

Fig. 3.

The functions $F(p_{\text{ren}}, n)$ and $f(p_{fs}, N)$ are presented in Figure 3 and provide information about the existence domain of the solution to Equation (6). The values of the right-hand side of Equation (6) belong to the interval

$$0 \leqslant f(p_{fs}, N) \leqslant f^{\max} = \frac{N-1}{N}\left(\frac{1}{N+1}\right)$$

that determines the domain of the p_{ren} values for the whole range of p_{fs}

$$1 \leqslant p_{\text{r en}}^{\min} \leqslant p_{\text{ren}} \leqslant p_{\text{r en}}^{\max} = \left(\frac{3+n}{2+n}\right)^{3+2n} \tag{7}$$

From (5) the range of a variation along the envelope $0 \leqslant a < \infty$ follows, i.e., $p_r = p_{\text{r en}}^{\max}$ corresponds to the limit point on the envelope $g(r_s)$ $p_{fs} \to 1$, $a \to 0$, and $p_r = p_{\text{r en}}^{\min}$ corresponds to $p_{fs} \to \infty$, $a \to \infty$.

It is seen from (7) that the domain of p_{ren} values on $g(r_s)$ for all admissible k, γ is bounded from above by

$$\lim_{n \to \infty} \left(\frac{3+n}{2+n}\right)^{3+2n} = e^2 = 7.39.$$

In every concrete case the interval of p_{ren} values is rather narrow, for example, $2.94 \leqslant p_{\text{ren}} \leqslant 4.21$ for $k = \gamma = 5/3$ (see Figure 3).

The left-hand side of Equation (6) increases monotonously when p_{ren} decreases and for $p_{\text{ren}} = 1$ is equal to $(1 + 2n)/(3 + 2n) = 1/k$. Thus the condition of envelope existence reads as

$$\frac{N-1}{2N}\left(1 - \sqrt{\frac{1}{N+1}}\right) \leqslant \frac{1}{k}$$

Since the left-hand side of inequality (8) does not exceed 0.5, the condition is satisfied for any γ if $k \leqslant 2$; $n = 2 + 2n$ for $k = \gamma$ and inequality (8) becomes

$$\frac{N+1}{2N}\left(1 - \sqrt{\frac{1}{N+1}}\right) = \frac{1}{2} + \frac{1 - \sqrt{N+1}}{2N} \leqslant 1,$$

that is always satisfied. Evidently, for $k < \gamma$ (8) is all the more valid.

Thus a small region $2 < k \leqslant 3$ remains where envelope existence is not evident for large p_{fs} and $\gamma \lesssim 1.17$. As an example, the maximum value

p_{fs} above which there is no envelope for $k = 3$ ($n = 0$), $\gamma = 1.1$ ($N = 21$) is shown in Figure 3.

Let us compare the values of p_{fs} and p_f on the envelope and on the point-explosion curve in the region of high values. In that region p_{ren} asymptotically approaches p_{ren}^{min} on the envelope and

$$\frac{r_{sen}}{r_d} \rightarrow l_{en}^{min} = \lim_{p_{fs} \rightarrow \infty} \left[1 + \frac{2M}{1 - M_{cc}} \frac{1}{a} p_{ren}^{\frac{1+n}{3+2n}} \right] .$$

Excluding a with the use of Equation (5) and calculating a limit we obtain

$$l_{en}^{min} = 1 + \frac{2N}{N - 1} \left[1 - \sqrt{\frac{1}{N + 1}} \right]^{-1} \frac{1 + 2n}{2 + n} \left(p_{ren}^{min} \right)^{\frac{n}{3+2n}} - \frac{3 + 2n}{2 + n} \tag{9}$$

For the strong stage of the point explosion the similarity condition yields (in dimensioned form [3]),

$$p_{fp} = \frac{8E_0}{9(\gamma + 1)\alpha^0 R} . \tag{10}$$

The ratio p_{fen} to p_{fp} for the same $R = R_s$, the same energy E_0 and $k = \gamma$ can be written as

$$\frac{p_{fen}}{p_{fp}} = \frac{g(\gamma - 1)\alpha^0 l_{en}^{min}}{16 p_{ren}^{min}} . \tag{11}$$

The results of calculation using Equation (11) are presented in Table 1. The envelope $g(r_s)$ lies higher than the point explosion curve for small n and lower for large n, but the difference is small.

The subsequent analysis is hampered by the necessity to solve Equation (6) relating p_{ren}, p_{fs}, n, N.

The slow variation of p_r along the envelope allows us to consider the constant p_r solution.

First we treat the case

$$p_r = p_{ren}^{max} = \left(\frac{3 + n}{2 + n} \right)^{3+2n} .$$

It is of interest because the relative energy $\varphi_{cs} = E_{cs}/E_0$ transferred through the contact surface during the time τ_{cs} needed for the rarefaction wave to catch up with the contact surface turns out to be maximum. Indeed, for the plane explosion

$$\varphi_{cs} = \frac{2P_{cs} \cdot U_{cs} \cdot \tau_{cs}}{E_0} = \frac{4P_{cc} \cdot U_{cc} \cdot \tau_c}{E_0} = 4\left(1 - p_r^{-\frac{1}{3+2n}}\right) p_r^{-\frac{2+n}{3+2n}} \qquad (12)$$

it results in

$$\varphi_{cs} = \varphi_{cs}^{\max} = \frac{4}{3+n}\left(\frac{2+n}{3+n}\right)^{2+n} \text{ at } p_r = p_r^{\text{opt}} = \left(\frac{3+n}{2+n}\right)^{3+2n} \qquad (13)$$

All the admissible values $\varphi_{cs}^{\max}$, p_r^{opt} are in the ranges $0 \leqslant \varphi_{cs}^{\max} \leqslant 0.592$, $3.375 \leqslant p_r^{\text{opt}} \leqslant 7.39$.

The simple expression for p_r^{opt} enables us to examine this regime more thoroughly. To begin with we note its main features:

1) p_r^{opt} depends only on the slab gas specific heat ratio;

2) the contact surface velocity is close to the sound velocity in the slab:

$$U_{cs}^{\text{opt}} = \frac{1+2n}{3+n} a_1;$$

for $n = 2$ $(k = 1.4)$ $U_{cs}^{\text{opt}} = a_1$;

3) the rarefaction wave sound velocity decreases $(3 + n)/(2 + n)$ times, i.e., 1.5 times at most;

4) the density decreases $\left(\dfrac{3+n}{2+n}\right)^{1+2n}$ times, i.e., from 1.5 to e^2 times;

5) the Mach number after the rarefaction wave flow acceleration

$$M_r = \frac{1+2n}{2+n}, \quad \text{i.e., } 0.5 \leqslant M_r < 2;$$

for $n = 1$, $M_r = 1$?

6) the shock wave Mach number is defined by the expression

$$\frac{M^2 - 1}{M} = \frac{N}{N - 1}\frac{1+2n}{3+n} a,$$

from which it follows that for high pressures p_{fs} the shock wave propagation velocities in different gases is $(k = \gamma)$

$$\frac{2}{3} Q_1 \leqslant D < 2a_1$$

for $n = 1$, $D = a_1$, i.e., shock wave velocity equals the sound velocity in the slab (velocity of energy releasing the rarefaction wave propagation).

From Equation (4) the expression relating r_s and p_{fs} follows for $p_r = p_r^{opt}$ regime

$$r_{sopt} = r_d\left\{1 + \frac{2(1 + 2n)}{3 + n}\left(\frac{3 + n}{2 + n}\right)^{1+n} \frac{Np_{fs} + 1}{(N - 1)(p_{fs} - 1)}\left[1 - \sqrt{\frac{p_{fs} + N}{p_{fs}(N+1)}}\right]^{-1}\right\}$$

$$= r_d[1 + Y(n)Z(N, p_{fs})]. \tag{14}$$

The results of calculations using this formula for $k = \gamma = 1.4$, 1.1, 1.66 are shown by solid ·circles in Figure 4. The difference with the point explosion solution for p_f in the whole range $r = r_s$ considered does not exceed 7% and is usually smaller.

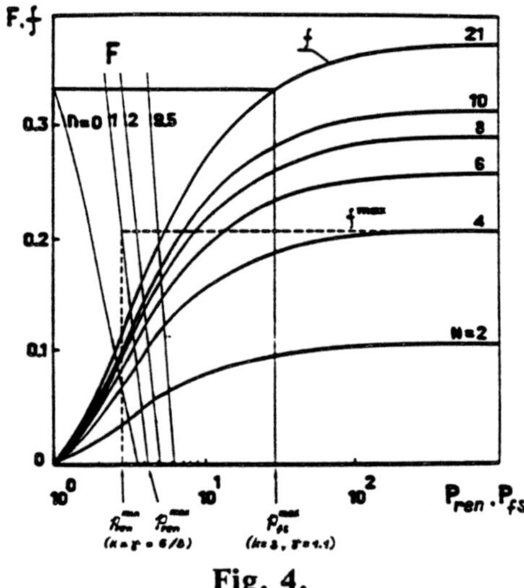

Fig. 4.

When $p_{fs} \to \infty$ $r_{sopt} \to 0$, but

$$\frac{r_{sopt}}{r_d} \to l_{opt}^{min} = 1 + \frac{1 + 2n}{2 + n}\left(\frac{3 + n}{2 + N}\right)^n \frac{2N}{N - 1}\left(1 - \sqrt{\frac{1}{N + 1}}\right)^{-1}$$

We employ this limiting value to compare the $p_r = p_r^{opt}$ regime to the point explosion at high pressure. To this end we substitute l_{opt}^{min} and p_r^{opt} in Equation (11) instead of l_{en}^{min} and p_{ren}^{min}. The results are presented in Table 1. As expected, for the regime p_r^{opt} the pressure p_{fs}^{opt} is lower than the envelope and for small n is closer to the point explosion curve.

Table 1

n	$\dfrac{P_{fs}^{opt}}{P_{fp}}$	$\dfrac{P_{fs}^{son}}{P_{fp}}$	$\dfrac{P_{fen}}{P_{fp}}$	α^0
1	1.064	1.064	1.092	0.603
2	1.042	1.064	1.065	1.077
9.5	0.965	0.954	0.977	4.506

For $p_{fs} \to 1$ Equation (14) reduces to an asymptotic solution

$$
\underset{rs \to \infty}{p_{fs} \to} 1 + 2(N + 1)\sqrt{\frac{\alpha^0}{2 + n}\left(\frac{2 + n}{3 + n}\right)^{3+n}\frac{1}{N(N - 1)}\frac{1}{r\, r_s}}
$$

$$
= 1 + \frac{4\gamma}{r\,2(\gamma + 1)}C\left(k, \frac{E_0}{P_0}\right)\frac{1}{r R_s}
$$

$$(16)$$

The structure of Equation (16) is similar to Sedov's asymptotic solution [11]:

$$
p_f \simeq 1 + \frac{2}{\gamma + 1}\sqrt{\frac{const}{R}},
$$

however, it is necessary to keep in mind a principle distinction in the nature of the process to be described. Sedov's formula describes an attenuation of the point explosion shock wave at a large distance while Equation (16) gives the asymptotic decrease of the initial pressure for a shock wave arising due to slab explosion with initial parameters providing sufficiently large coordinate shock wave catching by the rarefaction wave. Moreover,

$$
C\left(k, \frac{E_0}{P_0}\right) = \sqrt{\left(\frac{2 + n}{3 + n}\right)^{3+n}\frac{1}{2 + n}\frac{E_0}{P_0}}
$$

is not constant since R_s depends on E_0 and P_0. Nevertheless, Equation (16) at $k = \gamma$ provides good accord with the point explosion pressure for $r = r_s \geq 2$ (see Table 2) where p_{fk} is the solution for the point explosion in homogeneous atmosphere [3]).

Table 2

$n = 1$		$n = 2$		$n = 9.5$	
r, r_s	$\dfrac{p_{fs}}{p_{fk}}$	$r, _r s$	$\dfrac{p_{fs}}{p_{fk}}$	r, r_s	$\dfrac{p_{fs}}{p_{fk}}$
1.792	0.982	1.698	0.956	1.606	0.949
4.804	0.989	4.504	0.988	4.172	0.993
7.671	0.994	7.195	0.995	5.599	0.995
41.87	0.999	38.72	1.003	35.00	1.005

Evidently, the asymptotic solution (16) is valid for an envelope with $p_{\text{ren}} \to p_r^{\text{opt}}$ at $p_{fs} \to 1$ as well.

Thus the $p_r = p_r^{\text{opt}}$ regime gives the relation $p_{fs}(r_s)$ which is close to the point explosion for all $0 < r_s < \infty$.

Another regime for which the pressure p_{fs} is close to the point explosion solution for $r = r_s$ occurs when gas flow is accelerated in the rarefaction wave to the sonic velocity at the early stage of slab explosion. In that case,

$$\varphi_{cs}^{\text{son}} = \frac{2}{1 + n}\left[\frac{1 + 2n}{2 + 2n}\right]^{2+n}, \quad p_r^{\text{son}} = \left(\frac{2 + 2n}{1 + 2n}\right)^{3+2n} \tag{17}$$

$$r_s^{\text{son}} = r_d\left[1 + 2\left(\frac{2 + 2n}{1 + 2n}\right)^n Z(N, p_{fs})\right] \tag{18}$$

The main features of the sonic regimes are:

1) the sonic and optimal regimes are identical for $n = 1$ ($k = 5/3$);

2) $p_r^{\text{son}} \leqslant p_r^{\text{opt}}$ for $n \geqslant 1$, for $n < 1$ $p_r^{\text{son}} > p_r^{\text{opt}} = p_{r\,\text{en}}^{\text{max.}}$,

3) the rarefaction wave sound velocity decrease is $(2 + 2n)/(1 + 2n)$, i.e., at most twofold;

4) the rarefaction wave density decrease is $\left(\dfrac{2 + 2n}{1 + 2n}\right)^{1+2n}$, i.e., at least twofold;

5) the shock wave Mach number is defined by the relation

$$\frac{M^2 - 1}{M} = \frac{N}{N - 1} \frac{1 + 2n}{2 + 2n} \, a,$$

Table 3

r, r_s	p_{fk}	p_{fs}^{opt}	δ_{opt}	p_{fs}^{son}	δ_{son}
$N = 4 \quad n = 1 \ (k = \gamma = 5/3)$			$p_r^{opt} = 4.214 \quad p_r^{son}$		
0.003415	99.47	106.25	0.068		
0.02135	17.75	18.96	0.068		
0.1078	5.08	5.38	0.059		
3.222	1.425	1.452	0.019		
11.89	1.214	1.225	0.009		
41.87	1.113	1.116	0.003		
$N = 6 \ n = 2 \ (k = \gamma = 1.4) \quad p_r^{opt} = 4.768$				$p_r^{son} = 3.583$	
0.003423	109.9	115.3	0.049	117.5	0.069
0.04862	9.72	10.28	0.058	10.39	0.069
0.1071	5.45	5.77	0.059	5.79	0.062
0.4493	2.53	2.65	0.047	2.64	0.045
0.9473	1.94	2.01	0.036	2.00	0.031
38.72	1.119	1.127	0.007	1.126	0.006
$N = 21 \ n = 9.5 \ (k = \gamma = 1.1) \quad p_r^{opt} = 6.261$				$p_r^{son} = 2.925$	
0.003406	124.4	122.3	-0.017	121.6	-0.022
0.006297	68.49	67.64	-0.012	67.10	-0.020
0.01003	43.90	43.63	-0.006	43.20	-0.016
0.02111	22.03	22.23	0.009	21.87	-0.007
0.04840	10.79	11.12	0.031		
0.1050	6.012	6.295	0.047		
0.2541	3.489	3.685	0.056	3.550	0.017
0.4341	2.687	2.834	0.055		
0.9026	2.030	2.128	0.048		
1.606	1.717	1.787	0.041	1.739	0.013
2.916	1.503	1.552	0.033		
4.172	1.409	1.449	0.028		
35.00	1.129	1.142	0.011		

which reduces to $(M^2 - 1)/M = a$ for $k = \gamma$, i.e., for $p_{fs} \to \infty$ and any n $D \to a_1$.

The results of calculation using (18) are presented by open circles in Figure 4. Table 1 contains the ratio P_{fs}^{son}/P_{fp} for $k = \gamma$ and

$$l_{s\,on}^{min} = 1 + 2 \left[\frac{2 + 2n}{1 + 2n} \right]^n \frac{N}{N - 1} \left(1 - \sqrt{\frac{1}{N + 1}} \right)^{-1}, \quad p_r = p_r^{son}$$

To demonstrate agreement between the function $p_{fs}(r_s)$ and the function $p_f(r)$ correspondent to the point explosion problem, Table 3 presents a comparison of the results obtained by Formulae (14) and (18) with the results of numerical calculation p_{fk} of the point explosion in the finite pressure ambient medium [3].

The relative error δ is defined as $(p_{fs} - p_{fk})/p_{fk}$.

Thus the conditions for slab explosion realization considered provide the dependence of the shock wavefront pressure p_{fs} on the coordinate r_s of the shock wave by catching by the rarefaction wave which is close enough to the point explosion one $p_f(r)$ for the whole range of p_f values. These regimes are either optimal or close to optimal with regard to explosion energy transfer to the ambient gas during the time needed by the rarefaction wave to catch the contact surface. Apparently, it is this property that predetermined the closeness to point explosion solutions, since the last model provides the ideal (optimal) energy transfer to the ambient medium.

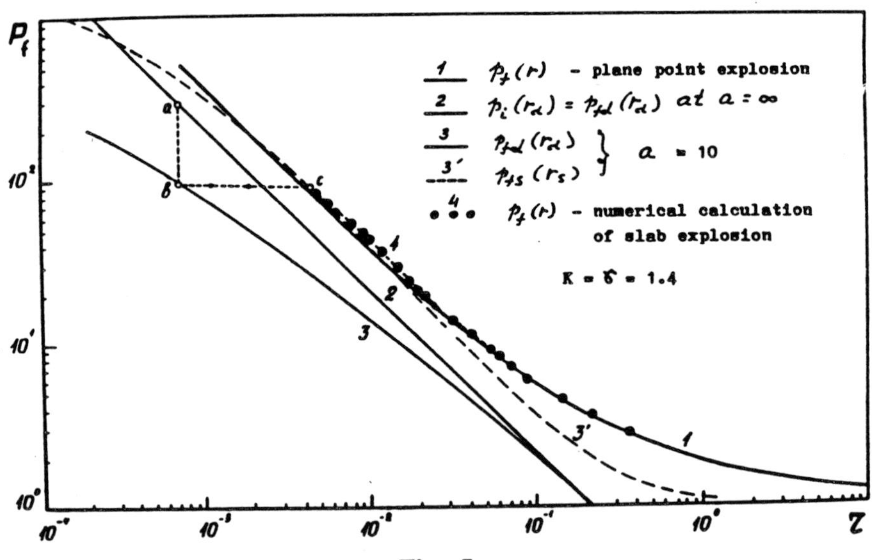

Fig. 5.

For other values of determinant parameters the shock wavefront pressure at $r = r_s$ is less than the point explosion one and approaches it only at $r \gg r_s$.

As selected numerical computations reveal, the flow after the catching of the shock wave by the rarefaction wave evolves in such a way that the shock wavefront pressure follows the point explosion dependence also for $r > r_s$ if the initial conditions are in accordance with the envelope or close to it. The example of computations ($a = 10$, $r_d = 7.18 \cdot 10^{-4}$, $p_i = 300$, $k = \gamma = 1.4$) is presented in Figure 5.

The established properties of the slab explosion solution appear to be useful when analyzing explosion processes in real media or choosing the conditions of explosion simulation.

References

1. L. I. Sedov, *Prik. Matem. i Mekh.*, vol. 9, no. 4, 1945, pp. 35-39 (in Russian).
2. L. I. Sedov, *Dokl. AN SSSR*, vol. 52, no. 1, 1946, pp. 7-9 (in Russian).
3. Kh. S. Kestenboim, G. S. Roslyakov, and L. A. Chudov, *Point Explosion*, Moscow: Nauka, 1974 (in Russian).
4. V. P. Korobeinikov, *Point Explosion Theory Problems*, Moscow: Nauka, 1985 (in Russian).
5. G. Broud, *Explosions Calculations on Computers. Gasdynamics of Explosions*, Moscow: Mir, 1976, pp. 96-159 (in Russian).
6. A. S. Fonarev and S. Yu. Chernyavskii, *Izv. AN SSSR, Mekhanika Zhid. n Gaza*, no. 5, 1968, pp. 169-174 (in Russian).
7. P. C. Chou and S. L. Huang, *J. Appl. Phys.*, vol. 40, no. 2, 1969, pp. 752-759.
8. M. Lutzky and D. L. Lehto, *Ibid.*, vol. 41, no. 2, 1970, pp. 844-846.
9. E. V. Nazarov and M. P. Syschikova, *Zh. Tekhn. Fiz.*, vol. 56, no. 11, 1986, pp. 2262-2265 (in Russian).
10. K. P. Stanjukovich, *Nonstationary Motion of Continuum*, Moscow: Nauka, 1971 (in Russian).
11. L. I. Sedov, *Similarity Methods in Mechanics*, Moscow: Nauka, 1965 (in Russian).

CALCULATION OF SHOCK WAVE-MOVING BODY INTERACTION

V. P. Goloviznin, I. V. Krasovskaya

*Abstract:*An effective method is proposed for calculation of nonstationary supersonic plane and axisymmetric inviscid blunt body flows in nonuniform flowfields of the explosion type, thermal nonuniformity, etc. The method is based on an explicit predictor-corrector MacCormack scheme for Euler equations written in the form of integral conservation laws using Cartesian coordinates. A computational grid is adapted to the motion of the bow shock wave that is treated as a discontinuity bounding the computational domain.

Calculation results are presented on the axisymmetric interaction of a plane shock wave with constant concurrent flow parameters and a blast shock wave from instantaneous point explosion in a medium with finite pressure with a blunted body moving at supersonic speed.

1. Introduction

Nonstationary effects arising with supersonic flight in nonuniform media essentially change the shock layer structure and aerodynamic coefficients of the object. Demand for the determination of such flow parameters has stimulated constant interest in the calculation methods. Inviscid flow calculation results provide the data necessary for convective and radiative heat transfer evaluation and the outer boundary conditions for boundary layer equations.

Most numerical methods for supersonic flow calculations using bow shock fitting procedures (treating the bow shock as a discontinuity bounding the computational domain) exploit coordinate transformations. The reason for using coordinate transformation is that, for the problems considered, boundary condition implementation often turns out to be more complicated than the basic equations. Boundary conditions are easily applied in a transformed computational domain because of its simple shape. In the present paper, however, the basic equations are written in integral form using the

Cartesian coordinate system and their solution is sought in the physical flow domain. A simple way to satisfy all boundary conditions is demonstrated. The use of the integral form of the basic equations implies that gas discontinuities in the shock layer (if there are any) will be manifested as the steep gradients of gasdynamic parameters.

2. Formulation of the Problem

A blunted axisymmetric body is supposed to move at supersonic speed V_∞ in quiescent uniform isotropic medium. Speed V_∞ is constant and parallel to the body symmetry axis. Stationary flow with a detached bow shock wave occurs near the body under these conditions. Suppose that there is a spherical domain initiated by an instantaneous point explosion at finite ambient pressure ahead of the body. The explosion center does not move relative to the medium and is located at the body symmetry axis. The problem is to calculate nonstationary flow parameters during body motion within the explosion domain. Figure 1 shows a scheme for axisymmetric blast shock wave interaction with a body flying at supersonic speed. Here OX is the body and flow symmetry axis; FED is the body surface generatrix, PBC is the bow shock wave before interaction; O' is the explosion center, HP is part of the blast shock wave of radius $X_1(t_0)$ at moment t_0 before interaction; GB is part of the blast shock wave of radius $X_1(t)$ at moment $t > t_0$; AB is part; of the refracted bow shock wave; XOY is the Cartesian orthogonal coordinate system bound up with the body. Three regions may be considered in the gas flow: 1) Explosion region unperturbed by the body; 2) Shock layer near the body; 3) Still gas. Region 2 is bounded by the body surface, FED; flow axis segment AF; part of the bow shock wave ABC; and segment CD which closes the region 2 boundary. Segment CD is located in the shock layer in such a way that the normal gas velocity component at this boundary exceeds the local sound velocity value. Within region 3 flow parameters, i.e., gas velocity relative to the body $q_\infty = -V_\infty$ density ρ_∞ and pressure p_∞ are known. If gas parameters in region 1 are known at any moment as well, we can calculate the flow in region 2.

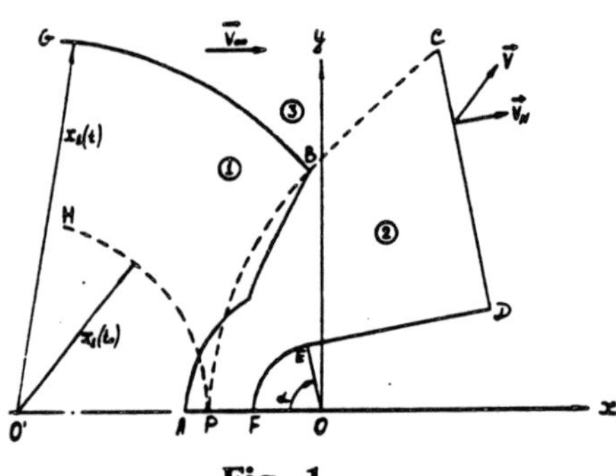

Fig. 1.

Gas parameters in regions 1 and 2 will be determined within the framework of an inviscid gas model described by Euler equations. The unsteady Euler equation for two-dimensional flows may be written with the use of a Cartesian orthogonal coordinate system (Figure 1) in the following differential form [1]:

$$\frac{\partial F_t}{\partial t} + \frac{\partial F_x}{\partial x} + \frac{\partial F_y}{\partial y} = 0,$$

(1)

where

$$F_t \equiv U = \begin{bmatrix} \rho \\ \rho u \\ \rho v \\ \rho e \end{bmatrix} \quad \vec{\mathcal{H}} = \begin{bmatrix} \rho \vec{q} \\ \rho u \vec{q} + p \vec{i}_x \\ \rho \vartheta \vec{q} + p \vec{i}_y \\ (\rho e + p) \vec{q} \end{bmatrix} \quad F_x = \vec{i}_x \cdot \vec{\mathcal{H}}, \quad F_y = \vec{i}_y \cdot \vec{\mathcal{H}},$$

(2)

$\vec{i}_t, \vec{i}_x, \vec{i}_y$ are the orths of coordinate axes t, x, and y, respectively; t, q, p, $\vec{q}$, and e are time, density, pressure, gas velocity, and total specific energy, respectively

$$e = \frac{1}{\gamma - 1} \frac{p}{\rho} + \frac{1}{2} \vec{q}^2,$$

(3)

γ is the gas specific heat ratio; u, ϑ are velocity vector $\vec{q}$ components with respect to the axes OX and OY:

$$u = \vec{q} \cdot \vec{i}_x, \quad \vartheta = \vec{q} \cdot \vec{i}_y.$$

Introduce the vectors

$$\vec{F} = F_t \vec{i}_t + F_x \vec{i}_x + F_y \vec{i}_y \quad \text{and} \quad \vec{\nabla} = \frac{\partial}{\partial t} \vec{i}_t + \frac{\partial}{\partial x} \vec{i}_x + \frac{\partial}{\partial y} \vec{i}_y$$

then equations (1) can be written as

$$\vec{\nabla} \cdot \vec{F} = 0,$$

(4)

Integrating (4) over volume Ω in (t, x, y) space and applying the Gauss theorem gives

$$\int_S \vec{F} \cdot \vec{N} dS = 0,$$

(5)

where S is the volume Ω boundary, $\vec{N}$ is the orth of the outer normal to the surface S.

3. Finite-Difference Grid Generation

Computational region 2 is divided into the mesh cells (i, j), $i = 1, 2,$ ..., I, $j = 1$, ..., J, as shown in Figure 2. The same indices (i, j), $i = 1, 2,$ ..., $I+1$, $j = 1, 2,$ ..., $J+1$ will be used to denote the nodes. For instance, node 1 of the cell (i, j) in Figure 2 is denoted by indices (i, j); node 2 by $(i, j+1)$; node 3 by $(i + 1, j + 1)$; node 4 by $(i + 1, j)$. Later comments will be made about whether the indices (i, j) relate to the mesh cell or to the node. The nodes $j = J + 1$ are always located at the bow shock wave, the nodes $j = 1$ are located at the body surface. The mesh segments with $i = $ const do not change their location in physical space. Thus as the computational grid is moving, its cells fill the shock layer whose thickness varies with time. This grid is rather general and applies to calculation of the flows around bodies of various shapes.

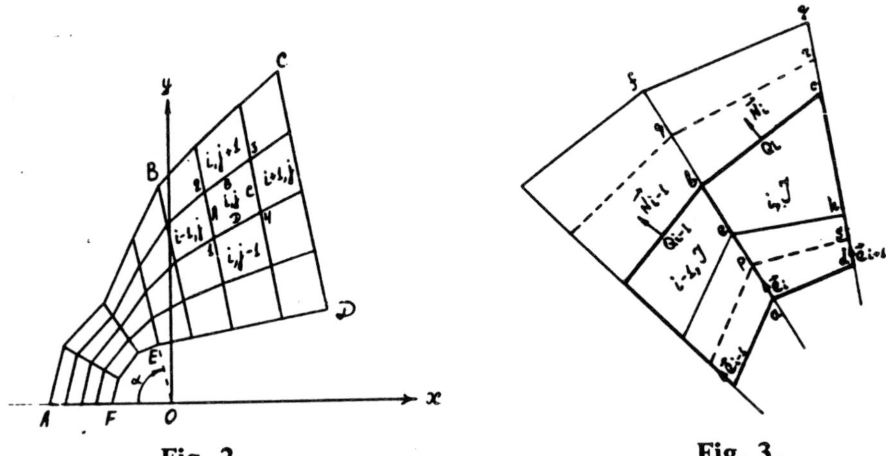

Fig. 2.   Fig. 3.

To describe the grid motion consider cell (i, J) adjacent to the shock wave (Figure 3) at moments $t^n > t_0$ (quadrangle *abcd*) and $t^{n+1} = t^n + \Delta t$, $\Delta t > 0$ (quadrangle *efgh*). Gas parameters in all cells of region 2 at moment $t = t^n$ are known. To calculate the node coordinates of cells (i, j) at the $(n+1)^{\text{th}}$ time step Rankine-Hugoniot relations are used at part *bc* of the bow shock wave

$$\rho_{2i}^n(q_{2Ni}^n - D_{i,J+1/2}^n) = \rho_{mi}^n(q_{mNi}^n - D_{i,J+1/2}^n),$$

$$p_{2i}^n + \rho_{2i}^n(q_{2Ni}^n - D_{i,J+1/2}^n)^2 = p_{mi}^n + \rho_{mi}^n(q_{mNi}^n - D_{i,J+1/2}^n)^2, \tag{6}$$

$$\frac{\gamma \rho_{2i}^n}{(\gamma - 1)\rho_{2i}^n} + \frac{(q_{2Ni}^n - D_{i\,J+1/2}^n)^2}{2} = \frac{\gamma \rho_{mi}^n}{(\gamma - 1)\rho_{mi}^n} + \frac{(q_{mNi}^n - D_{i\,J+1/2}^n)^2}{2},$$

where $D_{i\,J+1/2}^n$ is velocity of the bow shock wave part b_c along the outer normal $\bar{N}_i^n$ to the shock front,

$$q_{2Ni}^n = \bar{q}_{2i}^n \cdot \bar{N}_i^n,$$

$$q_{mNi}^n = \bar{q}_{mi}^n \cdot \bar{N}_i^n.$$

This direction corresponds to positive q_{2Ni}^n and $D_{i\,J+1/2}^n$ values. Oncoming stream parameters (regions 1 and 3 in Figure 1) are denoted by index $m = 1$, 3. Gas parameters just behind the bow shock front are denoted by index 2.

Equations (6) are easily transformed to

$$\rho_{2i}^n = \rho_{mi}^n \frac{(\gamma - 1)p_{mi}^n + (\gamma + 1)p_{2i}^n}{(\gamma + 1)p_{mi}^n + (\gamma - 1)p_{2i}^n},$$

$$D_{i\,J+1/2}^n = q_{mNi}^n + \left[\frac{p_{2i}^n - p_{mi}^n}{\rho_{mi}^n(1 - \rho_{mi}^n/\rho_{2i}^n)} \right]^{1/2}, \tag{7}$$

$$q_{2Ni}^n = D_{i\,J+1/2}^n + \rho_{mi}^n(q_{mNi}^n - D_{i\,J+1/2}^n)/P_{2i}^n.$$

If all oncoming stream parameters (regions 1 and 3) and pressure p_{2i}^n behind the bow shock front are known we can determine $D_{i\,J+1/2}^n$ from (7). Pressure p_{2i}^n at a point Q_i halving the segment bc is determined by a quadratic extrapolation with the use of pressure values $p_{i\,J}^n$, $p_{i\,J-1}^n$, $p_{i\,J-2}^n$ inside the shock layer [2]. In the case of uniformity in the j mesh we have, for instance,

$$p_{2i}^n = 0.125 \cdot (15p_{i\,J}^n - 10 \qquad p_{i\,J-1}^n + 3p_{i\,J-2}^n) \tag{8}$$

and obtain all values $D_{i\,J-1/2}^n$, $i = 1, 2, ..., I$ from Equations (7). Suppose $\bar{w}_{i\,j}^n$ is mesh node (i, j) velocity in the ray $i = $ const directions; l_i^n is the distance between the nodes $(i, J + 1)$ and $(i + 1, J + 1)$, $i = 1, 2, ..., I; e_i$ is the orth of the ray $i = $ const (Figure 3). Then we can write the following formulae for velocity

$$\bar{w}_{i\,j}^n = \frac{\dfrac{D_{i-1\,J+1/2}^n l_i^n}{(e_i \cdot N_{i-1}^n)} + \dfrac{D_{i\,J+1/2}^n l_{i-1}^n}{(e_i \cdot N_i^n)}}{(l_{i-1}^n + l_i^n)J} \; (j - 1) \cdot \bar{e}_i, \quad i \neq 1, i \neq I + 1;$$

$$\vec{w}^n_{1,j} = \frac{D^n_{i,J+1/2}}{(\vec{e}_1 \cdot \vec{N}^n_1)J} \, (j - 1) \, \vec{e}_1, \qquad (9)$$

$$\vec{w}^n_{I+1,j} = \frac{D^n_{I,J+1/2}}{(\vec{e}_{I+1} \cdot \vec{N}^n_I)J} \, (j - 1) \, \vec{e}_{I+1}, \qquad (9)$$

Denoting the radius-vector of the (i, j) node at the nth time step by $\vec{r}^n_{i,j}$ we have at $(n + 1)^{\text{th}}$ time step

$$\vec{r}^{n+1}_{i,j} = \vec{r}^n_{i,j} + \vec{w}^n_{i,j}\Delta t \qquad (10)$$

where $i = 1, 2..., I + 1, \ j = 1, 2, ..., J + 1$.

4. Calculation of Gas Parameters in the Inner Mesh Cells

We will write below the MacCormack scheme that approximates the set of Equation (5). The finite-difference representation of these equations corresponds to time $t = t^{n+1/2} = t^n + \Delta t/2$. Coordinates of the mesh nodes at this moment

$$\vec{r}^{n+1/2}_{i,j} = \vec{r}^n_{i,j} + \tfrac{1}{2}\vec{w}^n_{i,j}\Delta t \qquad (11)$$

where $i = 1, 2, ..., I + 1, \ j = 1, 2, ..., J + 1$.

Predictor stage:

$$\begin{aligned}
\bar{U}_{i,j} \, \mathrm{vol}^{n+1}_{i,j} = {} & U^n_{i,j} \, \mathrm{vol}^n_{i,j} - \Delta t[\vec{\mathcal{H}}^n_{i,j}\vec{S}^{n+1/2}_{i-1/2,j} + \vec{\mathcal{H}}^n_{i+1,j}\vec{S}^{n+1/2}_{i+1/2,j} \\
& + (\vec{\mathcal{H}}^n_{i,j+1} - U^n_{i,j+1}\bar{D}^n_{i,j+1/2})\vec{S}^{n+1/2}_{i,j1/2} \\
& + (\vec{\mathcal{H}}^n_{i,j} - U^n_{i,j}\bar{D}^n_{i,j-1/2})\vec{S}^{n+1/2}_{i,j-1/2} \\
& - G^n_{i,j}(\vec{S}^{n+1/2}_{i-1/2,j} + \vec{S}^{n+1/2}_{i+1/2,j} + \vec{S}^{n+1/2}_{i,j1/2} + \vec{S}^{n+1/2}_{i,j-1/2}) \cdot \vec{i}_y].
\end{aligned} \qquad (12)$$

Corrector stage:

$$\begin{aligned}
U^{n+1}_{i,j} \, \mathrm{vol}^{n+1}_{i,j} = {} & \tfrac{1}{2}\{U^n_{i,j}\mathrm{vol}^n_{i,j} + \bar{U}_{i,j}\mathrm{vol}^{n+1}_{i,j} - \Delta t[\vec{\mathcal{H}}_{i-1,j}\vec{S}^{n+1/2}_{i-1/2,j} \\
& + \vec{\mathcal{H}}_{i,j}\vec{S}^{n+1/2}_{i+1/2,j} + (\vec{\mathcal{H}}_{i,j} - \bar{U}_{i,j}\bar{D}^n_{i,j+1/2})\vec{S}^{n+1/2}_{i,j+1/2}
\end{aligned}$$

$$+ \tilde{\mathcal{H}}_{i,j-1} - \bar{U}_{i,j-1}\bar{D}^n_{i,j-1/2})\bar{S}^{n+1/2}_{i,j-1/2} - \tilde{G}_{i,j}(\bar{S}^{n+1/2}_{i-1/2,j}$$

$$+ \bar{S}^{n+1/2}_{i+1/2,j} + \bar{S}^{n+1/2}_{i,j+1/2} + \bar{S}^{n+1/2}_{i,j-1/2}) \cdot \bar{i}_y]\}.$$

Here $G = (o, o, p, o)$ is column-vector; $\text{vol}^n_{i,j}$ and $\text{vol}^{n+1}_{i,j}$ are the (i, j) cell volumes at n^{th} and $(n+1)^{\text{th}}$ time steps; $\bar{S}^{n+1/2}_{i-1/2,j}$, $\bar{S}^{n+1/2}_{i,j+1/2}$, $\bar{S}^{n+1/2}_{i+1/2,j}$, $S^{n+1/2}_{j,j-1/2}$ are vectors of verges area of the (i, j) mesh cell at $t = t^n + 1/2\Delta t = t^{n+1/2}$ (Figure 2); $\bar{D}^n_{i,j+1/2}$ and $\bar{D}^n_{i,j-1/2}$ are velocities of verges B and D of the mesh cell (i, j). Products $\bar{D}^n_{i,j+1/2}\bar{S}^{n+1/2}_{i,j+1/2}$ and $\bar{D}^n_{i,j-1/2}\bar{S}^{n+1/2}_{i,j-1/2}$ in Formulae (12) and (13) are replaced by $\text{vol}_{bfgc}/\Delta t$ and $\text{vol}_{aehd}/\Delta t$, respectively (see Figure 3). It allows us to avoid determination of the velocities $\bar{D}^n_{i,j+1/2}$ and $\bar{D}^n_{i,j-1/2}$ of the verges B and D. The volume of quadrangles $bfgc$ and $aehd$ are easily calculated.

Coordinates (x_k, y_k), $k = 1, 2, 3, 4$ (Figure 3) of the nodes of mesh cell (i, j) at n^{th} or $(n+1)^{\text{th}}$ time steps being known we can calculate the mesh cell volumes $\text{vol}_{i,j}$ using the following formulae:

for plane flow:

$$\text{vol}^n_{i,j} = \tfrac{1}{2}[|(x^n_2 - x^n_1)(y^n_4 - y^n_1) - (x^n_4 - x^n_1)(y^n_2 - y^n_1)| + \tag{14}$$

$$+ |(x^n_2 - x^n_3)(y^n_4 - y^n_3) - (x^n_4 - x^n_3)(y^n_2 - y^n_3)|],$$

for axisymmetric flow

$$\text{vol}^n_{i,j} = \tfrac{1}{3}[(x^n_3 - x^n_2)((y^n_2)^2 + (y^n_3)^2 + y^n_2 y^n_3) +$$

$$+ (x^n_2 - x^n_1)((y^n_2)^2 + (y^n_1)^2 + y^n_1 y^n_2) + (x^n_1 - x^n_4)((y^n_1)^2 + \tag{15}$$

$$+ (y^n_4)^2 + y^n_1 y^n_4) + x^n_4 - x^n_3)((y^n_4)^2 + (y^n_3)^2 + y^n_3 y^n_4)].$$

Verge area and outer normal for mesh cell (i, j) in plane flow are determined by formulae (superscripts n, $n+1/2$, or $n+1$ are omitted):

$$S_{i-1/2,j} = |\bar{S}_{i-1/2,j}| = ((x_2 - x_1)^2 + (y_1 - y_2)^2)^{1/2},$$

$$N_{xi-1/2,j} = \bar{N}_{i-1/2,j} \cdot \bar{i}_x = (y_1 - y_2)/S_{i-1/2,j},$$

$$N_{yi-1/2,j} = \bar{N}_{i-1/2,j} \cdot \bar{i}_y = (x_2 - x_1)/S_{i-1/2,j},$$

$$S_{i,j+1/2} = ((x_3 - x_2)^2 + (y_2 - y_3)^2)^{1/2},$$

$$N_{x,i,j+1/2} = (y_2 - y_3)/S_{i,j+1/2},$$

$$N_{y,i,j+1/2} = (x_3 - x_2)/S_{i,j+1/2},$$

$$S_{i+1/2,j} = ((x_4 - x_3)^2 + y_3 - y_4)^2)^{1/2},$$

$$N_{x,i+1/2,j} = (y_3 - y_4)/S_{i+1/2,j},$$

$$N_{y,i+1/2,j} = (x_4 - x_3)/S_{i+1/2,j}, \tag{16}$$

$$S_{i,j-1/2} = ((x_1 - x_4)^2 + (y_4 - y_1)^2)^{1/2},$$

$$N_{x,i,j-1/2} = (y_4 - y_1)/S_{i,j-1/2},$$

$$N_{y,i,j-1/2} = (x_1 - x_4)/S_{i,j-1/2}.$$

In an axisymmetric problem the outer normals to the mesh cell verges are determined by the same formulae as in the plane one. Verge areas are calculated in the following manner:

$$S_{i-1/2,j} = ((x_2 - x_1)^2 + (y_1 - y_2)^2)^{1/2}(y_1 + y_2),$$

$$S_{i,j+1/2} = ((x_3 - x_2)^2 + (y_2 - y_3)^2)^{1/2}(y_2 + y_3),$$

$$S_{i+1/2,j} = ((x_4 - x_3)^2 + (y_3 - y_4)^2)^{1/2}(y_3 + y_4), \tag{17}$$

$$S_{i,j-1/2} = ((x_1 - x_4)^2 + (y_4 - y_1)^2)^{1/2}(y_4 + y_1).$$

In the axisymmetric case values of $\bar{S}_{i-1/2,j}$ and $\bar{N}_{i-1/2,j}$ at $i = 1$, $j = 1, 2, ..., J$ are not calculated since this verge turns out to be the flow axis segment and the corresponding fluxes become equal to zero.

5. Calculation of Gas Parameters in the Boundary Cells

a) Body surface.

The verges D, of the cells $(i, 1)$, $i = 1, 2, ..., I$ are adjacent to the body surface. Boundary flux in (12) is

$$(\vec{\mathcal{H}}_{i,j}^n - U_{i,j}^n \bar{D}_{i,j-1/2}^n)\bar{S}_{i,j-1/2}^{n+1/2}|_{j=1} = \vec{\mathcal{H}}_{i,j}^n \cdot \bar{S}_{i,j-1/2}^{n+1/2}|_{j=1} \tag{18}$$

because the verges adjacent to the body surface are fixed $\bar{D}^n_{i,j-1/2}|_{j=1} = 0$. Since the vectors $\vec{q}^n_{i,j}|_{j=1}$ and $\vec{S}^{n+1/2}_{i,j-1/2}|_{j=1}$ are orthogonal

$$\vec{\mathcal{H}}^n_{i,j} \cdot \vec{S}^{n+1/2}_{i,j-1/2}|_{j=1} = \begin{bmatrix} 0 \\ p^*_i \cdot \vec{S}^{n+1/2}_{i,1/2} \cdot \vec{i}_x \\ p^*_i \vec{S}^{n+1/2}_{i,1/2} \cdot \vec{i}_y \\ 0 \end{bmatrix}. \tag{19}$$

Analytical solution of the Riemann problem [3] is used to determine p^*_i values at the body surface:

$$p^*_i = p^n_{i,1} \left[\frac{1 + \left(\dfrac{\gamma-1}{2} q^n_{Ni,1} \right)}{\left(\gamma \dfrac{p^n_{i,1}}{\rho^n_{i,1}} \right)^{1/2}} \right]^{2\gamma/(\gamma-1)} \tag{20}$$

$$p^*_i = p^n_{i,1} + \frac{\gamma+1}{4} \rho^n_{i,1}(q^n_{Ni,1})^2 + \left[\frac{1}{4}(2p^n_{i,1} + \frac{\gamma+1}{2} \times \right.$$
$$\left. \rho^n_{i,1}(q^n_{Ni,1})^2 \right]^2 - \left[(p^n_{i,1} - \frac{\gamma-1}{2} \rho^n_{i,1}(q^n_{Ni,1})^2 \, p^n_{i,1}) \right]^{1/2}. \tag{21}$$

The value of p^*_i is calculated according to (20) if normal to the body surface gas velocity component $q_{Ni,1}$ is directed outward and according to (21) in another case. The calculations analogous to (18)–(21) are carried out to obtain the term

$$(\vec{\mathcal{H}}_{i,j-1} - U_{i,j-1}\bar{D}^n_{i,j-1/2}) \cdot \vec{S}^{n+1/2}_{i,j-1/2}|_{j=1} = \vec{\mathcal{H}}_{i,j-1} \cdot \vec{S}^{n+1/2}_{i,j-1/2}|_{j=1} \tag{18'}$$

in Equation (13) ($i = 1, 2, ..., I$).

$$\vec{\mathcal{H}}^n_{i,j-1}\vec{S}^{n+1/2}_{i,j-1/2}|_{j=1} = \begin{bmatrix} 0 \\ \tilde{p}^*_i \vec{S}^{n+1/2}_{i,1/2} \cdot \vec{i}_x \\ \tilde{p}^*_i \vec{S}^{n+1/2}_{i,1/2} \cdot \vec{i}_y \\ 0 \end{bmatrix}. \tag{19'}$$

The values of $\tilde{p}^*_i$ are calculated according to Formulae (20) and (21) where p^*_i, $p^n_{i,1}$, $\rho^n_{i,1}$ and $q^n_{Ni,1}$ must be replaced by $\tilde{p}^*_i$, $\tilde{p}_{i,1}$, $\tilde{\rho}_{i,1}$ and $\tilde{q}_{Ni,1}$, respectively.

b) Outflow boundary, i.e., segment *CD* located in the supersonic flow area.

Verges *C* of cells (I, j) $j = 1, 2, ..., J$ are adjacent to this boundary. Correspondent fluxes in (12) and (13) are calculated by formulae

$$\vec{\mathcal{H}}^{n}_{i+1,j} \cdot \vec{S}^{n+1/2}_{i+1/2,j}\big|_{i=I} = \begin{bmatrix} (\rho\bar{q})^{n}_{I,j} \\ (\rho u\bar{q} + \bar{p}\bar{i}_x)^{n}_{I,j} \\ (\rho\vartheta\bar{q} + \bar{p}\bar{i}_y)^{n}_{I,j} \\ ((\rho e + p)\bar{q})^{n}_{I,j} \end{bmatrix} \cdot \vec{S}^{n+1/2}_{I+1/2,j},$$

$$\vec{\mathcal{H}}_{i,j} \cdot \vec{S}^{n+1/2}_{i+1/2,j}\big|_{i=I} = \begin{bmatrix} (\tilde{\rho}\tilde{q})_{I,j} \\ (\tilde{\rho}\tilde{u}\tilde{q} + \tilde{p}\tilde{i}_x)_{I,j} \\ (\tilde{\rho}\tilde{\vartheta}\tilde{q} + \tilde{p}\tilde{i}_y)_{I,j} \\ ((\tilde{q}\tilde{e} + \tilde{p})\tilde{q})_{I,j} \end{bmatrix} \cdot \vec{S}^{n+1/2}_{I+1/2,j}. \tag{22}$$

The "drift" of gas parameter values in (22) is substituted by supersonic flow velocity near this boundary.

c) Flow axis boundary.

Verges *A* of cells $(1, j)$, $j = 1, 2, ..., J$ are adjacent to this boundary. In axisymmetric flows the verges *A* degenerate to the flow axis segment, therefore correspondent fluxes in (12) and (13) are equal to zero

$$\vec{\mathcal{H}}^{n}_{i,j} \cdot \vec{S}^{n+1/2}_{i-1/2,j}\big|_{i=1} = \vec{\mathcal{H}}_{i-1,j} \cdot \vec{S}^{n+1/2}_{i-1/2,j}\big|_{i=1} = 0, \quad j = 1, 2, ..., J. \tag{23}$$

In plane flow segment *AF* is not part of the calculation domain boundary.

d) Bow shock wave boundary.

Verges *B* of cells (i, J) $i = 1, 2, ..., I$ coincide with the shock wave surface and move with velocity $D^{n}_{i,J+1/2}N^{n}_i$. It should be noted that $D^{n}_{i,J+1/2}$ is the velocity of the bow shock part bounded by the fixed rays i and $i + 1$. If the verge *B* center Q_i of the cell (i, J) is located at interval *BC* (see Figure 1) and does not coincide with point *B*, then calculating

$$(\vec{\mathcal{H}}^{n}_{i,j+1} - U^{n}_{i,j+1}\vec{D}^{n}_{i,J+1/2}) \cdot \vec{S}^{n+1/2}_{i,j+1/2}\big|_{j=J} \tag{24}$$

in (12) and

$$(\vec{\mathcal{H}}_{i,j} - U_{i,j}\vec{D}^{n}_{i,J+1/2}) \cdot \vec{S}^{n+1/2}_{i,j+1/2}\big|_{j=J} \tag{25}$$

in (13) unperturbed gas parameters from region $\underline{3}$, i.e., ρ_∞, $\bar{q}_\infty$, p_∞ are employed instead of $\rho^n_{i\,j+1}$, $\bar{q}^n_{i\,j+1}$, $p^n_{i\,j+1}$, $\rho_{i,j}$, $q_{i,j}$, and $p_{i,j}$, respectively. If the gas parameters of the oncoming stream ahead of the bow shock at points Q_i of region 1 (Figure 1) at moments t^n and t^{n+1} are known we can easily calculate the correspondent fluxes by substituting these parameters in (24)–(25). Scheme (12)–(13) along with boundary conditions (a)–(d) allows us to obtain the solution for region 2 at $t = t^n$ if the data on the solution at $t = t^{n+1}$ are disposed.

To provide stability, time increment value Δt of this scheme is limited by the following conditions [1, 2]:

1)

$$\Delta t \leqslant \min_{i,j} \frac{\text{vol}^n_{i,j}}{|\bar{q}^n_{i,j}\bar{S}^n_{i-1/2,j}| + |\bar{q}^n_{i,j}\bar{S}^n_{i,j-1/2}| + \left[\left(\gamma\dfrac{p^n_{i,j}}{\rho^n_{i,j}}\right)\left((\bar{S}^n_{i-1/2,j})^2 + (\bar{S}^n_{i,j-1/2})^2\right)\right]^{1/2}}$$

This inequality is the well-known KFL condition that is necessary for convergence of the numerical method as well.

2) Grid displacement bf at each time step must not exceed the (i, J) cell size, ab (Figure 3), $i = 1, 2, ..., I$.

3) Shock wave displacement GB (Figure 1) at each time step must not exceed bc (Figure 3). Numerical solutions obtained by means of Scheme (12)–(13) are smoothed at each time step with the use of a flux-correction procedure [4]. In the case of two-dimensional flow the smoothing algorithm is as follows

$$f^{n+1}_{i,j} = \tilde{f}_{ij} + (\varphi^*_{i+1/2,j} - \varphi^*_{i-1/2,j} + \varphi^*_{i,j+1/2} - \varphi^*_{i,j-1/2})/\text{vol}^n_{i,j},$$

$$\varphi^*_{i+1/2,j} = \begin{cases} \varphi_{i+1/2,j}, & \text{if } (\delta\tilde{f}_{i,j}\delta\tilde{f}_{i+1,j} < 0) \vee (\delta\tilde{f}_{i,j}\delta\tilde{f}_{i-1,j} < 0), \\ 0, & \text{in another case} \end{cases} \tag{27}$$

$$\varphi_{i+1/2,j} = Q\delta f^n_{i,j}\,\Psi(\text{vol}^n_{i,j}, \text{vol}^n_{i+1,j}),$$

$$\delta f_{i,j} = f_{i+1,j} - f_{i,j}.$$

Fluxes $\varphi^*_{i-1/2,j}$, $\varphi^*_{i,j+1/2}$, and $\varphi^*_{i,j-1/2}$ are determined by analogous formulae. Here $Q = \text{const}$ is the smoothing parameter, Ψ is a function that satisfies the condition

$$\Psi(a, b) = \Psi(b, a) \quad \Psi(\alpha, \alpha) = \alpha.$$

In the present paper we assume $\Psi(a, b) = 2ab/(a + b)$.

Vector $f = (\rho, \rho u, \rho \vartheta, \rho e)$ represents a smoothed solution; vector $\tilde{f}$ the solution before smoothing.

6. Calculation of Gas Parameters in Region 1

If region 1 (see Figure 1) is the explosion region the problem of instantaneous point explosion with finite ambient pressure is solved to obtain flux values at the AB boundary. Consider the algorithm applied to solve this one-dimensional problem for three types of symmetry: $\nu = 1$, plane explosion; $\nu = 2$, cylindrical explosion; and $\nu = 3$ spherical explosion. The coordinate system origin is placed in explosion center $0'$ and as noted earlier is fixed relative to the quiescent isotropic medium ($u_\infty = 0$, $\rho_\infty = $ const; $p_\infty = $ const). Let r be the distance from the explosion center, $r = X_1(t)$ corresponding to the front of the blast wave propagating with velocity D_g relative to the ambient medium. The following dimensionless gas dynamic parameters are used (dimensional quantities are denoted by traits):

$$t = \frac{t'}{t^0}\alpha_0^{1/\nu}, \quad r = \frac{r'}{r^0}\alpha_0^{1/\nu}, \quad \rho = \frac{\rho'}{\rho_\infty}, \quad p = \frac{p'}{p_\infty}, \quad u = \frac{u'}{c_\infty}\,\acute{r}\bar{\delta} \tag{28}$$

where

$$r^0 = \left(\frac{E^0}{p_\infty}\right)^{1/\nu}, \quad t^0 = r^0\left(\frac{\rho_\infty}{p_\infty}\right)^{1/2}, \quad c_\infty = \sqrt{\gamma\frac{p_\infty}{\rho_\infty}}$$

E^0 is the explosion energy that is referred to as unit area or unit length in the cases $\nu = 1$ and 2, respectively. The values of dimensionless factor $\alpha_0 = \alpha_{0(\gamma, \nu)}$ are presented in Ref. [5]. Since we apply the solution of the explosion problem to obtain the boundary conditions for calculating gas parameters in region 2 it is reasonable to use the same (formulae (28)) nondimensional gas parameters in this region. Since they are written for nondimensional quantities (28) Equations (5) do not change.

The computational domain in the explosion

$$G = \{0 \leqslant r \leqslant X_1(t), t_0 \leqslant t \leqslant T\}$$

is divided into two parts

$$G_0 = \{0 \leqslant r \leqslant X_0, t_0 \leqslant t \leqslant T\} \tag{29}$$

and

$$G_1 = \{X_0 \leqslant r \leqslant X_1(t), \ t_0 \leqslant t \leqslant T\}. \tag{30}$$

Asymptotic solutions [5] are valid in the central part (29)

$$
\begin{aligned}
u &= \alpha(t)r, \\
\rho &= b(t)r^s, \\
p &= p(t),
\end{aligned}
\tag{31}
$$

where $\alpha(t)$, $b(t)$ are some functions independent of r

$$S = S(\nu, \gamma) = \frac{\nu}{\gamma - 1}, \tag{32}$$

In addition

$$u|_{r=0} = 0, \quad \rho|_{r=0} = 0. \tag{33}$$

Rankine-Hugoniot relations are employed at the bow shock wave. Taking into account $u_\infty = 0$ these relations can be written for nondimensional quantities as follows

$$
\begin{aligned}
\rho &= \frac{(\gamma + 1)p + (\gamma - 1)}{(\gamma + 1) + (\gamma - 1)p}, \\
D_b &= \left[\frac{(\gamma + 1)p + (\gamma - 1)}{2} \right]^{1/2} \\
u &= D_b(1 - 1/\rho).
\end{aligned}
\tag{34}
$$

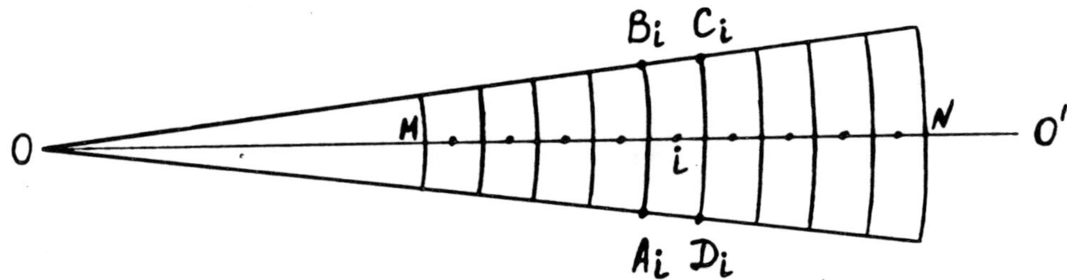

Fig. 4.

Gas flow in region 1 is described by Equations (5) written for two-dimensional (t, r) space. A computational grid in the case $\nu = 2$ is shown in

Figure 4. The grid is uniform and contains M mesh cells. Coordinates of nodes (cell centers) are r_m, $m = 1, 2, ..., M$, the coordinates of the cell verges are $r_{m-1/2}$, $m = 1, 2, ..., M, M+1$. Initial flow parameters are given by the solution of the explosion problem presented in Tables [5] for some moment $t = t_0$ at twenty points $R_1, R_2, ..., R_{20}$. It is assumed $r_1 = R_1$, $r_{M+1/2} = R_{20}$. Initial values of the gas parameters in the cells $i = 2, 3, ..., M$ are obtained by means of linear interpolation using the table values at points $R_1, R_2, ..., R_{20}$.

Geometrical grid parameters are calculated by the following formulae:

$$\text{vol}_i^n = (r_{i+1/2}^n)^\nu - (r_{i-1/2}^n)^\nu, \tag{35}$$

$$\text{vol}_i^{n+1} = (r_{i+1/2}^{n+1})^\nu - (r_{i-1/2}^{n+1})^\nu, \tag{36}$$

$$S_{i\pm1/2}^{n+1/2} = (r_{i\pm1/2}^{n+1/2} + r_{i\pm1/2}^n)^{\nu-1} - \tfrac{1}{2}(\nu - 1)(\nu - 2)r_{i\pm1/2}^{n+1}r_{i\pm1/2}^n, \tag{37}$$

$$W_{i-1/2}^n = D_b \frac{r_{i-1/2}^n - X_0}{X_1(t^n) - X_0}, \quad r_{i\pm1/2}^{n+1/2} = r_{i\pm1/2}^n + w_{i\pm1/2}^n\Delta t \tag{38}$$

where $W_{i-1/2}^n$ $(i = 1, 2, ..., M + 1)$ is the cell verges velocity.

The finite-difference MacCormack scheme for Equations (5) is

$$\tilde{U}_i\text{vol}_i^{n+1} = U_i^n\text{vol}_i^n - \Delta t[(\mathcal{H}_{i+1}^n - U_{i+1}^n W_{i+1/2}^n)S_{i+1/2}^{n+1/2} -$$

$$-(\mathcal{H}_i^n - U_i^n w_{i-1/2}^n)S_{i-1/2}^{n+1/2} - G_i^n(S_{i+1/2}^{n+1/2} - S_{i-1/2}^{n+1/2})], \tag{39}$$

$$U_i^{n+1}\text{vol}_i^{n+1} = \tfrac{1}{2}\{\tilde{u}_i\text{vol}_i^{n+1} + U_i^n\text{vol}_i^n - \Delta t[(\tilde{\mathcal{H}}_i - \tilde{U}_i w_{i+1/2}^n)S_{i+1/2}^{n+1/2}$$

$$-(\tilde{\mathcal{H}}_{i-1} - \tilde{U}_{i-1}w_{i-1/2}^n)S_{i-1/2}^{n+1/2} - \tilde{G}_i(S_{i+1/2}^{n+1/2} - S_{i-1/2}^{n+1/2})]\}. \tag{40}$$

Here

$$U = \begin{bmatrix} \rho \\ \rho u \\ \rho e \end{bmatrix} \quad \mathcal{H} = \begin{bmatrix} \rho u \\ \rho u^2 + p \\ (\rho e + p)u \end{bmatrix} \quad G = \begin{bmatrix} 0 \\ p \\ 0 \end{bmatrix} \tag{41}$$

Gas energy e is calculated by Formula (3). Boundary conditions would be prescribed if the fluxes through the verges $i - 1/2$ at $i = 1$ and $i + 1/2$ at $i = M$ were known.

a) $i = 1$. Using the asymptotic formulae (31) gives

$$u_{1/2}^n = u_1^n\frac{r_{1/2}^n}{r_1^n}, \quad \rho_{1/2}^n = \rho_1^n\left(\frac{r_{1/2}^n}{r_1^n}\right)^{\nu/(\gamma-1)}, \quad p_{1/2}^n = p_1^n. \tag{42}$$

Here $u^n_{1/2}$, $\rho^n_{1/2}$, $p^n_{1/2}$ are velocity, density, and pressure values at point $r = r^n_{1/2} = X_0$ (at the left verge of the cell $i = 1$). By the same method we find

$$\tilde{u}_{1/2} = \tilde{u}_1 \frac{r^{n+1}_{1/2}}{r^{n+1}_1}, \quad \tilde{\rho}_{1/2} = \tilde{\rho}_1 \left(\frac{r^{n+1}_{1/2}}{r^{n+1}_1} \right)^{\nu/(\gamma-1)}, \quad \tilde{p}_{1/2} = \tilde{p}_1. \tag{43}$$

Now we can write the formulae for correspondent fluxes through the left boundary of the computational domain, i.e., $A_1 B_1$ verge (see Figure 4). Note that in accordance with (38)

$$W^n_{i-1/2}|_{i=1} = 0. \tag{44}$$

Accounting for (44) we have

$$(\mathcal{H}^n_i - U^n_i W^n_{i-1/2}) S^{n+1/2}_{i-1/2}|_{i=1} = \begin{bmatrix} \rho^n_{1/2} u^n_{1/2} \\ \rho^n_{1/2} (u^2)^n_{1/2} + p^n_{1/2} \\ (\rho^n_{1/2} e^n_{1/2} + p^n_{1/2}) u^n_{1/2} \end{bmatrix} S^{n+1/2}_{1/2}, \tag{45}$$

$$(\tilde{\mathcal{H}}^n_{i-1} - \tilde{U}^n_{i-1} W^n_{i-1/2}) S^{n+1/2}_{i-1/2}|_{i=1} = \begin{bmatrix} \tilde{\rho}_{1/2} \tilde{u}^n_{1/2} \\ \tilde{\rho}_{1/2} (\tilde{u}^2)_{1/2} + \tilde{p}_{1/2} \\ (\tilde{\rho}_{1/2} \tilde{e}_{1/2} + \tilde{p}_{1/2}) \tilde{u}_{1/2} \end{bmatrix} S^{n+1/2}_{1/2}, \tag{45}$$

The values of $\rho^n_{1/2}$, $u^n_{1/2}$, $p^n_{1/2}$, $\tilde{\rho}_{1/2}$, $\tilde{u}_{1/2}$, $\tilde{p}_{1/2}$ in (45) and (46) are calculated according to the formulae (42) and (43) and the values $e^n_{1/2}$, $\tilde{e}_{1/2}$ according to Formula (3) with (42) and (43) taken into account.

b) $i = M$. The fluxes through the right boundary of the computational domain, i.e., verge $C_M D_M$ (Figure 4) of cell $i = M$ calculating we must substitute undisturbed gas parameters u_∞, ρ_∞, p_∞ in the correspondent fluxes of (39) and (40). In addition we must take into account that at the shock wave $w^n_{i+1/2}|_{i=M} = D^n_b$.

$$(\mathcal{H}^n_{i+1} - U^n_{i+1} w^n_{i+1/2}) S^{n+1/2}_{i+1/2}|_{i=M} = (\tilde{\mathcal{H}}_i - \tilde{U}_i w^n_{i+1/2}) S^{n+1/2}_{i+1/2}|_{i=M} =$$

$$\begin{bmatrix} \rho_\infty (u_\infty - D^n_b) \\ \rho_\infty u_\infty (u_\infty - D^n_b) + p_\infty \\ \rho_\infty e_\infty (u_\infty - D^n_b) + p_\infty u_\infty \end{bmatrix} S^{n+1/2}_{M+1/2}. \tag{47}$$

Time step Δt in (39) and (40) is chosen to satisfy stability condition [5].

$$\Delta t = \min \left[(r^n_{i+1} - r^n_i)/(|u^n_i| + \sqrt{\gamma p^n_i/\rho^n_i}) \right] \qquad (25)$$

$i = 1, 2, ..., M.$

In addition, the condition

$$D^n_b \Delta t < r^n_{M+1/2} - r^n_{M-1/2} \qquad (49)$$

must be satisfied.

Smoothing procedure (27) is used that is easily implemented for one-dimensional problems.

7. Test Problems

The numerical method described earlier was tested on the problem of stationary flow about a circular spherically blunted cone (see Figure 1). In this problem the oncoming stream parameters in regions 1 and 3 are constant: $p_\infty = 1$, $\rho_\infty = 1$, $u_\infty = 2.366$, $\vartheta_\infty = 0$, $\gamma = 1.4$. The origin of Cartesian coordinate system XOY is placed in the sphere center O. Angle α (spherical part of the body) is equal to 85°, $LOED = 90°$. The stationary solution was obtained by means of the time-asymptotic technique. Figure 5 presents density and pressure distributions along the bow shock wave (ρ_w, p_w), along the mid-line of the shock layer ρ_m, p_m, and along the body surface ρ_b, p_b. In all three cases the distance along the correspondent line is taken from the flow symmetry axis with the mesh coil taken as a unit of measure. Solid circles show the data [6]. Good agreement of the results is seen.

Figure 6 shows pressure, density, and both momentum component distributions along the shock layer segment bw of the symmetry axes at angle 1.42°. Point b is at the body surface, point w is at the shock wave. Solid circles as in the previous figure show the results [6] for gas parameters at the symmetry axes. Again we see good agreement of the results. Note that because of the use of different computational grids comparison in Figure 6 is carried out for close but different points of physical space.

Figure 7a shows the ispoycnics and sonic lines (bold curve) in the shock layer. Figure 7b depicts the shape of the bow shock. Filled circles denote the results [6]. The stationary flow field past a blunted cone (Figures 5 and 7) was calculated on uniform grid $I = 50$ (30 cells on the spherical part of the body), $J = 20$, with Courant number, $K = 0.9$, and smoothing parameter $Q = 0.1$. Test calculations were carried out on the grid $I = 20$, $J = 6$ as well. The results obtained on both grids were found to be the same.

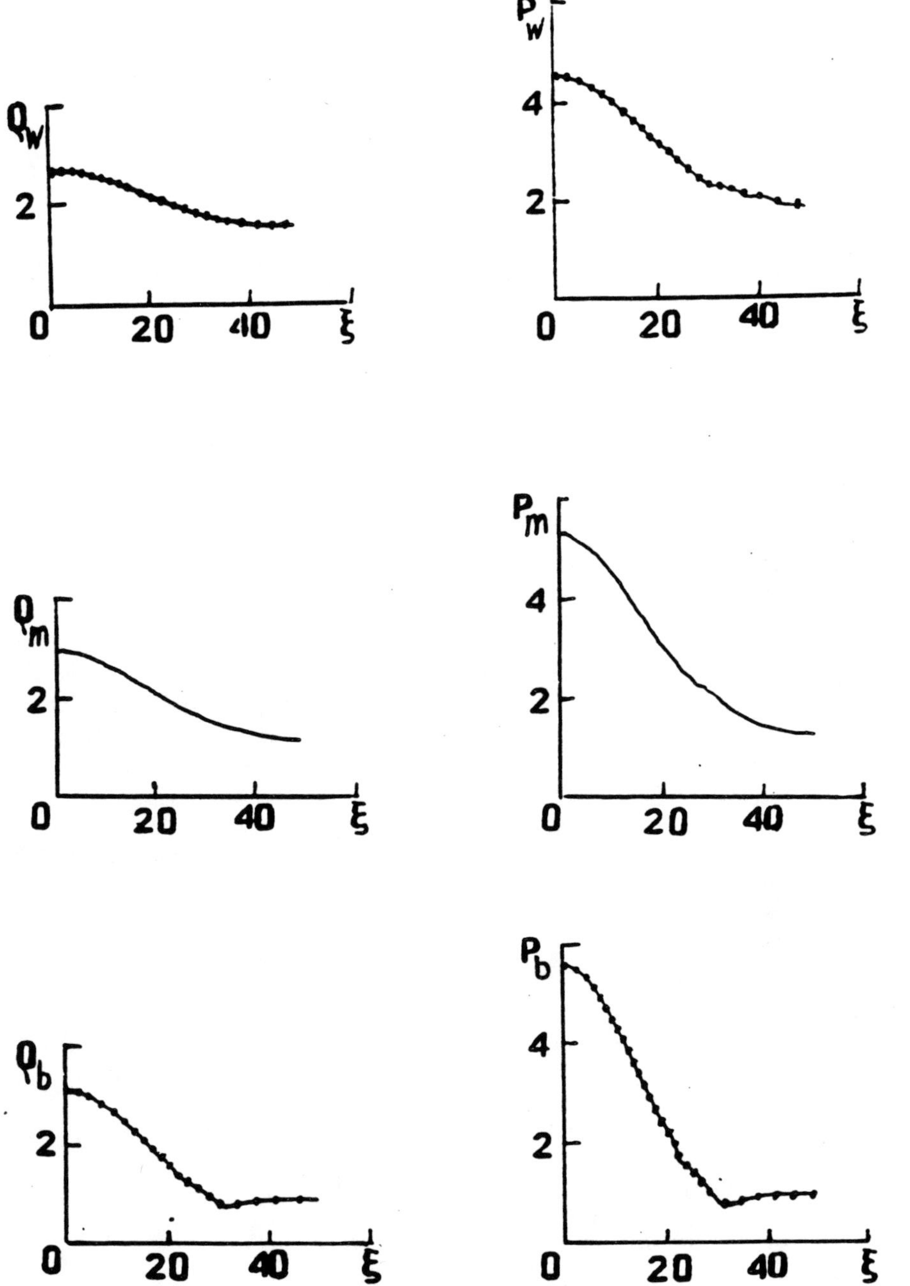

Fig. 5.

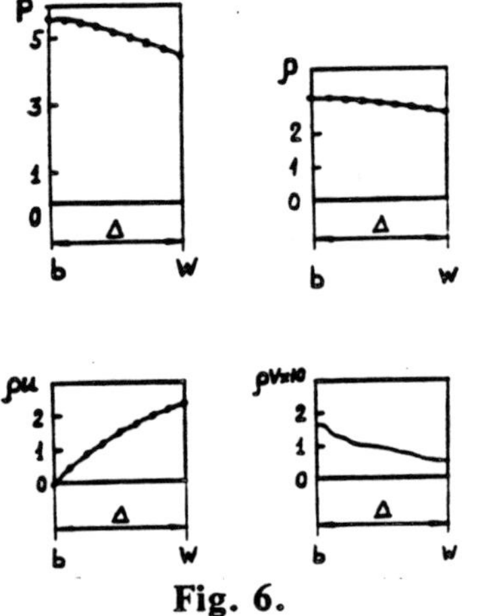

Fig. 6.

A stationary problem solving initial flow field and location of the bow shock wave were prescribed with low accuracy. Initial shock layer thickness was assumed to be constant along the body surface and equal to the body nose radius. Gas parameters in all mesh cells within the shock layer were assumed to be equal to constant oncoming stream parameters q_∞, ρ_∞, p_∞.

The problem of instantaneous point explosion with finite ambient pressure at $\nu = 3$, $\gamma = 1.4$ was solved on the grid with number of cells, $M = 99$; Courant number, $K = 0.8$; and smoothing parameter $Q = 0.1$). The data of tables [5] at nondimensional time $\tilde{t}_0 = 0.4969 \cdot 10^{-2}$ were used to generate initial flow field. Figure 8 shows the u, ρ, and p variation with a distance from the explosion center at two successive moments $t_1 = 0.6982 \cdot 10^{-2}$ and $t_2 = 0.9984 \cdot 10^{-2}$. These graphs represent calculation results obtained with the use of the scheme described previously. Solid circles denote the data [5] that agree well with the present calculation results. It should be noted here that the algorithm presented above is simpler and more efficient compared to the algorithm of Ref. [5].

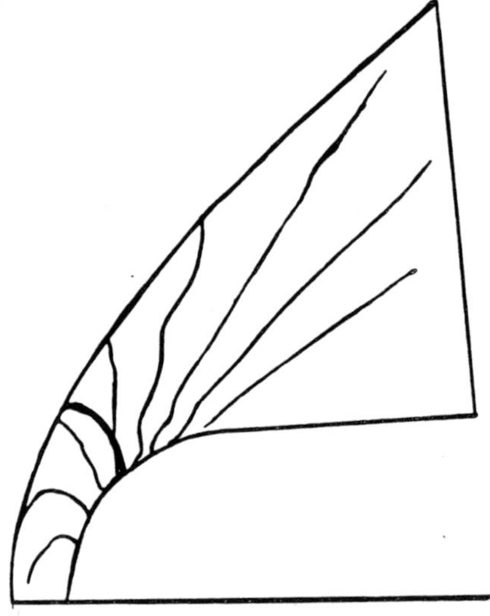
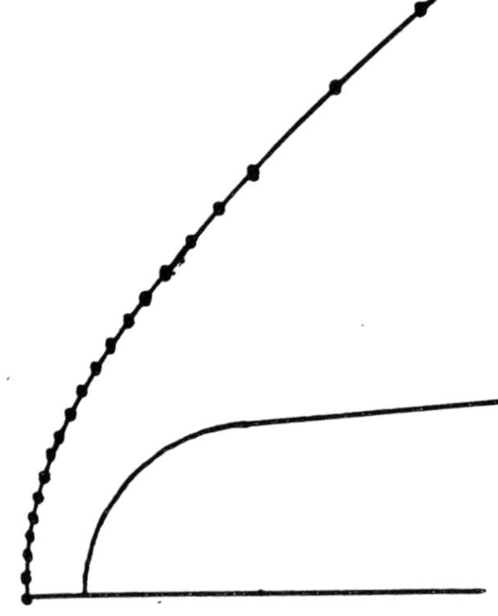

Fig. 7.

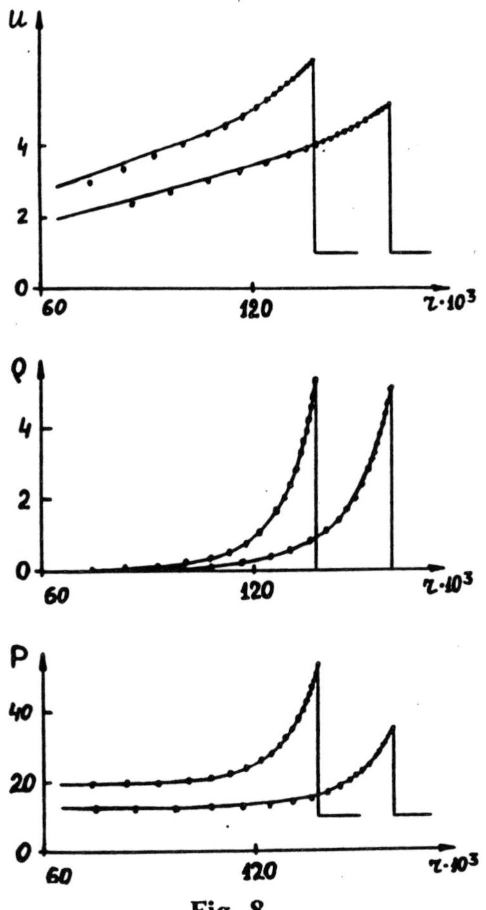

Fig. 8.

8. Shock-Wave-Body Interaction.
Calculation Results

Region 1 in Figure 1 may be the explosion region, a heated gas area, the region with constant gas parameters corresponding to concurrent flow behind plane shock wavefront, etc. In all cases it is important that the flow parameters ahead of the bow shock front are known. The numerical scheme described earlier is applied here to calculate non-stationary flow that occurs at the interaction of the body moving at supersonic speed with the shock wave propagating in the opposite direction [7-11]. Two cases will be considered. In the first case the incident shock wave is plane and concurrent flow parameters are constants; in the second the incident wave is initiated by instantaneous point explosion with finite ambient pressure and $\nu = 3$. Both problems are axisymmetric. A typical flow pattern is schematically shown in Figure 9. It contains a number of intersecting shock waves and slip-stream surfaces that form a complex flow field around the body. The following notation is employed in Figure 9: I — incident shock wave; B — part of the bow shock ahead of the body before interaction; II — refracted incident wave; BI — refracted bow shock wave; IR — reflected shock wave; M — Mach stem; BN — new bow shock forming ahead of the body. Dashed curves show the slip-stream lines, thin lines show the beginning and end of the expansion wave.

The results of stationary flow calculations presented previously were used to generate initial flow field parameters in region 2 (see Figure 1). At the moment of contact of the interacting wavefronts ($t = t_0$) gas parameters at the fronts of the incident plane and blase wave are assumed to be the same: Mach number, $M_1 = 4.37$, $p_1 = 22.12$, $\rho_1 = 4.755$. In the second case

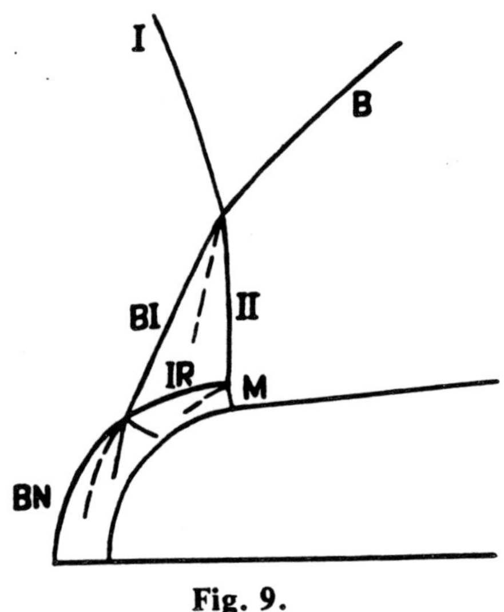

Fig. 9.

(blast incident shock wave) calculation of flow parameters in Region 2 and solution of the explosion problem were carried out simultaneously. time step Δt was chosen to be the same for both problems and to provide a stable solution for both regions.

Calculation results for each interaction variant are presented in the form of shock layer isopycnics at six successive moments and correspondent density $\rho(\xi)$ and pressure $p(\xi)$ distributions. Indices b, b and m refer to the values at the bow shock front, shock layer mid-line, and boundary stream-line, respectively. Distance ξ is taken from the symmetry axis in graphs $\rho_w(\xi)$, $p_w(\xi)$, $\rho_m(\xi)$, $p_m(\xi)$ and from the stagnation point in graphs $\rho_b(\xi)$ and $p_b(\xi)$ ($\xi < 0$) corresponds to flow axis, $\xi > 0$ to body surface).

At the beginning of the interaction process (Figures 10a, 11a relate to plane and blast incident shock waves) regular front collision of the incident and bow shocks occurs resulting in two refracted shocks and slip-stream surface formation. Refracted shock wave II is located within the shock layer. Correspondent density and pressure distributions are depicted in Figure 12 (plane incident shock wave) and Figure 13 (blast incident shock wave). Pressure and density distributions for both variants are similar. It proves that blast wave interaction may be considered as a quasistationary process at this stage. Density and pressure values in the second variant are somewhat less as compared to the first. This is explained by blast wave attenuation with time.

As the interaction process develops the refracted incident shock approaches the body surface and reflects. Regular reflection of the refracted incident shock on the spherical part of the body is seen in Figure 10b and 11b. Correspondent density and pressure distributions are presented in Figures 14 and 15. Nonuniformity of gas parameters in the explosion region begins to influence the variation of ρ_w, p_w, ρ_m, p_m, ρ_b, p_b and intensity of the slip-stream originating from the point of the incident and bow shock intersection. This is clearly seen in Figure 15 which presents dependence $\rho_m(\xi)$. Refracted shock wave reflection at the body surface is accompanied by significant ρ_b and p_b increase behind the reflection point $\xi \simeq 10$ (see Figures

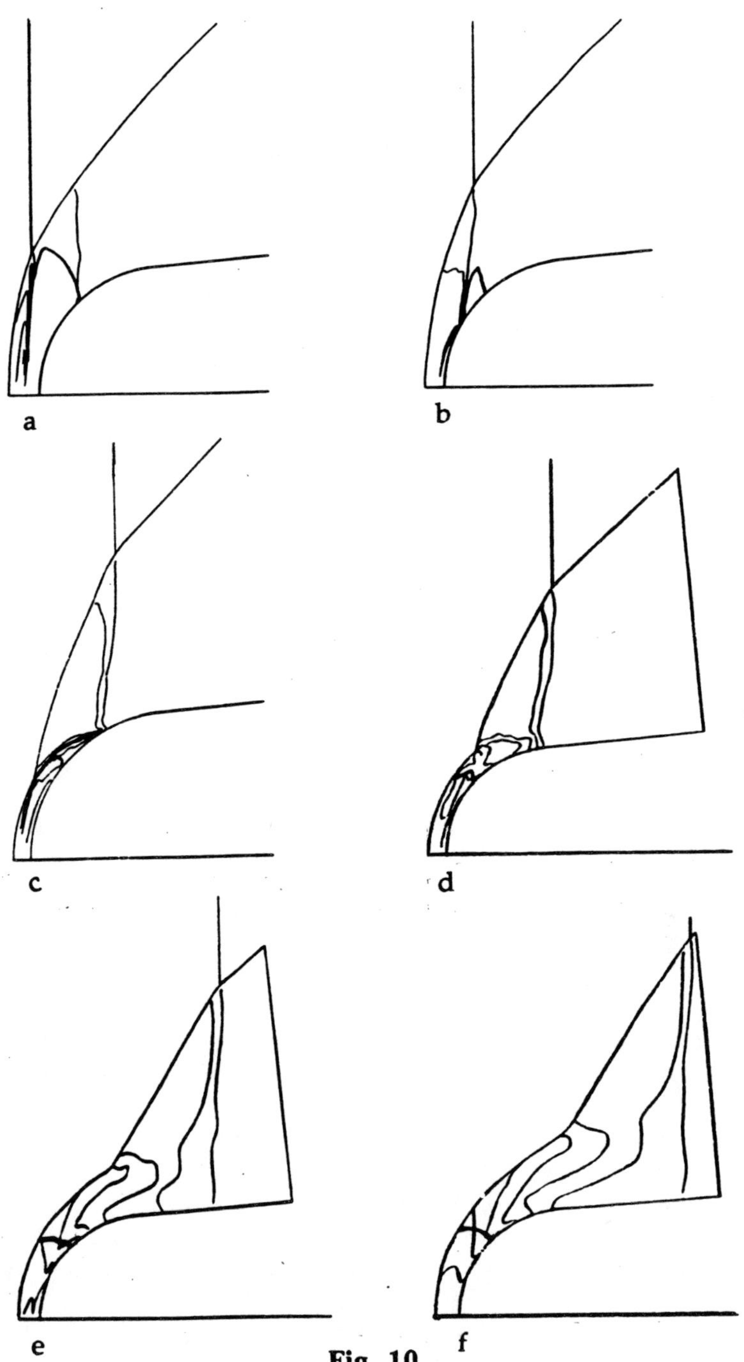

Fig. 10

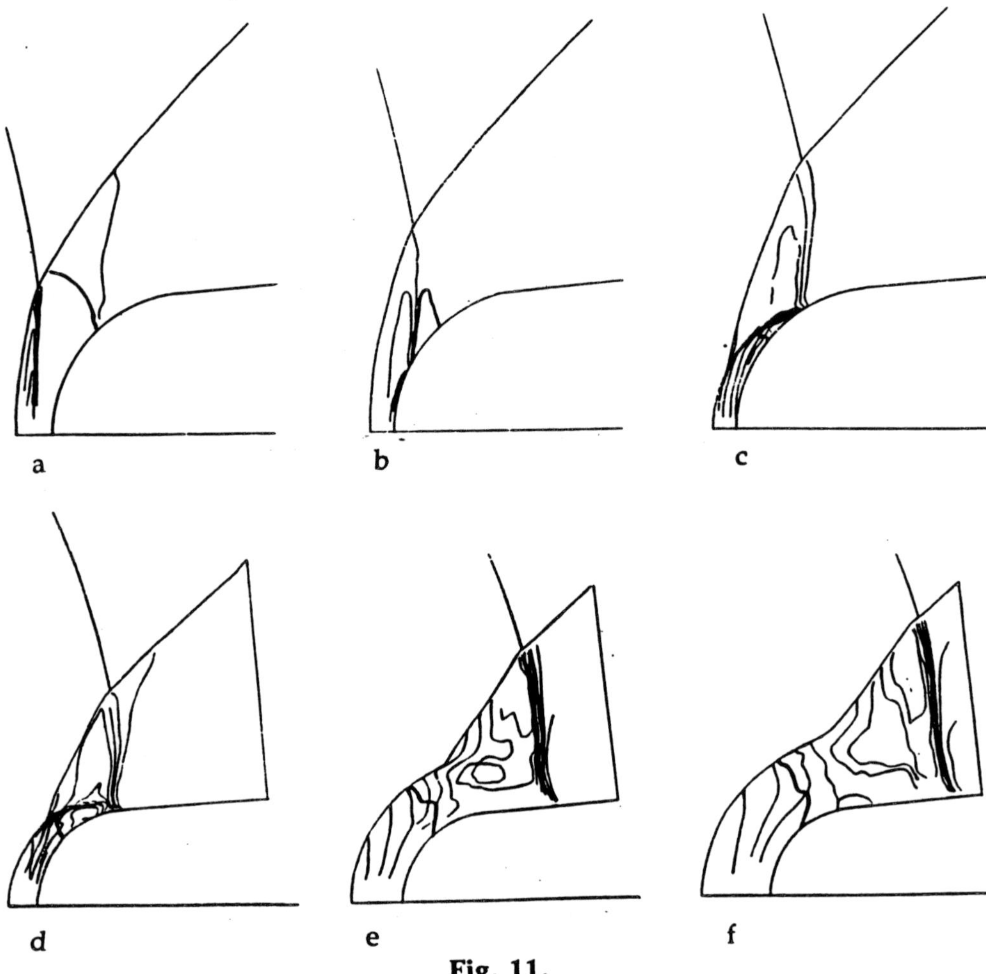

Fig. 11.

14 and 15) as compared with correspondent parameter behavior in Figures 12 and 13. Nonuniformity of ρ_b and p_b distributions along the part of body surface behind the reflection point in the case of blast incident shock waves is related to the behavior of gasdynamic parameters in the explosion region.

Later, as the refracted incident shock moves along the body surface, the reflection character changes from regular to the Mach type. The reflected shock wave propagates from the body surface, refracts at the slip-stream originating from the incident and bow shocks intersection point, and catches the bow shock wave *BI*. This is accompanied by formation of a resultant shock wave, slip-stream and expansion wave propagating to the

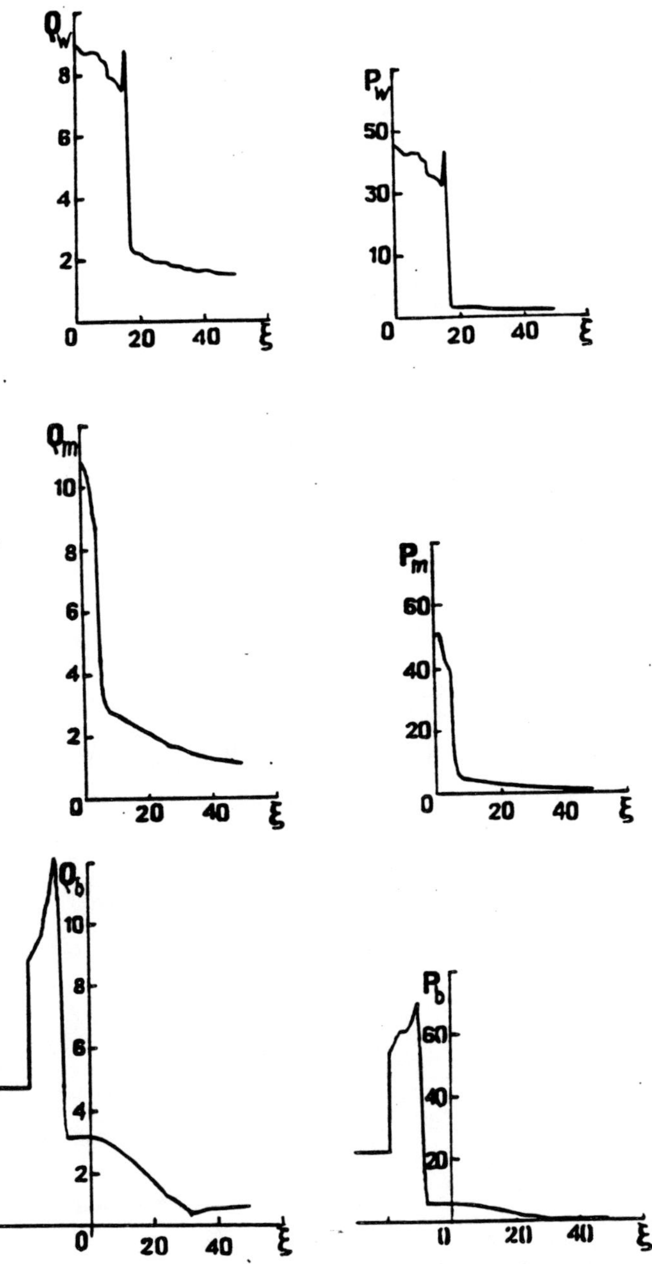

Fig. 12.

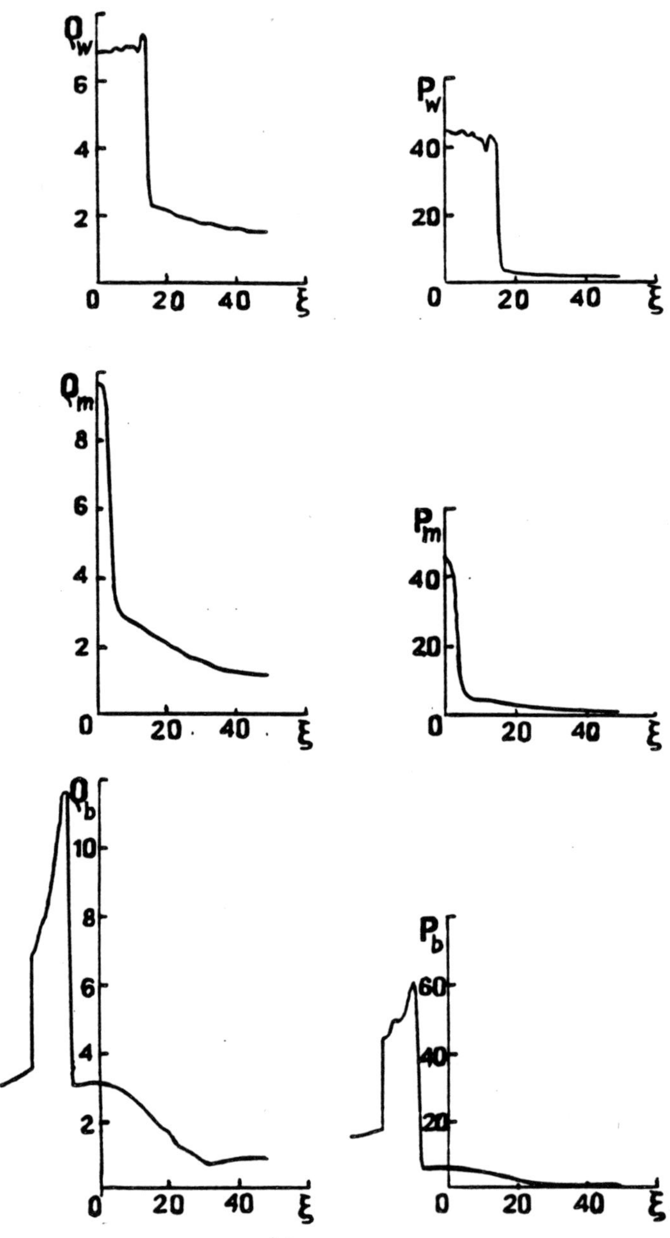

Fig. 13.

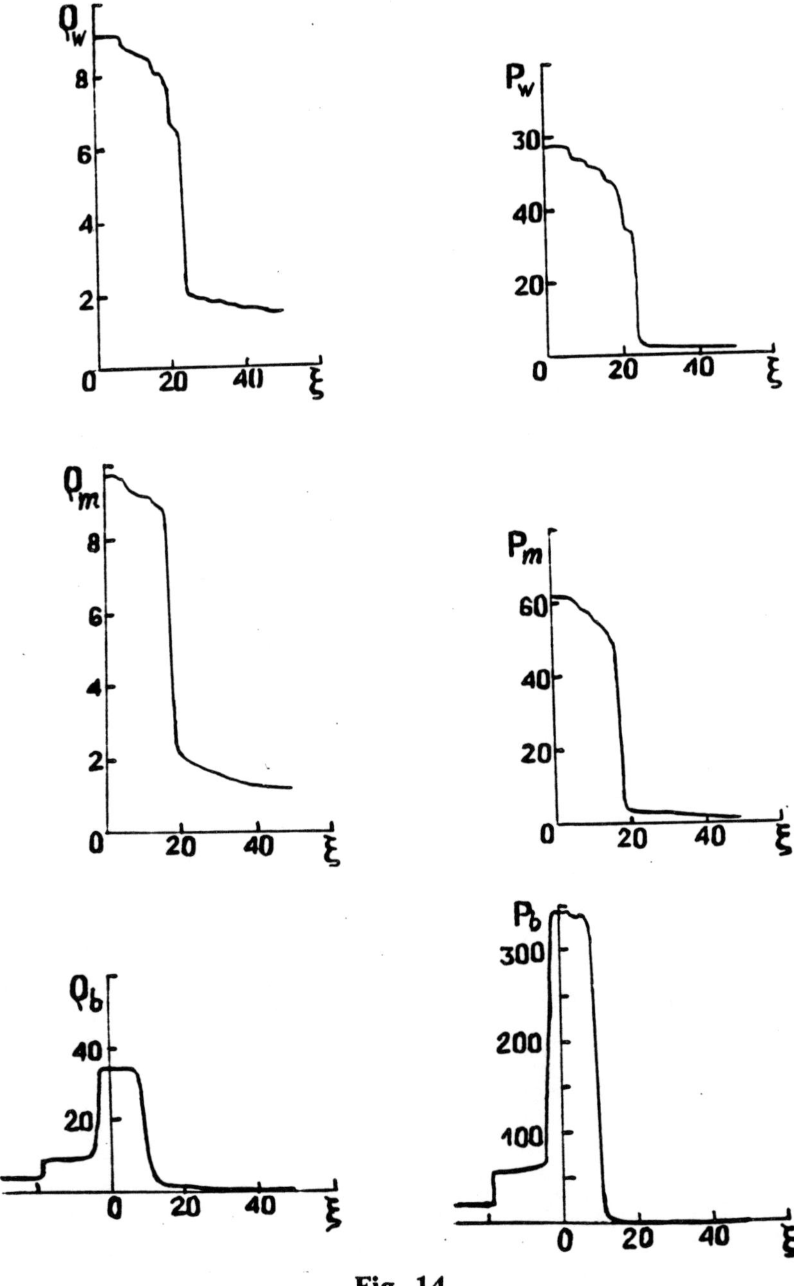

Fig. 14.

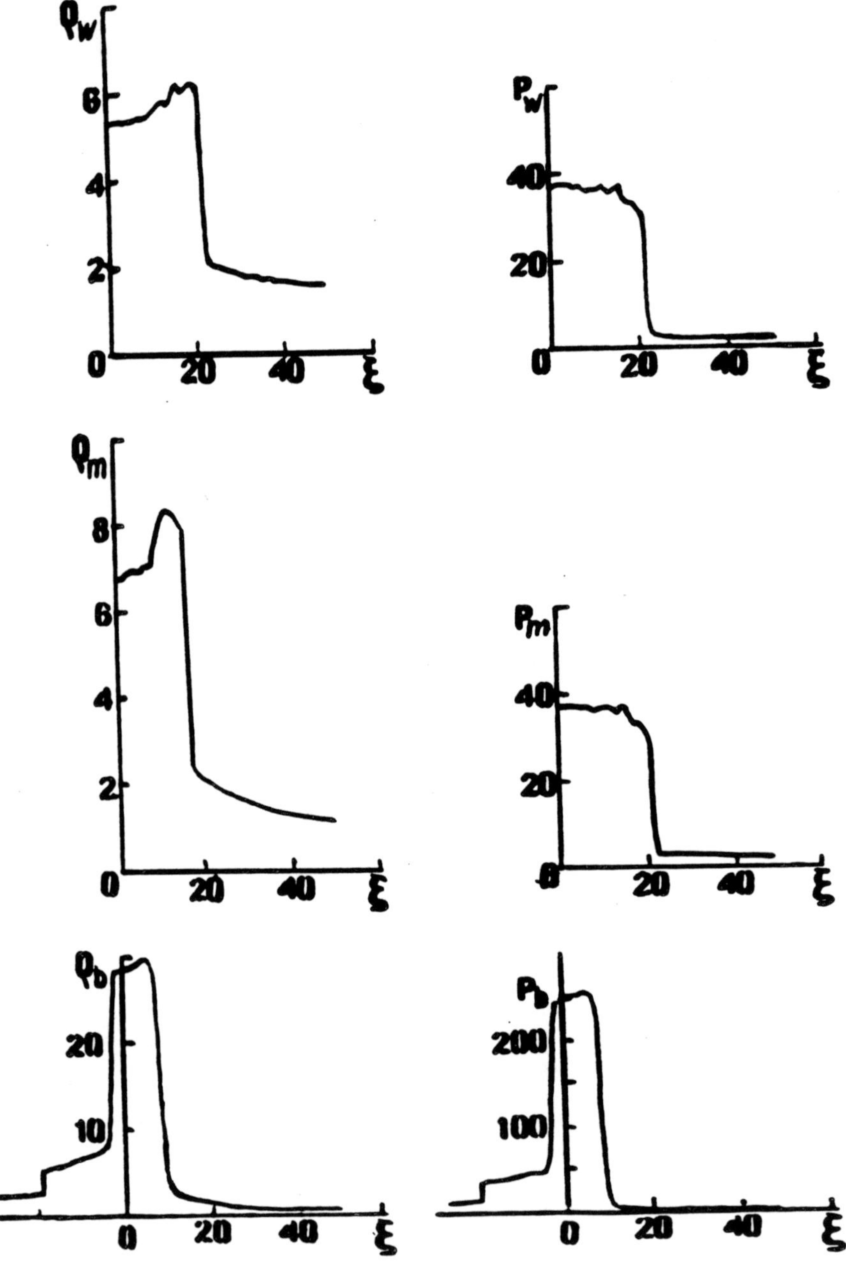

Fig. 15.

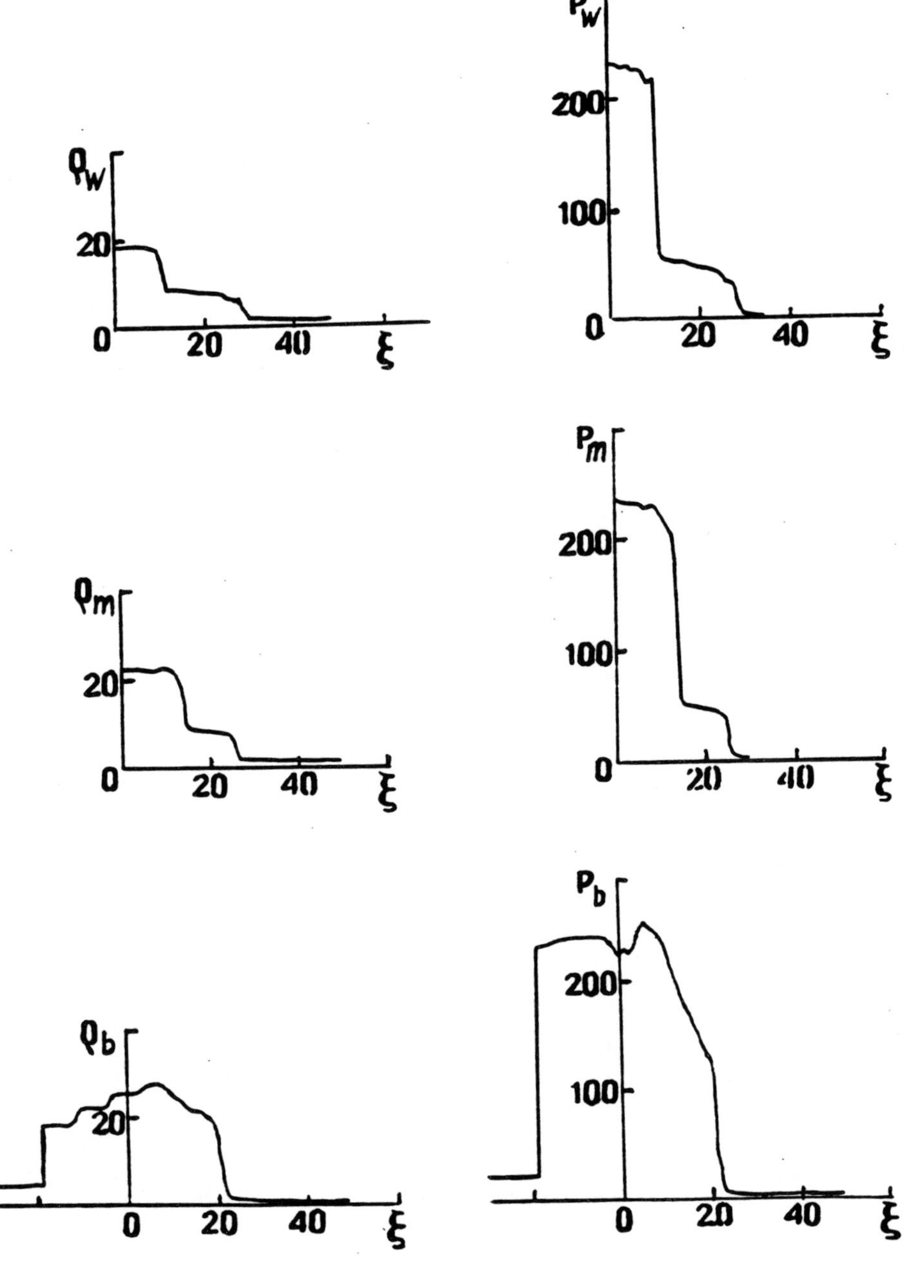

Fig. 16.

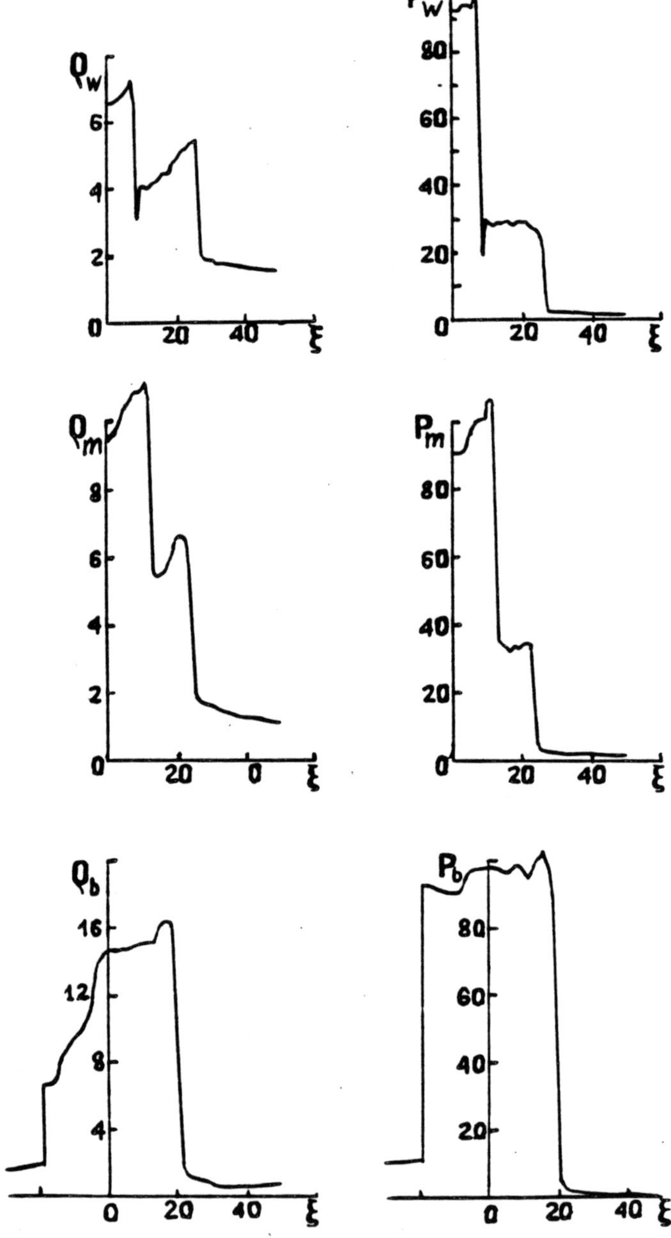

Fig. 17.

body surface. Having passed through the shock layer the expansion wave reflects from the body as the expansion wave as well. Isolines in Figures 10c and 11c demonstrate Mach configuration and triple shock configuration arising when catching shock wave interaction. Novel flow structure is clearly seen in density and pressure distributions presented in Figures 16 and 17. Pressure and density increase behind the novel shock wavefront *BN* as compared Figure 14 and 15. Since reflection type at the body has changed, gas pressure and density on the body surface immediately behind the Mach stem decrease as compared with the previous moment. Complexity of the interaction process including reflected shock wave *IR* propagation through the shock layer, its refraction at slip-stream surface, reflection of resultant expansion wave from the body as well as limited amount of data presented do not give satisfactory interpretation for density and pressure distributions along the boundary stream line. Additional calculations are necessary to demonstrate the flow structure details.

Later on (Figures d, e, f) novel stationary flow is formed near the body interacting with the plane incident shock wave with constant gas parameters of the concurrent flow. This is clearly seen in Figures 18, 20, 22. Density and pressure distributions in Figure 22 are similar to correspondent distributions for the stationary flow (Figure 5).

The process of blast incident shock wave interaction with the body remains nonstationary. When moving through the explosion region the body enters more heated gas areas with less density which results in an increase of the bow shock wave detachment (Figures 11 d, f, e). Figure 19, 21, and 23 show density and pressure distributions at the moments correspondent to Figures 11 d, f, e. A decrease of all gasdynamic discontinuity intensity with time is seen. The body entering the explosion region, the density and pressure profiles behind the reflected and refractive shock waves resemble to a greater extent the profiles of gas dynamic parameters behind a blast wave. Note that there is a rather extensive nearly isobaric area in the shock layer body entry in higher temperature and bow density area where the velocity of small perturbation propagation is much higher than flow velocity that results in pressure smoothing.

Bow shock wave detachment from the body is rather sensitive to oncoming stream parameter variation. The graphs illustrating bow shock wave detachment Δ variation at the stagnation line with x are presented in Figure 24. In the case of blast incident shock wave $x = X_1(t) - OO'$ (see Figure 1). In other words, x is the coordinate of the point undisturbed blast shock wavefront intersection with OX axis of Cartesian coordinate system XOY. In the case of a plane incident shock wave, x is the coordinate of the

ρ_W
20
0
20
40
ξ
P_W
200
100
0
20
40
ξ
ρ_m
20
0
20
40
ξ
P_m
200
100
0
20
40
ξ
ρ_b
20
0
20
40
ξ
P_b
100
0
20
40
ξ

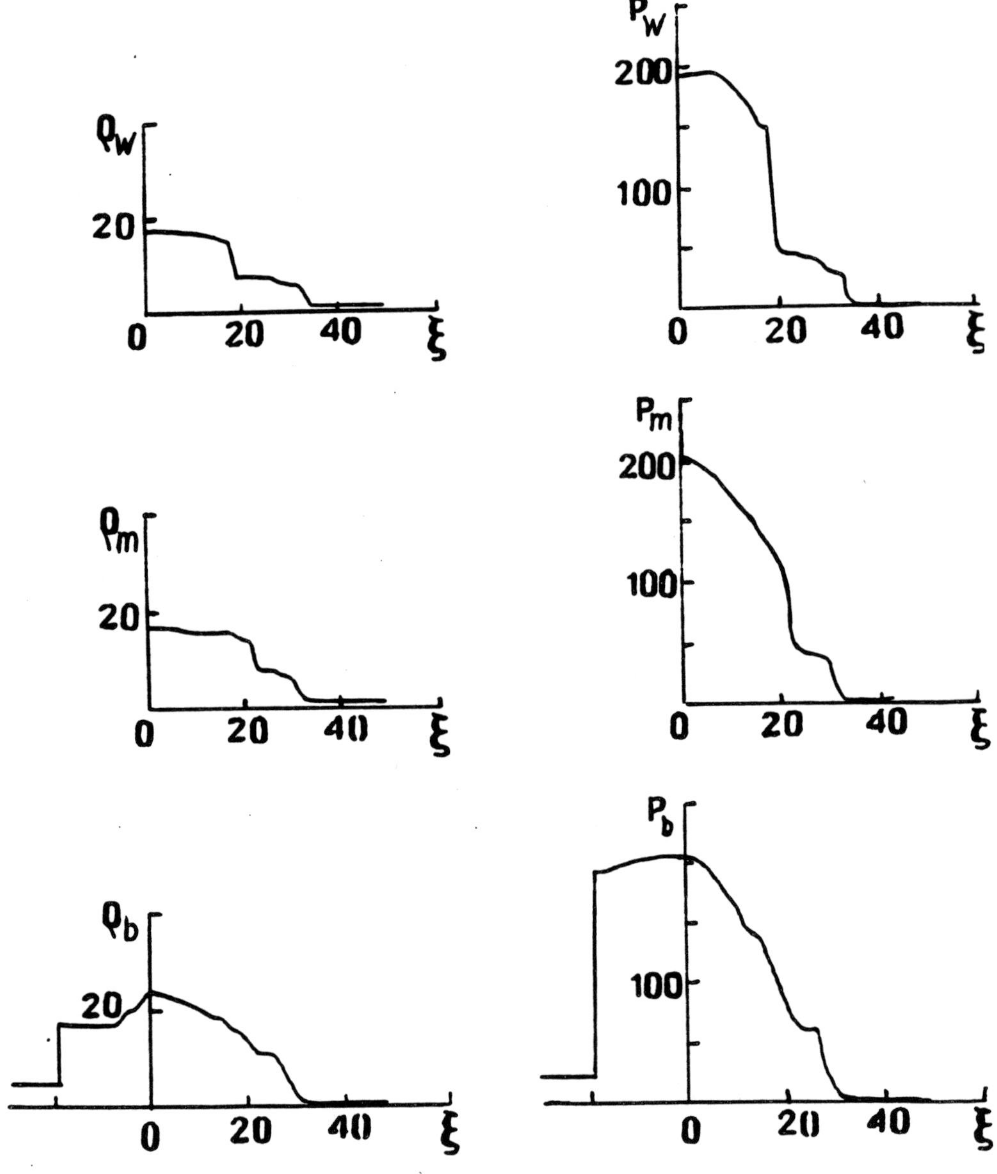

Fig. 18.

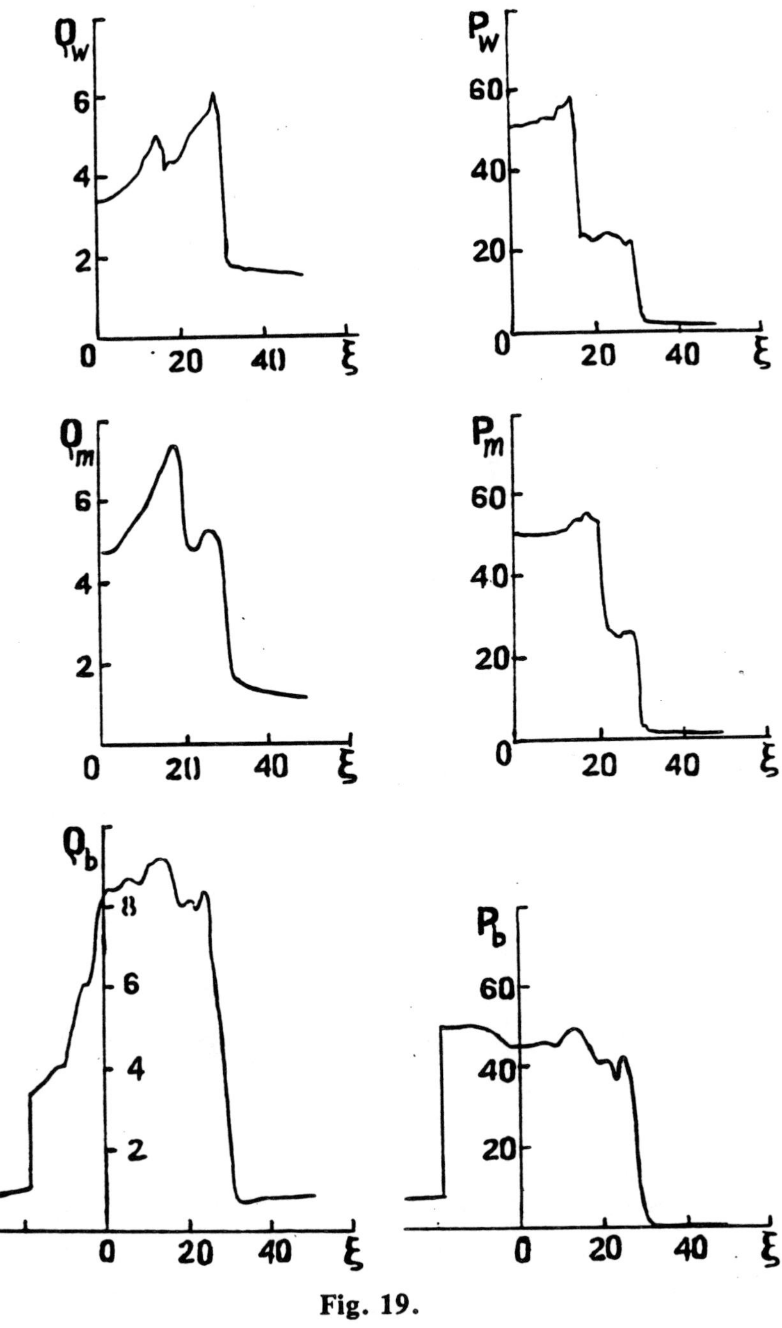

Fig. 19.

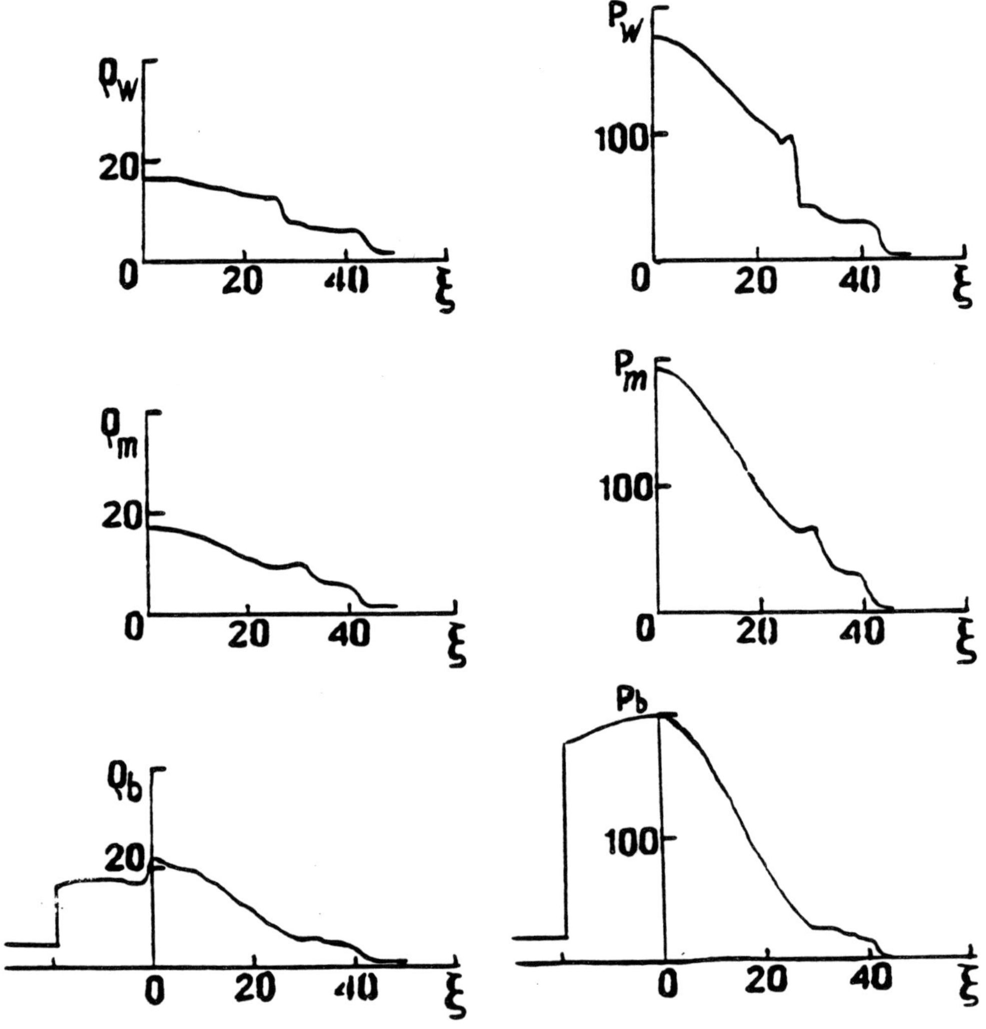

Fig. 20.

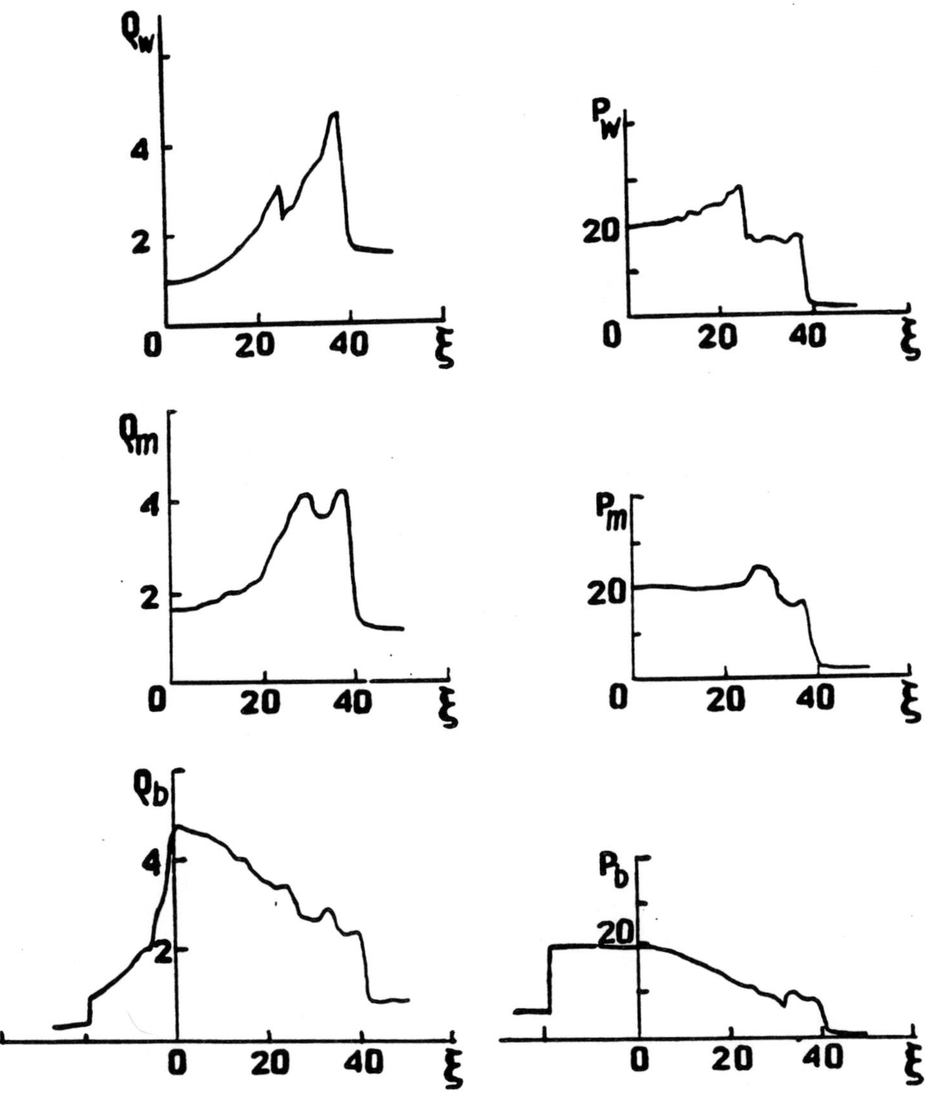

Fig. 21.

Q_w
20
0
20   40   ξ
P_w
100
0
20   40   ξ
Q_m
20
0
20   40   ξ
P_m
100
0
20   40   ξ
Q_b
20
0   20   40   ξ
P_b
100
0   20   40   ξ

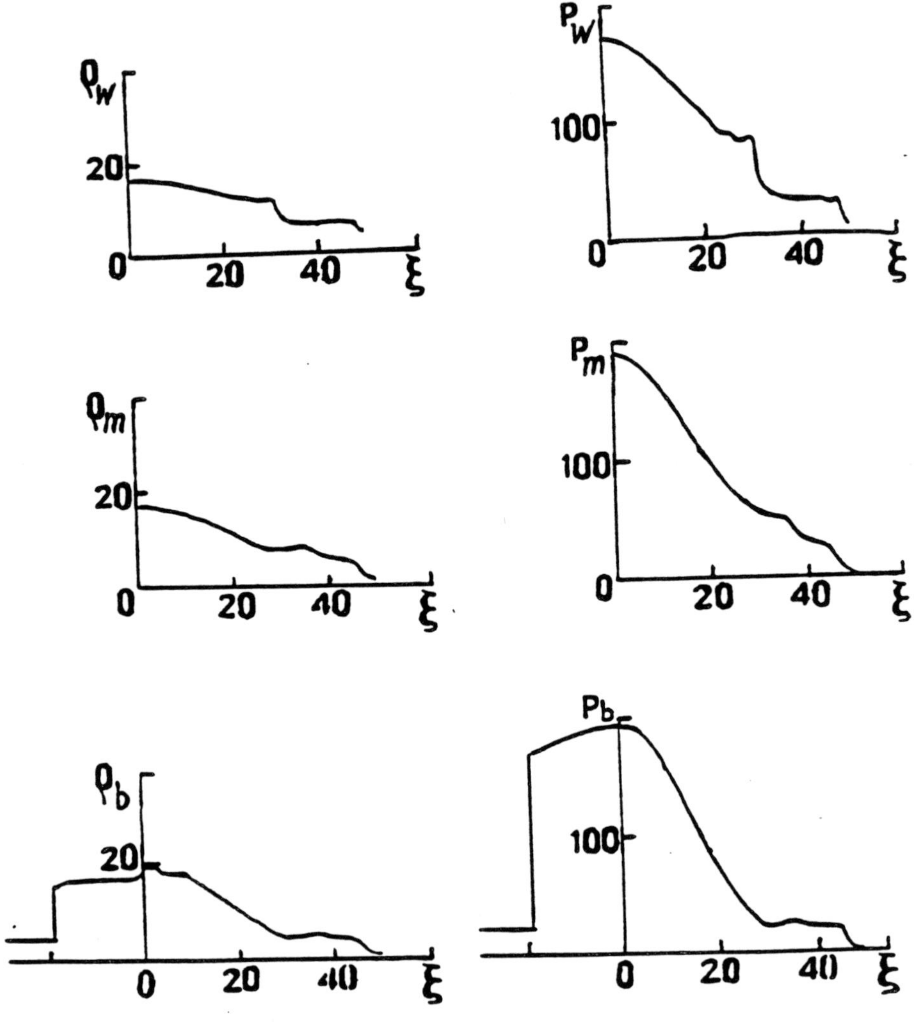

Fig. 22.

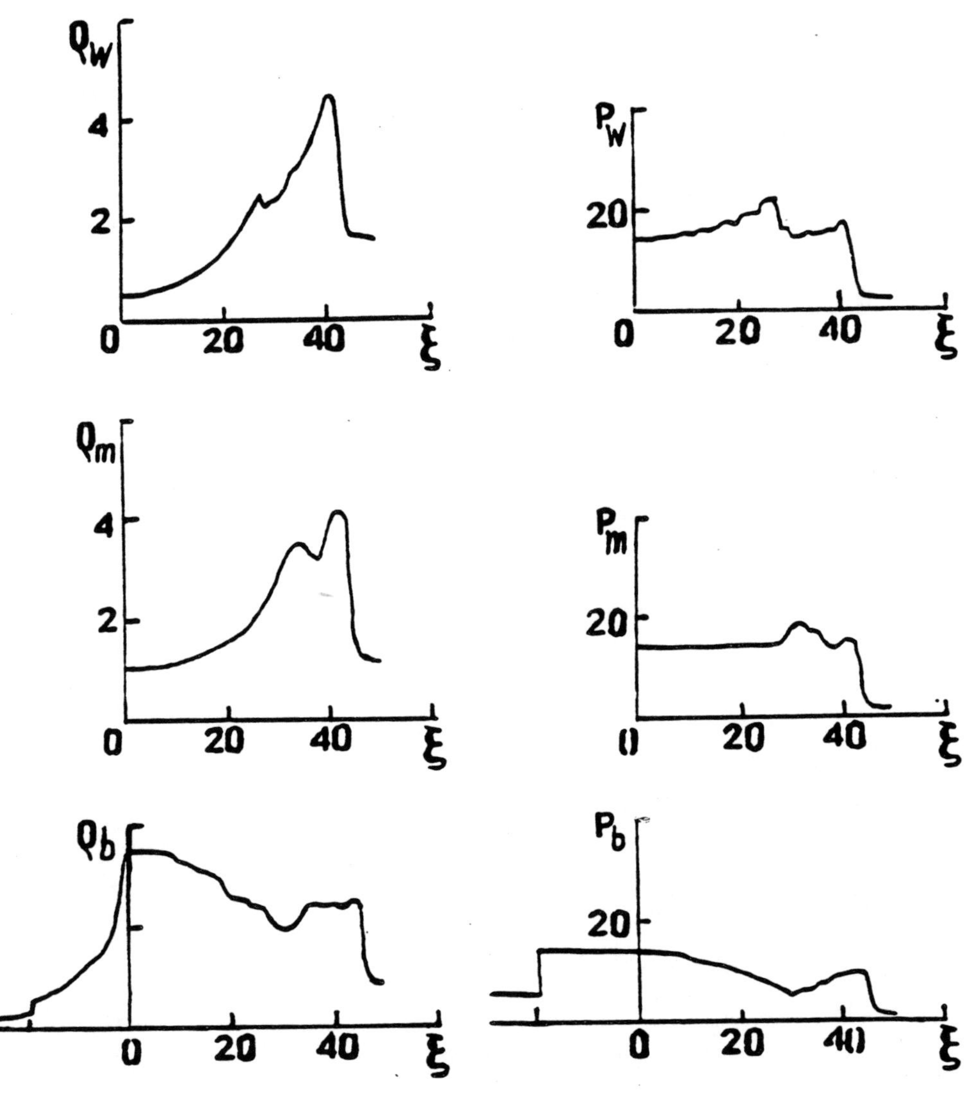

Fig. 23.

incident shock part located outside the shock layer in coordinate system *XOY*. In Figure 24 curve 1 is plotted using six points correspondent to Figures 10a-e, curve 2 is plotted using the data of Figures 11a-e, curve 3 presents calculation results for blast incident shock wave energy $E^0_{(3)}$ = 0.0086 $E^0_{(2)}$. In the last case, gas parameters at the incident shock front at the moment of its contact with the bow shock wave have the following values: M_1 = 1.205, p_1 = 1.527, ρ_1 = 1.35. The blast incident shock wave radius is the same as in the second case. As is seen in Figure 24, at the beginning of the interaction, detachment Δ decreases in all three cases considered. At this stage regular collision of incident and bow shocks is the most important process determining flow formation. Detachment Δ decreases until the shock wave reflected from the body surface catches the refracted bow shock front. Beginning from this moment the bow shock wave location is governed by catching the interaction of two shocks, and therefore, detachment Δ increases. During the body entry in the concurrent flow region behind the plane incident shock wave (curve 1) detachment increase diminishes and it

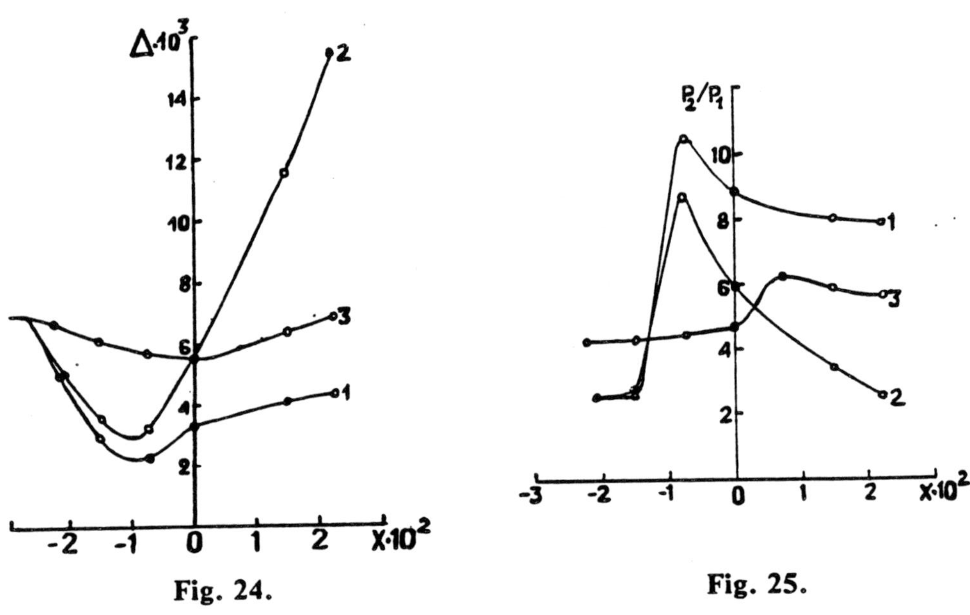

Fig. 24.   Fig. 25.

asymptotically approaches the novel stationary value. As the body enters the explosion region, gas temperature increases, and density decreases. Therefore bow shock wave detachment continues to increase with *x* (curves 2 and 3).

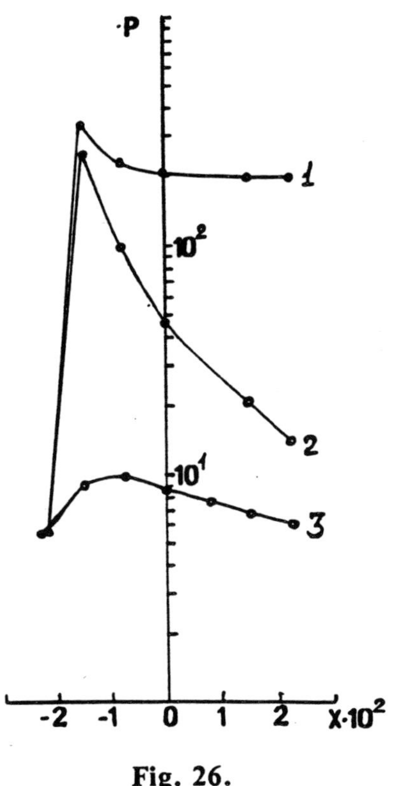

Fig. 26.

Figure 25 shows x-dependence of gas compression ratio at the bow shock front p_2/p_1 on the boundary stream line. Here x means the same quantity and curves 1, 2, 3 correspond to the same flows as in the previous figure. Figure 25 shows that maximum gas compression at the bow shock occurs when the shock wave reflected from the body catches the bow shock. Peak values of the ratio p_2/p_1 on curves 1 and 2 take place approximately at the same value $x \approx -0.01$. The maximum of curve 2 is somewhat less than curve 1 because of blast shock wave attenuation with time. The reflected shock wave having caught up with the bow shock, ratio p_2/p_1 decreases (curves 1 through 3 in Figure 25). In curve 1 the compression ratio p_2/p_1 approaches a novel stationary value. In the case of blast incident shock wave x value increasing or body approaching the center of the ratio $p_2/p_1 \rightarrow 1$ (curves 2 and 3).

Figure 26 shows stagnation point pressure dependence on parameter x. Here again curves 1 through 3 correspond to the same flows as in Figure 24. Maximum pressure values occur at the moment of refracted shock wave arrival to the body surface. In the case of plane incident shock wave and constant concurrent flow parameters (curve 1) stagnation point pressure approaches a new stationary value with time. Body entry into the explosion region is accompanied by stagnation point pressure decrease (curves 2 and 3).

9. Conclusions

Numerical results presented above prove the applicability of the method proposed to simulate nonstationary gas flows with a complicated structure of interacting discontinuities. The use of Cartesian coordinates in the physical space simplifies the form of basic Equations (5) and operations with vectors and vector products. Boundary conditions are taken into account in a rather simple way as well. The method allows us to use simple uniform orthogonal grids as well as general curvilinear computational grids with nonuniform spacing. Note that assuming all $D^n_{i\,J+1/2}$ $(i = 1..., I)$ in (9) to be equal to zero we receive the uniform shock-capturing finite-difference scheme. In this case the bow shock wave will be manifest in the solution on the fixed computational grid as a region of steep gradients of gasdynamic parameters. This scheme can be used to solve the problem of shock wave-immovable body interaction. Other advantages of the method proposed are simplicity of implementation, efficiency, and ability to provide accurate results on the course grids. For example, the flow around a sphere moving through a heated gas area was calculated in 12 on the grid 6 × 12. Calculation results obtained on this coarse grid are in rather good agreement with experimental data.

References

1. R. W. MacCormack, A. Rizzi, and M. Inouye, In: *Computational Methods and Problems in Aeronautical Fluid Dynamics*, New York: Academic Press, 1976, pp. 424-447.

2. A. Rizzi and M. Inouye, *AIAA J.*, vol. 11, no. 11, 1973, pp. 1478-1485.

3. S. K. Godunov, A. V. Zabrodin, M. Ya. Ivanov, A. N. Kriko, and G. P. Prokopov, *Numerical Solution of Multidimensional Gasdynamic Problems*, Moscow: Nauka, 1976.

4. V. P. Goloviznin, A. I. Zhmakin, and A. A. Fursenko, *Zh. Vychisl. Mat. i Mat. Fiz.*, vol. 22, no. 2, 1982, pp. 484-488 (in Russian).

5. Kh. S. Kestenboim, G. S. Roslyakov, and L. A. Chudov, *Point Explosion Calculation Method, Tables*, Moscow: Nauka, 1974.

6. A. N. Lubimov and V. V. Rusanov, *Gas Flows About Blunt Bodies*, vol. 2, Moscow: Nauka, 1970.

7. J. Champney, D. Chausse, and P. Kutler, *AIAA Paper*, no. 82-0227.

8. V. P. Goloviznin, A. I. Zhmakin, and A. A. Fursenko, *Dif. Uravneniya i Ikh Primeneniye*, Vilnus, no. 31, 1982, pp. 47-50 (in Russian).

9. V. P. Goloviznin, S. P. Goloviznin, and A. A. Fursenko, In: *Numerical Methods for the Problems of Mathematical Physics*, Lvov, 1983, p. 11.

10. V. F. Kamenetskii and L. I. Turchak, In: *Numerical Simulation in Aerohydrodynamics*, Moscow: Nauka, 1986, pp. 104-105.

11. V. P. Goloviznin, I. V. Krasovskaya, *Numerical Simulation of the Interaction Between Head-On Shock Waves and a Blunt Body Flying at Supersonic Speed*, Preprint No. 1281, Physical-Technical Institute, Leningrad, 1988.

12.   V. P. Goloviznin, G. I. Mishin, Yu. Serov, and I. P. Yavor, *Zh. Tekhn. Fiz.*, vol. 57, no. 7, 1987, pp. 1433–1435.

NUMERICAL SIMULATION OF TWO-PHASE SUPERSONIC FLOWS ABOUT BODIES

V. L. Belousov, Yu. P. Golovachev, M. S. Ramm,
D. M. Sharov, A. A. Schmidt

Abstract: This paper examines numerical simulation of two-phase supersonic flows about bodies under the following headings; Introduction, Eulerian-Lagrangian Approach to the Multiphase Medium Modeling, Numerical Results, Conclusion.

1. Introduction

Today the dynamics of heterogeneous systems is one of rapidly developing branches of continuous medium mechanics. Multiphase mixtures are found in many chemical and energy-related industries and in transportation. Some natural phenomena involve multiphase mixture processes as well. And, finally, the multiphase medium is an intriguing example of a non-equilibrium system possessing a number of bulk and surface relaxation processes. In a structure of heterogeneous media, flows significantly depend on mixture component properties and on their interaction with flow boundaries. A model describing the multiphase flows must take these phenomena into account.

Exploiting a widespread model of interpenetrating continua [1-3], or kinetic descriptions of multiphase transport processes [4, 5], we must use assumptions limiting the applicability of these approaches, otherwise unavoidable closing difficulties arise.

This paper presents some results of the numerical simulation of gas containing solid or liquid particle supersonic flows about blunted bodies. The efficient Eulerian-Lagrangian (EL) model describing both bulk and surface processes intrinsic to the particular flows, is used. In the framework of this approach phases are treated in the most natural manner, namely, the continuous models describe the carrying phase and the trajectory probe particle model describes the dispersed phase.

2. Eulerian-Lagrangian Approach to Multiphase Medium Modeling

The idea of EL description suggested in [6] is based on the fact that for moderate volume fraction of the particulate phase the stresses in the mixture are defined by the carrying phase stress tensor and the action on the dispersed particles is determined by the gas-particle interaction force [3]. Deformation of interphase surfaces causes, in this case, a modification of the gas phase transport coefficients. Thus, the effect of the dispersed particles results in mass, momentum, and energy coupling terms and modified transport coefficients in the gas phase governing equations. The algorithm of the EL approach for steady multiphase flows is as follows:

- at the first Eulerian step, the carrying phase flow parameters are computed assuming the zero fraction of the particles;
- at the second, Lagrangian step, the first step data are employed to compute the dispersed phase parameter distribution and the coupling terms. Then the first step is repeated incorporating the coupling source terms obtained, etc. Although there are no *a priori* estimates of convergence for such a procedure, the computational experience has shown that the solution exists in a wide range of mixture parameters. It should be mentioned that the unsteady problem requires modification to the algorithm discussed below.

Let us consider each EL stage in more detail.

At Eulerian step Euler of Navier-Stokes, equations with source of sink coupling terms are solved for the carrier phase. Generally this phase is a multicomponent mixture consisting of a neutral gas and the dispersed phase material vapor.

The Lagrangian step is based on the substitution of all the dispersed particles by a number of probe particles and calculation of their trajectories through the carrying phase flow field within the domain under study. The probe particle represents a parcel of the dispersed particles, having approximately the same initial conditions. The data obtained are averaged over segments of the trajectories within each Eulerian cell used at the first step; and then yield the particulate phase parameter distribution and the coupling transport terms for a new Eulerian step.

The average value of a certain function Φ within an (i, j) Eulerian cell can be written as:

$$\langle \Phi \rangle_{ij} = \frac{\sum_{l \in (ij)} (\Phi\varphi)_{lij} \Delta n_{lij}}{\sum_{l \in (ij)} \tilde{\varphi}_{lij} \Delta n_{lij}}, \tag{2.1}$$

where Δn_{lij} is the number of the dispersed particles corresponding to the l-probe particle within the (i, j)-cell, φ is a weight function. The sums in (2.1) are calculated over all the probe particles belonging to the (i, j)-cell. The time-averaged functions are defined by:

$$\tilde{\Psi}_{lij} = \frac{1}{\tau_{lij}} \int_0^{\tau_{lij}} \Psi_{lij} dt, \tag{2.2}$$

where τ_{lij} is the residence time of the l-probe particle within the (i, j)-cell, Ψ is a certain trajectory functional. The number of dispersed particles and their concentration within the (i, j)-cell can be obtained from

$$\Delta n_{lij} = \eta_l \tau_{lij},$$

$$n_{pij} = \frac{1}{V_{ij}} \sum_{l \in (ij)} \Delta n_{lij}, \tag{2.3}$$

where η_l is the number flux of the dispersed particles along the l-probe particle trajectory, V_{ij} is the volume of the (i, j)-cell. It is suggested here that η_l depending on the initial conditions is constant, but a variation of η_l can easily be taken into account. Weight function $\varphi = 1$ corresponds to unchangeable, monodispersed particles, $\varphi = m_p$ corresponds to the case of interphase mass transfer, here m_p is the mass of the particle.

Relations (2.1)-(2.3) provide the averaging procedure completing the Lagrangian step.

Further description of the EL approach requires detailed models of the carrying phase and the probe particle motion. It is assumed below that the following set of inequalities is valid

$$\tau_{gg} \ll \tau_{gp}, \quad \tau_w^0 \ll \tau_{gp}, \quad \tau_{gp} \leqslant \tau_{pp}, \quad \tau_{gp} \gtrless \tau_g,$$
$$\tau_{gg} \ll \tau_g, \quad \tau_w^0 \ll \tau_g, \tag{2.4}$$

where τ_{gg}, τ_{pp}, are the characteristic time of the relaxation process in the carrying and dispersed phases, respectively; τ_g is the flow characteristic time; τ_w^0 is the particle internal relaxation characteristic time; τ_{gp} is the interphase transport relaxation time. Investigations [7, 8] have shown that inequalities (2.4) are satisfied in a broad range of mixture parameters. Relations (2.4) allow consideration of the carrying phase as a multicomponent mixture possessing a single temperature and a single velocity. In the framework of the Euler model, taking into account interphase mass, momentum and energy transfer in the mixture of air, water vapor, and water droplets, the governing equations for the carrying phase can be written as follows:

$$\frac{\partial \rho_g}{\partial t} + \nabla \cdot (\rho_g \vec{V}_g) = -\langle J_{gp}\rangle n_p,$$

$$\rho_g \frac{dC_1}{dt} = C_1 \langle J_{gp}\rangle n_p,$$

$$\rho_g \frac{d\vec{V}_g}{dt} = -\nabla p_g - \langle (\vec{F}_{gp} + J_{gp}(\vec{V}_p - \vec{V}_g)))\rangle n_p, \tag{2.5}$$

$$\rho_g \frac{dh_g}{dt} = \frac{dp_g}{dt} - \left[\left(Q_{gp} + \vec{F}_{gp} \cdot (\vec{V}_p - \vec{V}_g) + \right. \right.$$
$$\left. \left. + J_{gp}(h_2(T_p) - h_g) + J_{gp}\frac{(\vec{V}_p - \vec{V}_g)^2}{2} \right) \right] n_p,$$

and the equations of state are:

$$p_i = p_i(\rho_i, T_g), \quad h_i = h_i(T_g), \quad h_g = \sum_{i=1}^{2} c_i h_i, \tag{2.6}$$

where ρ is the density; $\vec{V}$ is the velocity vector; h is the enthalpy; C is the mass concentration; p is the pressure; F_{gp}, Q_{gp}, J_{gp} are the coupling terms accounting for mass, momentum, and energy interphase transfer; indices 1 and 2 denote air and water vapor; index g denotes the carrying phase; $\langle\rangle$ denotes time-space averaging. The sets of Equations (2.5) and (2.6) are obtained under the following assumptions: evaporation is a spherically symmetric process, the evaporated mass velocity is equal to that of the droplet, the energy due to interphase momentum transfer dissipates in the carrier phase, the interphase mass transfer occurs at the droplet temperature, the evaporating droplets lose the latent phase transition heat. These assumptions allow us to write the equations of the probe particle motion in the form:

$$\frac{d\vec{r}_p}{dt} = \vec{V}_p,$$

$$\frac{dm_p}{dt} = J_{gp},$$

$$m_p \frac{d\vec{V}_p}{dt} = \vec{F}_{gp}, \qquad (2.7)$$

$$C_p^0 m_p \frac{dT_p}{dt} = Q_{gp} + J_{gp} \cdot l$$

where C_p^0 is the specific heat of the dispersed phase material, l is the latent phase transition heat, r_p represent the radius-vector of the probe particle, index p denotes the dispersed phase.

In a number of problems it is necessary to study the heat transfer and friction on surfaces of bodies immersed in multiphase flows. The boundary approach fails to give reliable information because of intensive interphase transport, which changes the flow structure. In this case the Navier-Stokes model for carrying phase is suitable. The equations accounting for momentum energy and moment interphase transfer can be written as:

$$\frac{\partial \rho_g}{\partial t} + \nabla \cdot (\rho_g \vec{V}_g) = 0,$$

$$\rho_g \frac{d\vec{V}_g}{dt} = \text{Div}\, \vec{P}_g - \langle \vec{F}_{gp}\rangle n_p,$$

$$\rho_g \frac{dh_g}{dt} = \frac{dP_g}{dt} + 2\mu S^2 - \frac{2}{3}\mu(\text{div}\vec{V}g)^2 + \text{div}(\mu\,\text{grad}\,\frac{h_g}{P_r})$$
$$+ \langle \vec{F}_{gp} \cdot (\vec{V}_g - \vec{V}_p)\rangle n_p - \langle Q_{gp}\rangle n_p + \langle \vec{M}_{gp} \cdot (\tfrac{1}{2}\,\text{rot}\,\vec{V}_g - \vec{\omega}_p)\rangle n_p,$$

and the equations of state are

$$h_g = h_g(T_g), \quad p_g = p_g(\rho_g, T_g), \tag{2.9}$$

where $\overset{\leftrightarrow}{P}$ is the stress tensor; $\overset{\leftrightarrow}{S}$ is the deformation tensor; $\mu = \mu(T_g)$, C_p and λ are the viscosity, specific heat at constant pressure, and the heat conductivity for the carrier phase, respectively; $\mathrm{Pr} = C_p\mu/\lambda$ is the Prandtl number; M_{gp} is the term accounting for interphase moment transfer; ω_p is the angular velocity vector. Equations (2.8) involve the dispersed particle rotation in a steep velocity gradient flow field of the carrying phase. A detailed analysis of this oriented rotation of the particles [9, 10] shows that in a wide range of parameters the carrying phase stress tensor remains symmetrical.

Motion and heating of the probe particles can be written as follows:

$$\frac{d\vec{r}_p}{dt} = \vec{V}_p,$$

$$m_p \frac{d\vec{V}_p}{dt} = \vec{F}_{gp},$$

$$m_p \theta_p \frac{dx_p}{dt} = \vec{M}_{gp}, \tag{2.10}$$

$$C_p^0 m_p \frac{dT_p}{dt} = Q_{gp},$$

where θ_p is the particle inertial moment.

To close sets (2.5)–(2.7) and (2.8)–(2.10) it is necessary to introduce an explicit dependence of $\vec{F}_{gp}$, J_{gp}, Q_{gp}, and $\vec{M}_{gp}$ on the mixture parameters and flow regimes. For the mixtures under study it has been shown (see, for example, Ref. [11]) that the interphase momentum transfer can be determined by:

$$\vec{F}_{gp} = \vec{F}_{ST} + \vec{F}_M + \vec{F}_{SAF},$$

$$\vec{F}_{ST} = \frac{\Pi d_p^2}{2} \rho_g C_D (\vec{V}_g - \vec{V}_p)|\vec{V}_g - \vec{V}_p|, \tag{2.11}$$

$$\vec{F}_M = \frac{\pi d_p^3}{6} C_M (\vec{V}_g - \vec{V}_p) \times \vec{\omega}_p,$$

$$\vec{F}_{SAF} = 1.615\, d_p^2 \sum_m \mathrm{sign}\left(\frac{\partial V_{gn}}{\partial x_m}\right) \sqrt{\left|\frac{\partial V_{gn}}{\partial x_m}\right|}\,(V_{gn} - V_{pn})\vec{n}_m,$$

where $\bar{n}$ is the unit vector, m, n are the coordinate indices $m \neq n$. The second and third terms in the right-hand side of (2.11) are of importance only in the steep velocity gradient flows (if there is no external rotation of the particles).

Interphase mass transfer can be obtained by phenomenological relation, accounting for the diffusion and boiling processes [12]

$$J_{gp} = \pi dp D_2 \, \text{Sh}(\rho_2 - \rho_{2s}(T_p))(1 - \delta) - \frac{Q_{gp}}{l}\delta,$$

$$(2.12)$$

$$\delta = p_{2s}/p_g,$$

Neglecting the radiative transfer, interphase energy transport can be written in the form:

$$Q_{gp} = \pi \lambda d_p \text{Nu}(T_g - T_p). \tag{2.13}$$

The moment on the particle in a gradient flow field, is [13]:

$$\bar{M}_{gp} = \pi d_p^3 \mu (\tfrac{1}{2} \, \text{rot} \, \bar{V}_g - \bar{\omega}_p). \tag{2.14}$$

In relations (2.11)-(2.14) C_D is the drag coefficient, Nu is the Nusselt number, Sh is the Sherwood number, D is the diffusion coefficient, d_p is the diameter of the spherical monodispersed particle. Thus, in order to close the sets of governing equations appropriate critical relations for $C_D = C_D(\text{Re}_p, \text{M}_p, \text{Kn}_p, ...)$, $\text{Nu} = \text{Nu}(\text{Re}_p, \text{M}_p, \text{Kn}_p, ...)$, $\text{Sh} = \text{Sh}(\text{Re}_p, \text{M}_p, \text{Kn}_p, ...)$ are to be used here; Re_p, M_p, and Kn_p are the Reynolds, Mach, and Knudsen numbers for the particle moving through the carrying phase. There are a number of such criterial relations available for C_D, Nu, and Sh (see, for example, Refs. [14-18]). In the present study the data from Ref. [16] are used for solid particles and the data from Refs. [14, 17, 18] are used for droplets.

Equations (2.5)-(2.7) and (2.8)-(2.10), and Relations (2.11)-(2.14) along with criterial relations for C_D, Nu, and Sh represent the closed EL models. These models must be supplemented by appropriate boundary conditions discussed below.

One of the principal multiphase flow problems is that of the dispersed particle flow boundary interaction. During the interaction both the dispersed particle and boundary properties could change. It is assumed in this study that the state vectors of the incident and rebounded probe particle are related by

$$\bar{\Psi}_{lr} = k\bar{\Psi}_{li} \qquad (2.15)$$

where $\Psi = \{\bar{V}_p, T_p, \bar{\omega}_p, d_p\}$, k is the interaction coefficient matrix, indices i and r denote incident and rebounded particles, respectively. The number of probe particles may remain constant or change during the interaction, the former case is called a single-ray model and the latter a multi-ray model.

Body-particle interaction results in erosion of the body. Qualitative analysis of this process is performed on the basis of a phenomenological model which involves the idea of the erosion effective enthalpy H_{er} [20]:

$$H_{er} = \frac{G_p V_{pw}^2}{2 G_{er}}$$

where G_p is the incident particle mass flux, G_{er} is the mass flux of the rebounded particles and the eroded fragments of the body, index w denotes the body surface parameters. In terms of the model considered the body surface front velocity can be written in the form

$$\rho_w^0 \frac{\partial Z}{\partial t} = \frac{\rho_p V_{pw}^2 F(\alpha)}{2 H_{er}} \left[1 + \left(\frac{\partial Z}{\partial x} \right)^2 + \left(\frac{\partial Z}{\partial y} \right)^2 \right]^{1/2}, \qquad (2.16)$$

where $Z = Z(x, y, t)$ specifies the body surface, ρ_w^0 is the body material density, $F(\alpha)$ is a function of incident angle dependent on erosion regime. The averaged front velocity is defined by:

$$\left\langle \frac{\partial Z}{\partial t} \right\rangle = \frac{\int_{\Delta F} \left(\frac{\partial Z}{\partial t} \right) D\sigma}{\int_{\Delta F} d\sigma},$$

in this procedure ΔF is the area of the Eulerian boundary cell face.

The EL algorithm must be modified in the case of unsteady flow. The modification implies time discretization of the dispersed particles along with spatial discretization, namely, parcels of the probe particles successively entering the flow domain are introduced. An interphase transport is contributed by all the probe particles of all parcels within the domain. In an unsteady EL algorithm Lagrangian stage is performed at each step of the Eulerian stage. If an unsteady regime is due to external variation of dispersed phase parameters, the interval between the probe particle parcels is determined by the rate of this variation and the time step of the Eulerian stage is determined by the carrier phase response time to this variation. The response time can be estimated as

$$T \sim \mathrm{Sk}_g T_{\text{ext}}$$

where T_{ext} is the characteristic time of dispersed phase external variation, $\mathrm{Sk}_g - \tau_{gp}/(\tau_g \cdot \beta)$ is the Stokes number calculated taking into account the dispersed phase loading β.

Thus, the EL models described above allow investigation of the steady and unsteady multiphase flow structure including the interphase mass, momentum, energy and moment transfer, as well as the erosion of the body immersed in this flow.

3. Numerical Results

To illustrate the capabilities of the EL models some problems of axisymmetric particulate and droplet supersonic flows about a sphere are considered. The domain under study lies between the body surface, detached shock wave, stagnation streamline, and a certain line located in the supersonic region of the flow. Boundary conditions for the carrier phase are as follows: behind the shock wave the flow parameters are determined by normal Renkin–Hugonion relations for an interphase relaxation time much greater than the time required for the particle to pass through the shock wave. The conditions at the body surface take into account the body contour change due to the erosion process. The symmetry conditions are used on the flow axis. In droplet flow the vapor concentration is additionally assumed to be conserved within the shock.

Initial positions of the probe particles are distributed uniformly on the shock wave and in the free stream there is interphase temperature and velocity equilibrium. To describe the rebounded particles the interaction coefficient matrix is set. In spite of a number of studies of solid or liquid particle-solid surface interaction (for example, [17, 21, 22]), there is a shortage of quantitative data on this phenomenon. In the framework of ray model (2.15) the k-matrix components are chosen to agree with the numerical predictions of the ballistic range data [21, 23].

The number of probe particles was varied. In most of the problems it was sufficient to use up to 1000 probe particles in the parcel.

A body fitted coordinate system is used. The physical space is transformed into a rectangular computational space by normalizing the normal coordinate to the shock layer thickness. A computational mesh considered has up to 50 cells in each direction.

At the Eulerian stage of the EL algorithm implicit finite-difference schemes ([24] for the Euler carrier phase model and [25] for Navier-Stokes) are used. At each step the nonlinear set of finite-difference equations is solved using the Newton iterative technical. At the Lagrangian stage the ordinary differential equations for the probe particles (2.7) or (2.10) are solved using the technique providing accuracy correspondent to one of the Eulerian stage.

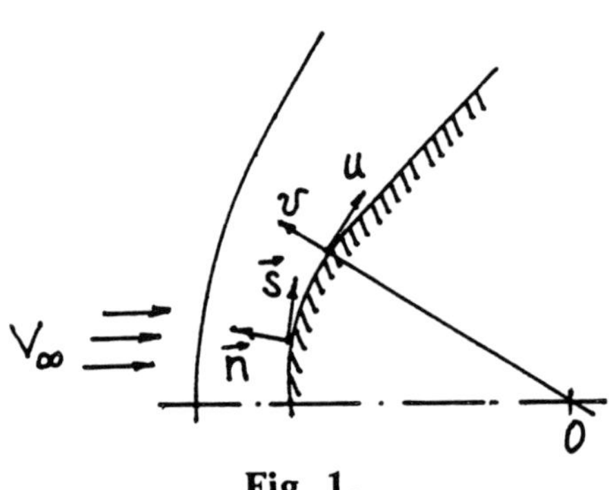

Fig. 1.

Figure 1 represents the flow pattern and the coordinate system. The effect of the relaxation parameter $Sk = \tau_{gp}/\tau_g$ and the k-matrix on the gas-solid particles shock layer structure is illustrated in Figures 2 and 3 for the free stream mixture velocity $V_\infty = 1000$ m/s, carrier phase density and temperature $\rho_\infty = 0.414$ kg/m^3, $T_\infty = 223$ K, respectively; dispersed phase material density and specific heat $\rho_p^0 = 2.6 \cdot 10^3$ kg/m^3, $C_p^0 = 741$ J/kgK, dispersed phase volume fraction in free stream $\varphi_\infty = 2 \cdot 10^{-5}$. It is assumed that Relation (2.15) is reduced to $v_{pr} = -K_v v_{pi}$. Here v_p is the normal velocity component. This is a rather crude assumption but comparison with experimental data presented below show that the approximation adopted takes into account the main features of the phenomenon.

Figure 2 shows the dispersed phase density profiles along the stagnation streamline. Curves 1, 2, 3, and 4 correspond to $Sk = 1.52$, 0.39, 0.15, and 0.08, respectively. Coefficient K_v is equal to 0.2. At $Sk \geqslant 1$ the flow is nonequilibrium, at $Sk \ll 1$ the flow becomes equilibrium. The structure of nonequilibrium flow is characterized by the essential role of the rebounded particles. Accumulation of these decelerated particles results in a nonmonotonic profile (curve 1). Decreasing the Stokes number the density profile maximum approaches the body surface because of more intensive interphase relaxation (curve 2). Further decrease of the Stokes number leads to predominant particle migration along the body surface. These particles are nearly in equilibrium with the carrier gas (curves 3 and 4).

The effect of the interaction coefficient K_v on the flow structure is illustrated in Figure 3, representing the density profiles of the dispersed phase along the stagnation streamline. Curves 1, 2, 3, and 4 correspond to $K_v = 0.2$, 0.15, 0.10, and 0.05, respectively. The Stokes number is equal to 1.52. It is shown that the K_v decrease is accompanied by a decrease of the rebounded region and increase of the density within this region. Thus, the

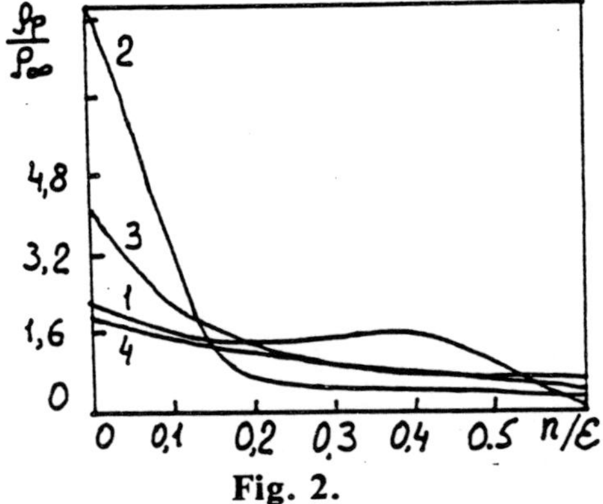

Fig. 2.

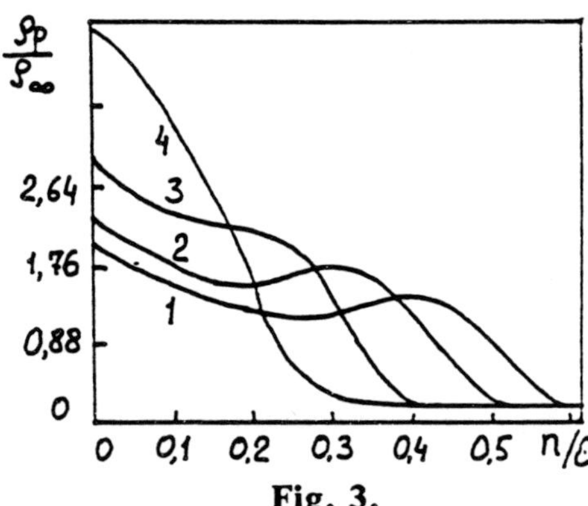

Fig. 3.

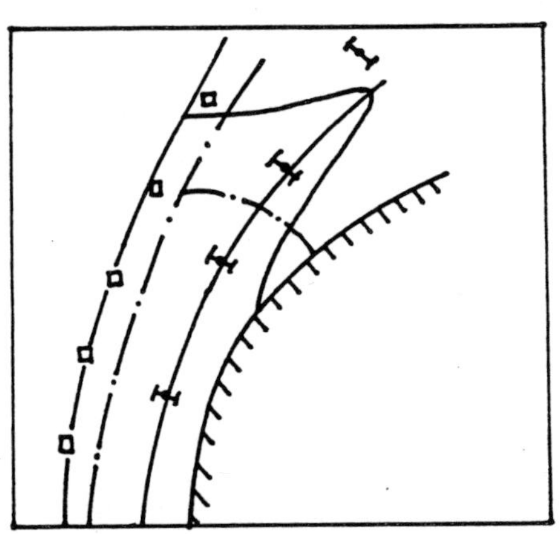

Fig. 4.

choice of appropriate value of K_v is one of the main problems of multiphase flows. Experimental data [23, 26] show that there is a high dispersed phase density region near the body flying in the gas-solid particle mixture at supersonic speed. Formation of this region is determined, on the one hand, by accumulation of the rebounded particles and, on the other hand, by dramatic change of the relaxation parameter during the particle-body surface interactions. The latter is due to fragile solid particle disintegration occurring in high speed impact. The direct measurements of the dispersed phase size distribution [23] before and after the body flight confirm this fact. Therefore, in addition to K_v it is necessary to introduce K_d which determines the shape and size of the rebounded particles. Study [23] carried out at $V_\infty = 850$ m/s, $\rho_\infty = 1.223$ kg/m^3, $T_\infty = 293$ K, $a_0 = 5$ mm, $d_{p\infty} = 120$ μm, $\rho_p^0 = 4 \cdot 10^3$ kg/m^3, $C_p^0 = 741$ J/kgK, $\varphi_\infty = 10^{-4}$ reveals that $K_v = 0.4$, and $K_d = 0.008$. Both shock layer and disperse phase high concentration layer thickness are presented in Figure 4 for comparison with experimental data. Solid lines show the number predictions, dots indicate experimental data, the dashed line indicates the sonic line, the broken curve corresponds to one-phase flow.

The effect of interphase mass transfer on the shock layer structure is investigated for the droplet flow about the sphere. It is assumed that $a_0 = 0.3$

m, V_∞ = 1000 m/s, ρ_∞ = 1 kg/m^3, T_∞ = 275 K, $d_{p\infty}$ = 500 μm, φ_∞ = 1.5·10^{-4}, and the liquid properties correspond to water. According to Ref. [21] a three-ray model for droplet-body surface interaction is used.

The fragments of the rebounded droplet are considered to be spherical and monodispersed. The interaction coefficients are K_v = 0.7, K_d = 0.03. Incident dispersed phase flow parameters are "frozen" and the corresponding Stokes number is equal to 7.5. The post first-impact history of motion, heating, and condensation-evaporation of the probe particle entering the shock layer near the stagnation line is presented in Figure 5. Two stages can be distinguished. The first stage includes the history of the rebounding droplet and the second includes the history of the turned back droplet. The former is characterized by velocity, temperature, and diameter increase; the latter by intensive evaporation and nonmonotonic variation of temperature due to interplay of interphase mass and energy transfer.

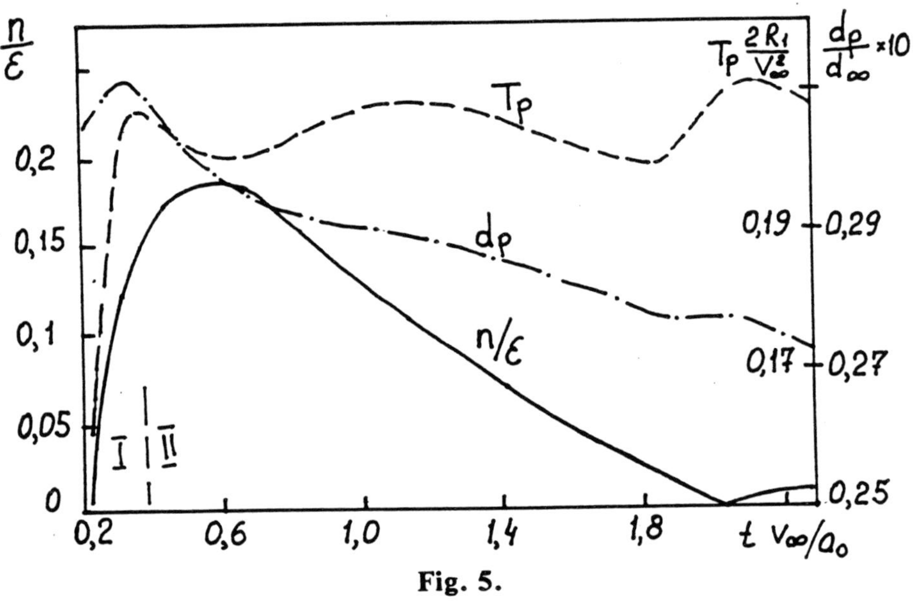

Fig. 5.

An accumulation of the decelerated rebounding droplets leads to formation of the dispersed phase high concentration regions. It is illustrated in Figure 6. The nonmonotonic vapor concentration profile across the shock layer presented in Figure 7 can be explained by intensive evaporation within these regions. One, 2, and 3 correspond to the profiles for various values of the distance along the body surface S/a_0 = 0, 0.524, and 0.786, respectively.

Thus, evaporating droplets significantly change the flow structure, and therefore, thermal and dynamic effects on the body surface.

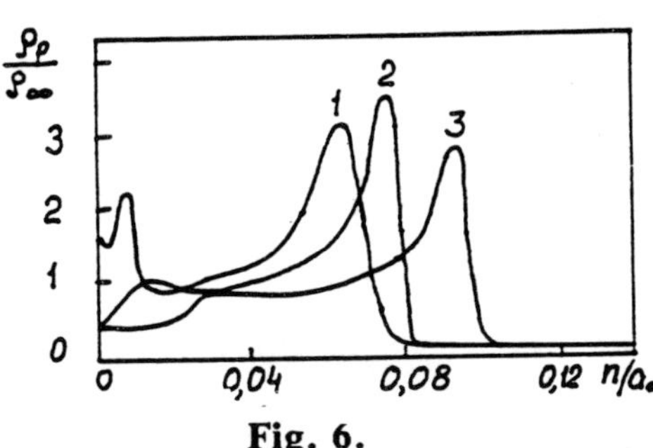

Fig. 6.

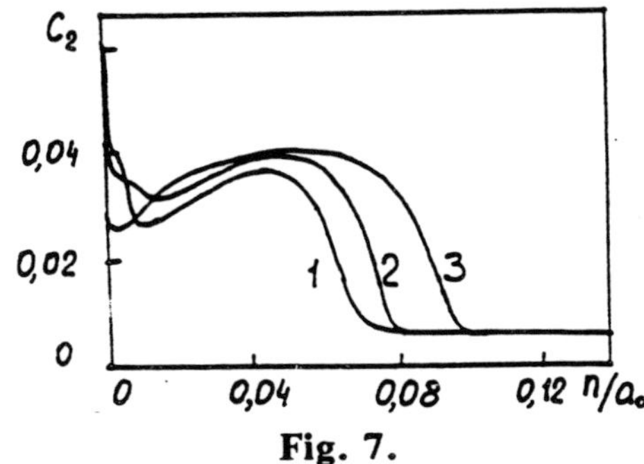

Fig. 7.

In a number of practical applications the problem of erosion is of great importance. The EL model is used to study erosion of the body immersed in the gas-solid particle flow. It is assumed that $a_0 = 0.3$ m, $V_\infty = 1000$ m/s, $\rho_\infty = 0.414$ kg/m^3, $T_\infty = 223$ K, $\rho_p^0 = 2.6 \cdot 10^3$ kg/m^3, $C_{p_g}^0 = 741$ J/kgK, $d_p = 30$ μm, $\rho_w^0 = 2.6 \cdot 10^3$ kg/m^3, $K_d = 0.25$, $K_v = 0.2$, $H_{er} = 10^6$ J/kg. Experimental data [20] show the effective enthalpy of erosion to be constant in a wide range of impact parameters. Estimates of the characteristic times of the flow and the erosion process yield:

$$\frac{\tau_g}{\tau_{er}} = \frac{\varphi_\infty V_\infty^2}{2H_{er}} \ll 1,$$

therefore, at the parameters under study the erosion process is a quasisteady one which can be taken into account as follows: steady heterogeneous shock layer flow field is computed, then Equation (2.16) can be used to give the surface front velocity and a new shape of the body surface is defined by

$$\langle Z \rangle|_{t+\Delta \tau_{er}} = \langle Z \rangle|_t + \left\langle \frac{\partial Z}{\partial t} \right\rangle\bigg|_t \cdot \Delta \tau_{er},$$

where $\Delta \tau_{er}$ is the time step of the erosion process. The above-mentioned estimate provides $\Delta \tau_{er} \gg \tau_g$ ($\Delta \tau_{er} = 10^4 \tau_g$ is chosen here). Figure 8 represents the change of the body surface shape and the detached shock wave shape occurring in the erosion process. Figure 8a corresponds to the initial state, Figures 8b and c correspond to the moment $t = 0.750 \, H_{er} \, a_0/\varphi_\infty/V_\infty^3$ for plastic and fragile materials. Solid lines show the body and shock wave contours, dashed curves show the sonic lines, and the broken lines denote the separatrixes of the rebounded particles. Layers with a particle concentration of not less than 0.85 of the maximum value are shaded. The plastic

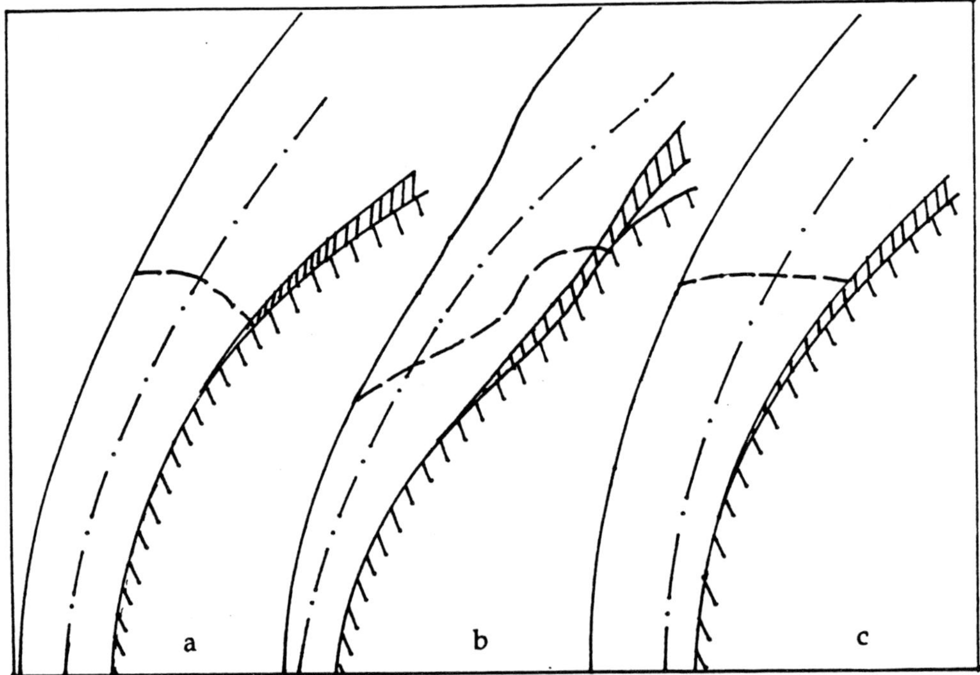

Fig. 8.

material is mainly eroded by the particles incident on the surface at an angle approximately equal to 30°, a circular hollow appears on the body surface, the body becomes more "sharp", the shock distance decreases. The circular hollow affects the flow structure qualitatively, namely, the maximum concentration layer separates, the sonic line is strongly deformed. The fragile material is mainly removed by the particles incident on the body normally, hence, the body becomes more "blunt", the shock distance increases."

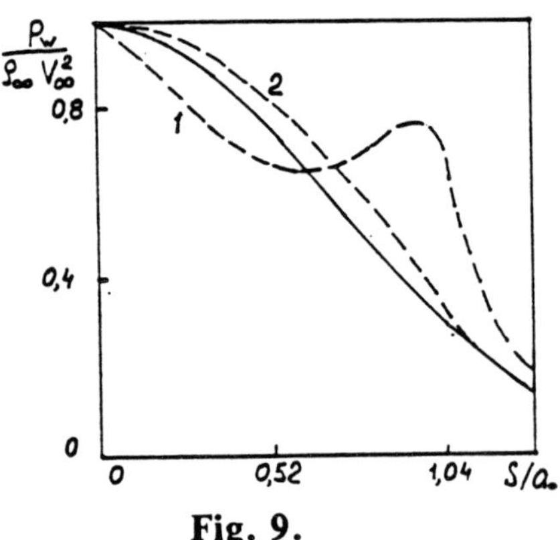

Fig. 9.

Figure 9 shows the pressure profiles along the body surface. Curves 1 and 2 correspond to the plastic and fragile materials. The solid line denotes the initial state. Plastic erosion leads to redistribution of the pressure, decreasing at the stagnation region due to the body nose "sharpening" and detached wave contour change. At the circular hollow the pressure increases and a trend to flow separation appears. The flow structure change during erosion of the fragile material leads to a pressure increase over the entire body contour.

Computations have shown that the erosion is a self-consistent process with strong feedback.

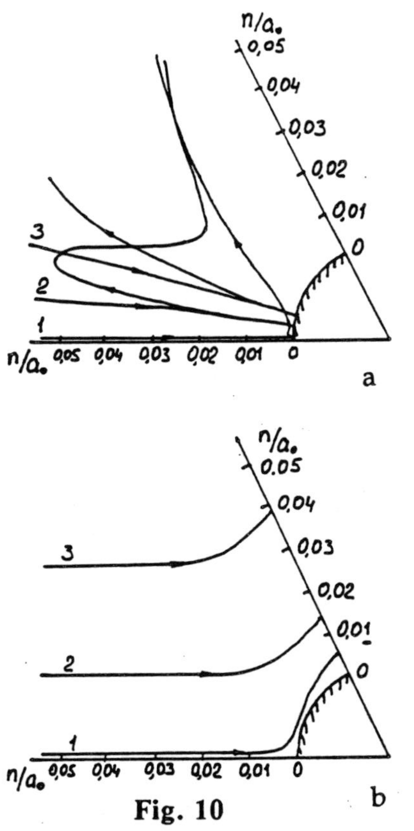

Fig. 10

To investigate the effect of viscosity and heat conductivity of the carrier phase on the particulate shock layer structure as well as on friction and heat transfer on the sphere surface the Navier-Stokes type EL model is used. Calculations are performed for a_0 = 0.43 m, $M_\infty \doteq 2$, ρ_p^0 = $3.7 \cdot 10^3$ kg/m^3, C_p^0 = 1250 J/kgK, $T_w = T_\infty$ = 300 K. The Reynolds number Re$_\infty$, loading ratio β = $\rho_{p\infty}/\rho_\infty$, particle diameter d_p, and components of the k-matrix are varied. It was shown that there are sedimentational and nonsedimentational regimes depending on the flow parameters. Figures 10a and b represent the detailed trajectories of the probe particles near the sphere surface in flows with various densities of the carrier phase (Figure 10a corresponds to Re$_\infty$ = $1.64 \cdot 10^4$, Figure 10b to Re$_\infty$ = $1.64 \cdot 10^6$). The dispersed particle diameter and the component of the k-matrix are considered to be d_p = 0.5 μm, K_v = 0.25, respectively. If the Reynolds number is increased, the multiphase flow passes to the nonsedementation regime, accompanied by stratification of the flow. This phenomenon significantly changes the whole flow structure.

It has been noted [27] that even at low loading ratios (β = 0.02-0.05) in hypersonic multiphase flows (Re$_\infty \sim 10^7$, $M_\infty \gg 1$) the heat flux on the body surface increases dramatically. This effect has been explained by dissipation of kinetic energy of incident particles within the boundary layer. The present analysis, accounting for the rebounded particles, gives the effect at moderate Mach and Reynolds numbers. Figure 11 shows St and C_f profiles along the body surface:

$$St = \frac{q_w}{\rho_\infty V_\infty C_p(T_s^0 - T_w)}, \quad C_f = \frac{\left(\mu\dfrac{\partial u}{\partial n}\right)_w}{\rho_\infty V_\infty^2},$$

here q_w is the heat flux on the body surface, indices S and O denote the parameters behind the shock wave and on the axis, respectively. The flow regime is as follows: $Re_\infty = 1.64\cdot10^4$, $K_v = 0.25$, $\beta_\infty = 0.05$; the dashed curves correspond to one-phase flow. For weak interphase interactions ($d_p = 1\ \mu$m or Sk = 10) the multiphase heat flux and friction on the body surface are nearly the same as for one-phase flow (curve 1). Then the interphase transfer becomes intensive ($d_p = 0.2\ \mu$m, Sk ≈ 0.4), heat flux and friction increase significantly due to accumulation of rebounded particles in the thin region within the boundary layer dissipation of their kinetic energy during migration along the body surface.

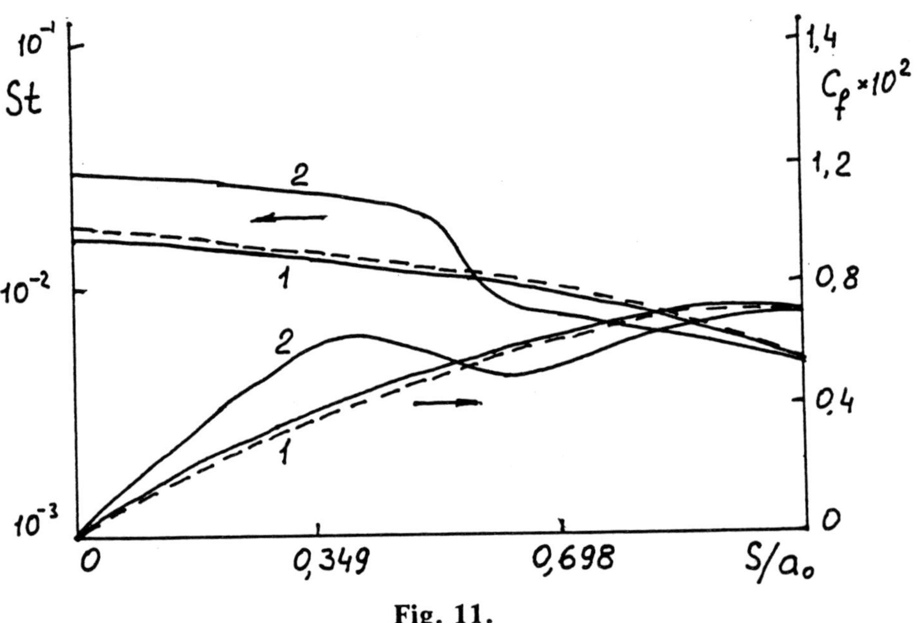

Fig. 11.

The effect of the dispersed phase of heat flux and skin friction can be amplified by the unsteady regime of the multiphase flow. To illustrate this phenomenon the supersonic motion of the sphere through the dusty layer is considered. The problem parameters are as follows: $Re_\infty = 1.64\cdot10^4$, $M_\infty = 2$, $d_p = 1\ \mu$m, $\rho_p^0 = 2.6\cdot10^3$ kg/m^3, $C_p^0 = 741$ J/kgK, $T_w = T_\infty = 300$ K, $\varphi_\infty = 2\cdot10^{-6}$, $\Delta L = 0.5\ a_0$, where ΔL is the dust layer thickness. It is assumed that the loading ratio jumps from $\beta = 0$ outside the layer to $\beta = 4$ inside the layer (the numerical method applied smears the front on the computational cell). Figures 12a and b represent isopycnics of the dispersed phase at moments $t = 0.24\ a_0/V_\infty$ and $t = 0.76\ a_0/V_\infty$, respectively. Figures 13a and b represent isobars at the same moments. Figures 12b and 13b correspond to the moment the particles rebounded by the stagnation region turn back. A complicated creasy structure of the dispersed phase is formed within the shock layer. This structure gives rise to discontinuities in the carrier phase flow. All these processes lead to unsteady dynamic and thermal effects on the body surface.

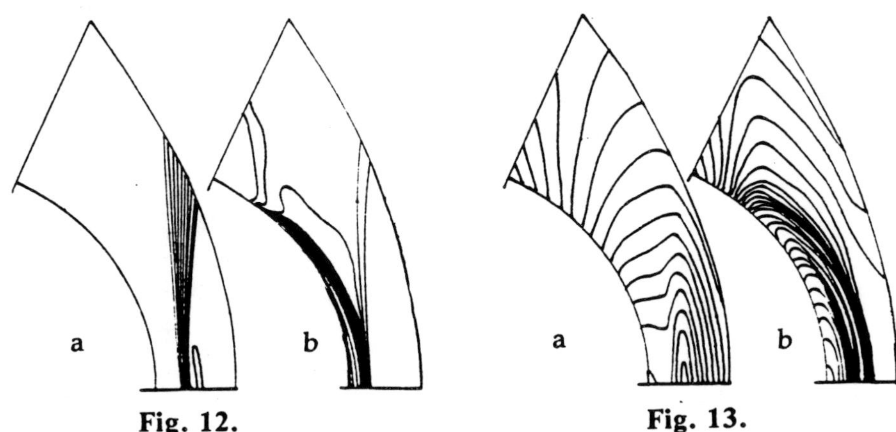

Fig. 12.   Fig. 13.

4. Conclusion

Thus, the set of models formulated in Section 2 allows us to treat multiphase flow problems uniformly. The software designed is efficient and universal. The data obtained demonstrate the significant effect of the dispersed phase on various aspects of the processes involving multiphase flows.

References

1.   Kh. A. Rakhmatulin, *Prikl. Mat. i Mekh.*, vol. 20, no. 2, 1959, pp. 184–195 (in Russian).
2.   F. E. Marbe, In: *Ann. Rev. Fluid Mech.* vol. 2, 1970, pp. 397–446.
3.   R. I. Nigmatulin, *The Foundations of Heterogeneous Media Mechanics*, Moscow: Nauka, 1978, 336 pp. (in Russian).
4.   V. P. Myasnkikov, *Zh. Prikl. Mat. i Tekhn. Fiz.*, no. 2, 1967, pp. 58–67 (in Russian).

5. Ya. D. Yankov, *Izv. AN SSSR, Mekh. Zhidk. i Gaza.*, no. 1, 1980, pp. 82-90 (in Russian).

6. C. T. Crowe and D. E. Stock, *Int. J. for Num. Meth. in Eng.*, vol. 10, 1976, pp. 185-196.

7. S. A. Senkovenko and A. L. Stasenko, *Relaxation Processes in Gas Supersonic Jets*, Moscow: Energoatomizdat, 1985, 118 pp. (in Russian).

8. V. V. Struminskii, *Prikl. Mat. i Mekh.*, vol. 38, no. 2, 1975, pp. 203-210 (in Russian).

9. H. Immich, *Int. J. Multiphase Flow*, 1980, vol. 6, pp. 441-471.

10. V. N. Nikolaevskii, In: *Modern Problems of Theoretical and Applied Mechanics*, Kiev: Naukova dumka, 1978, 128 pp. (in Russian).

11. R. G. Boothroyd, *Flowing Gas-Solids Suspensions*, London: Chapman & Hall, 1971.

12. H. Y. Lu and H. H. Chin, *AIAA J.*, vol. 4, no. 6, 1966, pp. 1008-1016.

13. S. I. Rubinov and J. B. Keller, *J. of Fluid Mech.*, vol. 11, pt. 3, 1961, pp. 447-459.

14. A. I. Ivandaev, A. G. Kutushev, and R. I. Nigmatulin, In: *Itogi Nauki i Tekhn. Mekh. Zhidk. i Gaza*, vol. 16, 1981, pp. 209-287 (in Russian).

15. N. N. Yanenko, et al., *Supersonic Two-Phase Flows in the Presence of Particle Velocity Nonequilibrium*, Novosibirsk: Nauka, 1980 159 pp. (in Russian).

16. D. Carlson and R. F. Hoglund, *AIAA J.*, no. 11, 1964, pp. 1980-1984.

17. A. L. Gonor and V. Ya. Rivkind, In: *Itogi Nauki i Tekhn. Mekh. Zhidk. i Gaza.*, vol. 17, 1982, pp. 86-159 (in Russian).

18. M. Renksizbulut and M. C. Yuen, *J. of Heat Transfer*, vol. 105, no. 5, 1983, pp. 384-388.

19. M. Renksizbulut and M. C. Yuen, *J. of Heat Transfer*, vol. 105, no. 5, 1983, pp. 389-397.

20. Yu. V. Polezhaev, et al., *Inzh. Fiz. Zh.*, vol. 37, no. 3, 1979, pp. 9-13 (in Russian).

21. I. A. Dukhovskii, P. I. Kovalev, and A. A. Schmidt, *Pisma v Zh. Tekhn. Fiz.*, vol. 10, no. 11, 1984, pp. 649-652 (in Russian).

22. W. Tabakoff, M. F. Malak, and A. Hammed, *AIAA J.*, vol. 25, no. 5, 1987, pp. 721-726.

23. I. A. Dukhovskii, et al., *Experimental and Numerical Study of Particle Disintegration Effect on Supersonic Dusty Gas Flow About Blunted Body*, Preprint FTI AN SSSR, no. 1300, 1988, 21 pp. (in Russian).

24. A. N. Lubimov and V. V. Rusanov, *Flows About Blunted Bodies*, vol. 1, Moscow: Nauka, 1970, 357 pp. (in Russian).

25. Yu. P. Golovachov and F. D. Popov, *Zh. Vych. Matem. i Matem. Fiz.*, vol. 12, no. 5, 1972, pp. 1292-1303 (in Russian).

26. I. A. Dukhovskii, et al., *Optika i Spektroskopiya*, vol. 63, no. 5, 1987, pp. 1105-1108 (in Russian).

27.   A. N. Osinov, E. G. Shapiro, *Izv. AN SSSR Mekh. Zhodk. i Gaza*, no. 5, 1986, pp. 55-62 (in Russian).

NUMERICAL INVESTIGATION OF SUPERSONIC VISCOUS BLUNT BODY FLOWS WITH THE USE OF ADAPTIVE GRIDS

V. L. Belousov, Yu. P. Golovachov

Abstract: Adaptive grids are generated within the framework of a well-known approach based on some function equidistribution along the coordinate lines in one dimension. Linear combinations of gas parameter gradients or truncation error of the finite-difference scheme are used as the weight functions governing grid point distribution. Various methods for constructing an adaptive grid are considered: time-discrete sequence of grids, periodic grid generation, application of moving coordinates. A strategy for constructing an optimal adaptive grid is discussed. Examples of viscous shock layer calculations are presented.

1. Adaptive Grid Generation

The most general approach for adaptive grid generation is based on variational principles [1, 2]. within the framework of this approach the adaptive grid is generated by minimizing a linear combination of the integrals representing grid properties, i.e., orthogonality, smoothness of cell size variation, deviation from Lagrangian coordinates, clustering of the grid points in the regions of steep gradients, etc. If we prescribe the coefficients of the above-mentioned linear combination we can construct grids with desirable properties.

The equidistribution schemes imply that the product of the mesh spacing and the positive weight function remains constant along the coordinate line. Such schemes can be derived by minimizing the integral representing the measure of grid point clustering. The main problem with these schemes is that of determining a suitable weight function and method for grid generation.

251

1.1. Choice of the Weight Function

Grid adaptation is carried out in order to minimize the truncation error of the finite-difference scheme and to provide its even distribution in the physical domain. Accounting for the fact that the truncation error is usually large in the region of steep gradients, combinations of gas parameter gradients are often used as the weight function W [3]. The use of truncation error Ψ itself seems more reasonable [4, 5]. However, it is rather difficult to ensure sufficient accuracy of this quantity evaluation when solving the problem.

Applying Taylor-series expansion to analyze the truncation error of the difference operator L_h we can obtain the formulas for the main part of the truncation error $\tilde{\Psi}(h, \varphi)$ containing higher derivatives of the unknown function φ. Practical application of these formulas requires numerical differentiation to be carried out that leads to significant error in $\tilde{\Psi}$ evaluation and therefore hampers the use of the result for constructing an adaptive grid.

There is another way to evaluate the truncation error of the difference operator free of numerical calculation of the higher derivatives of grid function φ_h. In this way the main part of truncation error on the grid with spacing h is given by

$$\tilde{\Psi}(h, \varphi) \approx \Psi_h(h, \varphi_{h/n}) = L_h \varphi_{h/n}$$

Here index h/N relates to the grid in which spacing is N times less than that of the initial grid; $\varphi_{h/n}$ is the solution of equation $L_{h/n}\varphi_{h/n} = 0$. The truncation error thus obtained $\tilde{\Psi}$ was employed in Ref. [6] to improve the accuracy of numerical solution of equation $L\varphi = 0$ by means of its injection with opposite sign into difference equation $L_h\varphi_h = 0$. The disadvantage of this approach is that we must obtain a solution on the finer grid as compared to the initial one.

Let us consider another method for evaluating $\tilde{\Psi}_h$ which incorporates the same idea. It is evident that having determined $\Psi_{Nh} = L_{Nh}\varphi_h$ at some values of N, for instance, at $N = 2$, we can obtain Ψ in the correspondent grid points by means of extrapolation in N. In this way Ψ values are calculated repeatedly on the grids displaced relative to each other, allowing us to obtain Ψ values in all points of the initial grids.

In the present work the linear combinations of gas parameter gradients as well as the truncation error of the finite-difference scheme were employed as weight functions governing the distribution of the grid points. The results of truncation error evaluations by means of a Taylor-series expansion and the extrapolation procedure described above are denoted with $\tilde{\Psi}_T$ and $\tilde{\Psi}_E$, respectively.

1.2. Grid Generation in Stationary and Nonstationary Problems

Solving the stationary gasdynamics problems by relaxation methods we can only reconstruct the computational grid from time to time and determine the function values in the points of a new grid by means of interpolation. Such a procedure will be further referred to as periodic grid generation (PGG).

Numerical investigations of the nonstationary flows are carried out with the use of a time-discrete sequence of grids (TDSG). In this case the grid is generated at each time step and this procedure is followed by interpolation of gasdynamic functions. Moving coordinates can be used for constructing an adaptive grid in the nonstationary problems as well. In this case we must prescribe or determine the node speeds in order to solve the problem. In the last case a complementary equation for the node speeds is formulated that is solved along with basic gasdynamic equations. Another (explicit) method of determining node speeds lies in numerical differentiation of the node coordinates at the adjacent time levels with respect to time. The use of moving coordinates allows us to avoid the interpolation procedure.

2. Solution of Stationary Problems

As an example of the stationary problems we will consider supersonic axisymmetric flow past a blunt body with mass injection. Gas specific heat ratio $\gamma = 1.4$, Prandtl number $\mathrm{Pr} = 0.7$, viscosity is determined by Sutherland's formula for air.

Bow shock wave is treated as the computational domain boundary using Rankine-Hugoniot relations with or without shock slip effects. Gas temperature, T_w; no-slip condition, $u = 0$; and injection velocity, ν_w are prescribed at the body surface. Stationary solution is obtained by means of a time-asymptotic technique using the finite-difference scheme [7].

Consider first, similar solutions of Navier-Stokes equations at the stagnation line. The results presented below correspond to free-stream Mach number $M_\infty = 6$, injection velocity $\nu_w = 0.09\,V_\infty$, body surface temperature $T_w = 0.35\,T^*$, where T^* is the free-stream stagnation temperature.

The physical domain is mapped into the computational domain in two stages: first, normalizing transformation $\xi = (r - 1)/\varepsilon$ is applied producing the segment of unit length in the computational plane (here r is radial coordinate, ε is bow shock wave distance from the body surface); then transformation $x = x(\xi)$ is applied permitting assignment of grid points in the physical plane where desired. Stretching function $x(\xi)$ is defined in analytic form or is determined by the equidistribution scheme. In the first case the computational grid is fixed in ξ-coordinate. In the second case gas temperature gradient or truncation error for energy equation are employed as the weight functions.

Truncation error $\tilde{\Psi}_T$ of the finite-difference scheme [7] is given by

$$\tilde{\Psi}_T = \frac{h^2}{x_\xi^2}\left[\frac{\rho C_p}{6\varepsilon}T^{\mathrm{III}}(\nu - \xi\varepsilon_t') - \frac{P^{\mathrm{III}}}{6\varepsilon}(2\nu - \xi\varepsilon_t') + \frac{8\mu\nu^{\mathrm{III}}}{9\mathrm{Re}}\left(\frac{u+\nu}{\xi} - \nu'\right)\right.$$

$$\left.- \frac{1}{\varepsilon^2\mathrm{Re}}\left[\frac{\eta T^{\mathrm{IV}}}{12} + \frac{\eta^{\mathrm{I}}T^{\mathrm{III}}}{6} + \frac{\eta^{\mathrm{II}}T^{\mathrm{II}}}{8} + \frac{\lambda^{\mathrm{III}}T^{\mathrm{I}}}{24}\right]\right] \tag{2.1}$$

Here h is uniform grid spacing in computational plane; ε_t' is bow shock wave velocity; ρ, T, and p are gas density, temperature, and pressure, respectively; ν is the radial velocity component, u is the derivative of the tangent velocity component in the angular coordinate; C_p is gas specific heat at constant pressure; μ, λ are viscosity and thermal conductivity; Re is the Reynolds number.

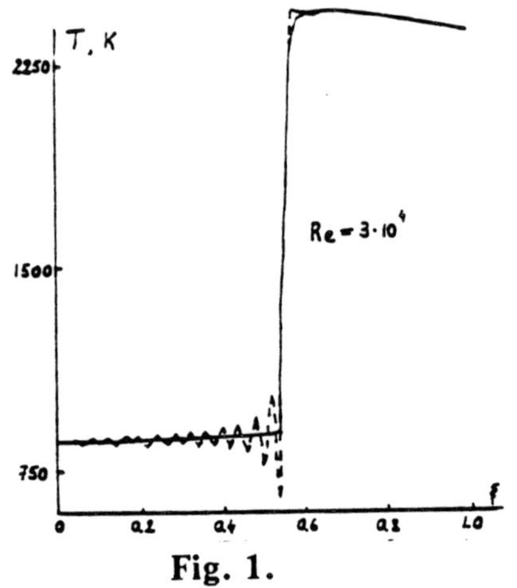

Fig. 1.

Calculations were carried out on grids containing 51 nodes between the bow shock wave and the body surface. At high Reynolds numbers the use of a uniformly spaced grid in the physical plane results in the appearance of oscillations in the numerical solution (dashed line in Figure 1). If the location of the mixing layer is known we can obtain monotonous solution by choosing appropriate analytic transformation $x - x(\xi, \xi^0, H)$, where ξ^0 is coordinate of the mixing layer center in the physical plane, H is the parameter that determines the measure of grid points clustering to point ξ^0. The solid line shows the temperature profile calculated with the use of such a grid generator [8]. This method of constructing the grid requires *a priori* information on parameters ξ^0 and H. Its application becomes difficult in the case of more than one dimension or when the flow structure varies with time.

In applying the equidistribution technique various weight functions were employed. Calculations show that the use of temperature gradient

$$W = |\nabla T/T^*|^\alpha + \delta \tag{2.2}$$

$$\alpha = 0.5, \quad \delta \lesssim 10^{-2}$$

or extrapolated truncation error

$$W = |\tilde{\Psi}_E| \tag{2.3}$$

permits generation of the adaptive grids which are better adjusted to the solution behavior than the use of truncation error $\tilde{\Psi}_T$.

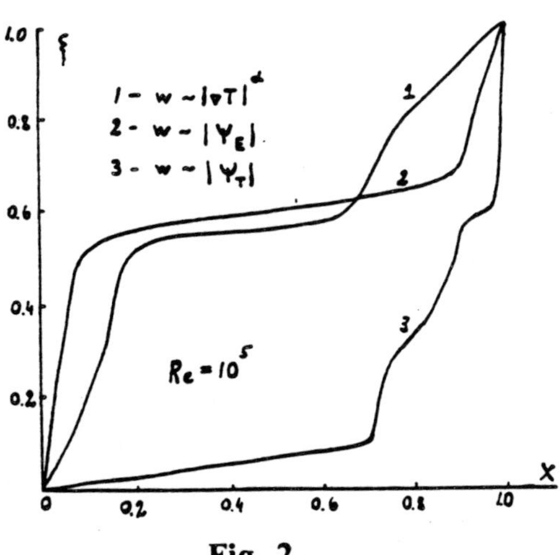

Fig. 2.

Transformations obtained with the use of the above-mentioned weight functions are presented in graphs in Figure 2. Gently sloping parts of the curves $\xi(x)$ correspond to grid point clustering, steep parts correspond to sparse grid point assignment. When equidistribution of the truncation error $\tilde{\Psi}_E$ is imposed the stationary solution can be obtained only if the correspondent grid generation procedure is applied once with preliminary use of other methods. Later weight function (2.2) is applied for grid generation in the problem considered.

Applying the TDSG method does not permit us to obtain the stationary solution because of violation of the iteration convergence by accumulation of the interpolation error. Various interpolation procedures were tested (linear, quadratic, and cubic spline interpolation) but the results were found to be nearly the same. Figure 3a shows time-evolution of the computational grid. Steady distribution of the grid points in the physical plane is obtained only if the Propagating technique is applied with period $\Delta t^a \gtrsim 10\tau$ where τ is the time increment in the finite-difference scheme (Figure 3b). The stationary solution was obtained at the least number of time steps with the use of a grid generated by means of PGG technique and fixed at the final stage of calculations (Figure 4).

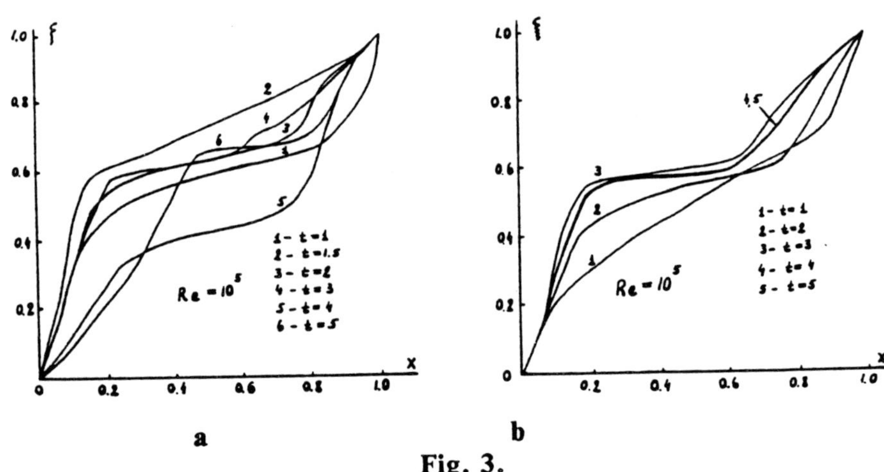

Fig. 3.

Fig. 4.

Consider generation of the two-dimensional adaptive grid using the problem of supersonic flow about the spherically blunted cone with mass injection at its spherical part as an example. Cone half-angle $\beta = 10°$, free-stream Mach number $M_\infty = 6$, body surface temperature $T_w = 0.078\ T^*$, injection velocity $\nu_w = 0.01\ V_\infty \cos s$ $(0 \leqslant s \leqslant \pi/2 - \beta)$, where s is a distance from the stagnation point referred to body nose radius; on the conical part of the body $\nu_w = 0$. The shock layer flow is described by reduced Navier-Stokes equations [7] written in the body-fitted curvilinear coordinate system. PGG method is applied along each ray $S = $ const to cluster the grid points in the region of steep gradients. Weight functions are determined by formula (2.2) and the formula

$$W = |\nabla T/T_s|^\alpha + |\nabla u/u_s|^\alpha + \delta \tag{2.4}$$

where $\alpha = 0.5$, $\delta = 10^{-2}$, index s refers to the values just behind the bow shock wave.

After each performance of the grid generation procedure function $\xi(x, s)$ is smoothed in the angular coordinate s direction.

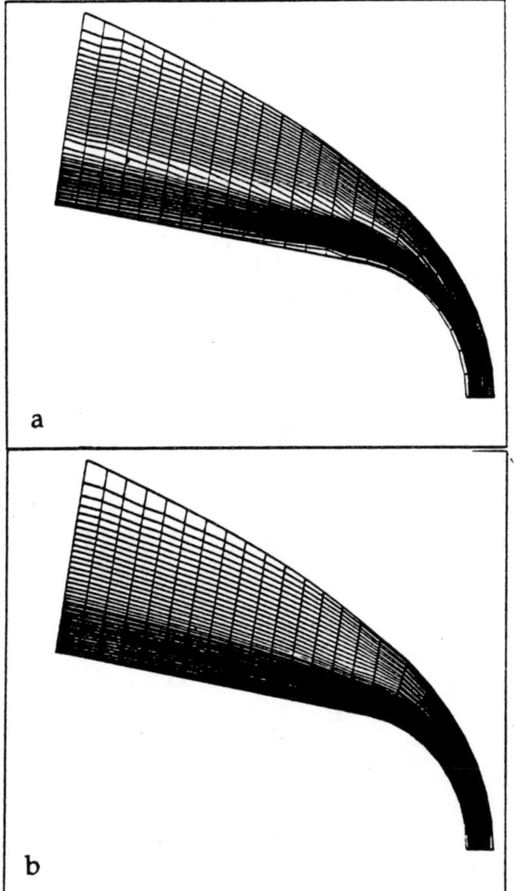

Fig. 5.

Figures 5a and b show the grids obtained by means of PGG method with weight functions (2.2) and (2.4). In the first case grid points are not clustered in the boundary layer near the cone base. In the second, the grid generation procedure ensures desirable grid point distribution at large s values as well. Temperature profiles in the shock layer are shown in Figure 6.

3. Solution of Nonstationary Problems

Application of the adaptive grids is considered using the problem of supersonic flow past a sphere moving at variable speeds as an example. It is supposed that the sphere moves with acceleration during time $t = T$. Time-dependence of the sphere acceleration W is shown in Figure 7. Mach number increases from $M_\infty = 4$ to $M_\infty = 6$, the Reynolds number under the initial flight conditions $Re = 2370$, sphere surface temperature $T_w = 0.35\ T^*$, gas properties are assumed to be the same as in the previous problem.

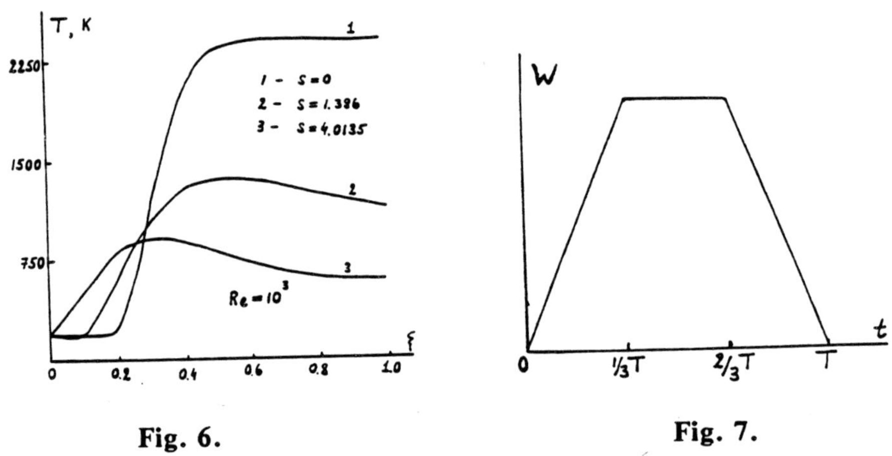

Fig. 6.   Fig. 7.

Nonstationary effects are determined by the Strouhal number $Sh = \tau_g/T$ where $\tau_g = R/V$ is characteristic flow time, R is sphere radius, V is body velocity before acceleration, T is acceleration time. The results discussed below correspond to $Sh = 5$, i.e., essentially nonstationary flow is considered.

Abrupt variation of the body motion speed can result in rise of intensive compression and expansion waves in the shock layer. Calculating this nonstationary flow on a fixed uniformly spaced grid we must significantly increase the number of grid points. Adaptive grid generation with the use of moving coordinates and both afore-mentioned methods for determining grid speeds were found to be applicable only for flows with Strouhal numbers $Sh \leqslant 1$. At higher Strouhal numbers this approach becomes ineffective because of the necessity to decrease the time increment value. Calculations proved the TDSG method to be the most effective. Weight function was determined by formula

$$W = |\nabla T/T_s|^\alpha + \delta$$

with $\alpha = 0.5$, $\delta = 0.2$.

Figure 8 shows pressure profiles on the stagnation line at some moments. The dashed line shows the calculation result at moment $t = T$ obtained on the fixed grid with the points clustered in the wall boundary layer.

It is seen that the use of an adaptive grid eliminates numerical oscillations. Figure 9 shows the computational grid at moment $t = 4/3\ T$. We can see grid points clustering near the body surface as well as in the vicinity of the inner shock.

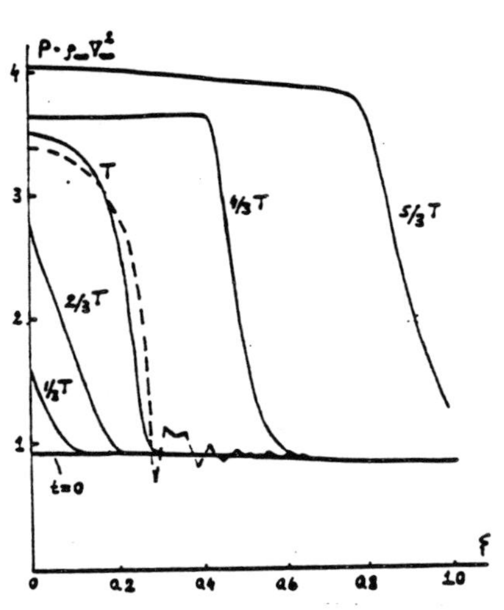

Fig. 8.

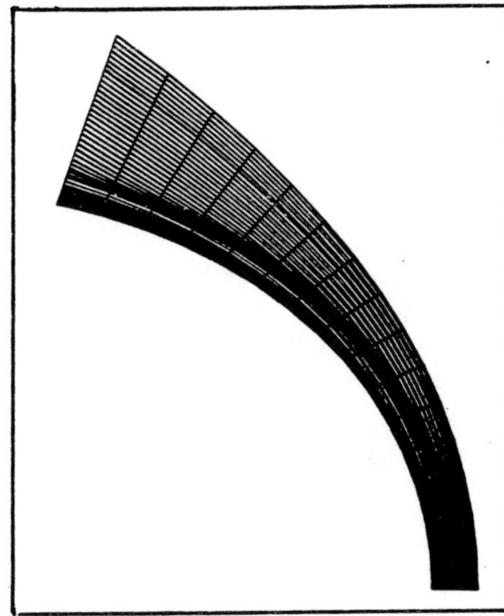

Fig. 9.

4. Conclusion

Some results of numerical experiments on the adaptive grid application in problems of supersonic flows around bodies are presented. The results show that the use of adaptive grids is an effective way to improve the accuracy of the numerical solution. The most consistent approach to the choice of weight function governing the grid point distribution is related to truncation error determination. The PGG method is most effective for stationary flow calculations. For nonstationary flows investigation we can apply the TDSG method or the moving coordinates. For essentially unsteady flows the use of moving coordinates becomes less effective because of the strong limitation of the time increment value.

References

1.   N. N. Yanenko, N. T. Danayev, and V. D. Liseikin, *Chisl. Meod. Mekh. Sploshn. Sred.*, vol. 8, no. 4, 1977, pp. 157–163 (in Russian).

2.   J. U. Brackbill and J. S. Saltzman, *J. Comput. Phys.*, vol. 46, 1982, pp. 342-368.

3.   H. A. Dwyer, R. J. Kee, and B. R. Sanders, *AIAA J.*, vol. 18, no. 10, 1980, pp. 1205-1212.

4.   L. M. Degtayrev, V. V. Drozdov, and T. S. Ivanova, *Adaptive Grid Method for One-dimensional Boundary Value Problems with the Boundary Layer*, Preprint No. 164, Inst. of Appl. Math., Moscow, 1986.

5.   B. L. Pierson and P. Kutler, *AIAA J.*, 1980, vol. 18, no. 1, pp. 49-54.

6.   K. Y. Fung, J. Tripp, and B. Goble, *Computer Meth. in Appl. Mech. and Eng.*, vol. 66, 1987, pp. 1-16.

7.   Yu. P. Golovachov and F. D. Popov, *Zh. Vychisl. Matem. i Mat. Fiz.*, vol. 12, no. 5, 1972, pp. 1292-1303 (in Russian).

8.   M. Vinokur, *J. Comput. Phys.*, vol. 50, no. 2, 1983, pp. 215-234.

9.   V. L. Belousov, Yu. P. Golovachov, and N. V. Leontieva, *Chisl. Metod. Mekh. Sploshn. Sred.*, vol. 2 (19), no. 2, 1988, pp. 31-38 (in Russian).

NUMERICAL METHOD FOR VISCOUS HEAT-CONDUCTING HYPOSONIC FLOW SIMULATION

L. A. Kadinskii, Yu. N. Makarov, A. I. Zhmakin

Abstract: An implicit numerical method for computation of hyposonic nonisothermal flow of viscous gas based on ideas similar to physical processes splitting is considered. Two modifications of the method which differ by way of the pressure equation solution are suggested. In the first version the generalized incompressibility condition is fulfilled by means of a modified Chorin-Hirt iteration procedure of the simultaneous pressure and velocity fields correction. The second version deals with the pressure equation solution by a conjugate gradient method with difference equations preconditioning by Poisson equation direct solving. Third-order upstream differences are used to approximate convective terms. Comparison of the methods for channel gas flow with inhomogeneous temperature boundary conditions is carried out.

1. Introduction

The present paper deals with the computation of the essentially nonisothermal flows of viscous heat-conducting gas under gravity at low Mach numbers. For the simulation of such flows Boussinesq approximation is inadequate because of the large density variations. On the other hand, it is known that computations of the low Mach subsonic flows based on a complete Navier-Stokes system for a compressible fluid encounter great difficulties [1, 2]. The efficiency of the finite-difference schemes, even implicit ones, drops sharply as the Mach number tends to zero. This degradation manifests itself in both execution time and accuracy of the computations. The problem of the formulation of the proper boundary conditions on the permeable boundaries and their numerical implementation is also very complicated [2, 3]. These difficulties do not arise when hyposonic flow equations [4-8] are used. For the flows in question hyposonic computations are identical to the full Navier-Stokes equations [6, 7, 9].

The hyposonic flow equations are implied by the full Navier-Stokes equations under two assumptions, namely: a) the characteristic gas velocity is

261

small compared to sound velocity; b) the thickness of the gas layer is small, i.e., the hydrostatic compressibility parameter $\varepsilon = gH/RT_e$ is infinitesimal, $\varepsilon \ll 1$.

The set of dimensionless equations for the case of the open domain may be written in the following manner:

$$\frac{\partial \rho}{\partial t} + \nabla \cdot (\rho V) = 0, \tag{1.1}$$

$$\rho \left[\frac{\partial V}{\partial t} + (V \cdot \nabla)V \right] = -\vartheta p^1 + \frac{\rho - 1}{Fr} j + \frac{1}{Re} \left[2 \, \text{Div} \, (\mu S) \right.$$
$$\left. - \frac{2}{3} \nabla (\mu \nabla \cdot V) \right] \tag{1.2}$$

$$\rho \left[\frac{\partial V}{\partial t} + (V \cdot \nabla)T \right] = \frac{1}{PrRe} \nabla \cdot (\lambda \nabla T), \tag{1.3}$$

$$\rho T = 1 \tag{1.4}$$

Here ρ, V, P^{-1}, T are the density, velocity vector, excess pressure, and temperature, respectively; μ and λ are dynamic viscosity and thermal conductivity, S is the deformation rate tensor, j is the unit vector in the direction of the gravity force; furthermore,

$$\text{Re} = \frac{\rho_0 V_0 H}{\mu_0}, \quad \text{Pr} = \frac{\mu_0 C_p}{\lambda_0}, \quad \text{Fr} = \frac{V_0^2}{gH}$$

are Reynolds, Prandtl, and Froude numbers, C_p is the specific heat at constant pressure, R is the gas constant, g is gravity acceleration, μ_0 and λ_0 are dynamic viscosity and thermal conductivity at temperature T_0.

Quantities ρ_0, V_0, $\rho_0 V_0^2$, T_0, H, H/V_0 are chosen as the scales of density, velocity, excess pressure, temperature, length, and time in Equations (1.1)–(1.4).

Note that in order to obtain the Boussinesq approximation equations we need an additional assumption for relative temperature variation to be small, $\Delta T/T_0 \ll 1$.

Equations set (1.1)–(1.4) similar to that for incompressible flow, does not describe acoustic wave propagation since P^1 is not related to density and temperature by the law of state. Differentiating Equation (1.4) with respect to time exploitation of Equations (1.1) and (1.3) yields a nonevolutional equation.

$$\nabla \cdot V = \frac{1}{PrRe} \nabla \cdot (\lambda \nabla T), \tag{1.5}$$

which generalizes the incompressibility condition $\nabla \cdot V = 0$ for hyposonic flow. Equation (1.5) can be used in computations instead of the continuity Equation (1.1) or the energy Equation (1.3).

Numerical methods for a hyposonic flow equation are based, as a rule, on the finite-difference schemes for incompressible viscous flow in primitive variables [1, 10]. The authors of [4] in order to solve Equations (1.1), (1.2), (1.4), and (1.5), have proposed a modification of the well-known MAC-method associated with the simultaneous pressure and velocity fields correction procedure generalized for stratified incompressible flows [11]. The same method has been used in [6] to solve Equations (1.2)-(1.5). An efficient implicit finite-difference scheme for the stationary flows is developed in [12] using the artificial compressibility concept [13] and both coordinate and physical process splitting. The high efficiency of this scheme is provided by means of the scalar factorization method. These schemes, as well as the implicit method for nonstationary equations described in the next section, exploit the MAC grid [1, 10].

2. The Difference Schemes

1. let the fields of velocity V^n, pressure P^n, temperature T^n, satisfying generalized incompressibility condition:

$$\nabla \cdot V^n = \frac{1}{PrRe} \nabla \cdot (\lambda^{n-1} \nabla T^n) \tag{2.1}$$

be known at some instant of time t^n. Then, for finding the unknown functions at the moment t^{n+1} we can introduce the following sequence of operational stages. First, the values of T^{n+1} and some intermediate velocity $\widetilde{V}$ are computed by the alternating direction method:

$$\frac{u - u^n}{\tau/2} + u^n \frac{\partial u}{\partial x} + v^n \frac{\partial u^n}{\partial y} = \frac{1}{\rho^n} \left[- \frac{\partial P^n}{\partial x} + \frac{\rho^n - 1}{Fr} j_x + \right.$$

$$+ \frac{1}{Re} \left\{ \frac{4}{3} \frac{\partial}{\partial x} \left(\mu^n \frac{\partial u}{\partial x} \right) - \frac{2}{3} \frac{\partial}{\partial x} \left(\mu^n \frac{\partial v^n}{\partial y} \right) + \frac{\partial}{\partial y} \left(\mu^n \frac{\partial u^n}{\partial y} \right) \right.$$

$$+ \frac{\partial}{\partial y}\left[\mu^n \frac{\partial v^n}{\partial x}\right]\Bigg\}\Bigg], \tag{2.2a}$$

$$\frac{v - v^n}{\tau/2} + u^n \frac{\partial v}{\partial x} + v^n \frac{\partial v^n}{\partial y} = \frac{1}{\rho^n}\left[-\frac{\partial P^n}{\partial y} + \frac{\rho^n - 1}{\mathrm{Fr}}\, j_y + \right.$$

$$+ \frac{1}{\mathrm{Re}}\left\{\frac{\partial}{\partial x}\left[\mu^n \frac{\partial u}{\partial y}\right] + \frac{\partial}{\partial y}\left[\mu^n \frac{\partial v}{\partial x}\right] + \frac{4}{3}\frac{\partial}{\partial y}\left[\mu^n \frac{\partial v^n}{\partial y}\right]\right.$$

$$\left.\left. - \frac{2}{3}\frac{\partial}{\partial y}\left[\mu^n \frac{\partial v^n}{\partial x}\right]\right\}\right], \tag{2.2b}$$

$$\frac{T - T^n}{\tau/2} + u^n \frac{\partial T}{\partial x} + v^n \frac{\partial T^n}{\partial y} = \frac{1}{\rho^n \mathrm{PrRe}}\left[\frac{\partial}{\partial x}\left(\lambda^n \frac{\partial T}{\partial x}\right) + \frac{\partial}{\partial y}\left(\lambda^n \frac{\partial T^n}{\partial y}\right)\right] \tag{2.3}$$

$$\frac{\tilde{\tilde{u}} - \tilde{u}}{\tau/2} + \tilde{u}\frac{\partial \tilde{u}}{\partial x} + \tilde{v}\frac{\partial \tilde{\tilde{u}}}{\partial y} = \frac{1}{\tilde{\rho}}\left[-\frac{\partial P^n}{\partial x} + \frac{\tilde{\rho} - 1}{\mathrm{Fr}}\, j_x + \right.$$

$$+ \frac{1}{\mathrm{Re}}\left\{\frac{4}{3}\frac{\partial}{\partial x}\left[\mu^n \frac{\partial \tilde{u}}{\partial x}\right] - \frac{2}{3}\frac{\partial}{\partial x}\left[\mu^n \frac{\partial \tilde{v}}{\partial y}\right] + \frac{\partial}{\partial y}\left[\mu^n \frac{\partial \tilde{\tilde{v}}}{\partial y}\right] + \right. \tag{2.4a}$$

$$\left.\left. + \frac{\partial}{\partial y}\left[\mu^n \frac{\partial \tilde{v}}{\partial x}\right]\right\}\right],$$

$$\frac{\tilde{\tilde{v}} - \tilde{v}}{\tau/2} + \tilde{u}\frac{\partial \tilde{v}}{\partial x} + \tilde{v}\frac{\partial \tilde{\tilde{v}}}{\partial y} = \frac{1}{\tilde{\rho}}\left[-\frac{\partial P^n}{\partial y} + \frac{\tilde{\rho} - 1}{\mathrm{Fr}}\, j_y + \right.$$

$$+ \frac{1}{\mathrm{Re}}\left\{\frac{\partial}{\partial x}\left[\mu^n \frac{\partial \tilde{u}}{\partial y}\right] - \frac{\partial}{\partial x}\left[\mu^n \frac{\partial \tilde{v}}{\partial x}\right] + \frac{4}{3}\frac{\partial}{\partial y}\left[\mu^n \frac{\partial \tilde{\tilde{v}}}{\partial y}\right] + \right. \tag{2.4b}$$

$$\left.\left. - \frac{2}{3}\frac{\partial}{\partial y}\left[\mu^n \frac{\partial \tilde{u}}{\partial x}\right]\right\}\right],$$

$$\frac{T^{n+1} - \tilde{T}}{\tau/2} + \tilde{u}\frac{\partial \tilde{T}}{\partial x} + \tilde{v}\frac{\partial T^{n+1}}{\partial y} = \frac{1}{\rho \mathrm{PrRe}}\left[\frac{\partial}{\partial x}\left(\lambda^n \frac{\partial \tilde{T}}{\partial x}\right) + \frac{\partial}{\partial y}\left(\lambda^n \frac{\partial T^{n+1}}{\partial y}\right)\right],$$

$$(2.5)$$

Third-order upstream differences are used for the convective terms [14], the others are approximated by central differences. Thus, the first stage arrives at a solution of the sequence of the scalar five-diagonal equation systems. The velocity fields obtained do not satisfy the generalized incompressibility condition and must be corrected at the third stage of the procedure, in the following way

$$V^{n+1} = \tilde{\tilde{V}} - \tau\frac{\nabla P^{n+1}}{\rho^{n+1}} + \tau\frac{\nabla P^n}{\tilde{\rho}} \tag{2.6}$$

where the pressure P^{n+1} at the $(n + 1)^{\mathrm{th}}$ level is defined by an elliptic type equation derived from (2.6) with the help of the divergence operator.

$$\nabla \cdot \left(\frac{\nabla P^{n+1}}{\rho^{n+1}}\right) = \frac{\nabla \cdot \tilde{\tilde{V}} - \nabla \cdot V^{n+1}}{\tau} + \nabla \cdot \left(\frac{\nabla P^n}{\tilde{\rho}}\right), \tag{2.7}$$

and

$$(\tilde{\rho})^{-1} = \tfrac{1}{2}[(\rho^n)^{-1} + (\tilde{\rho})^{-1}],$$

$$\nabla \cdot V^{n+1} = \frac{1}{PrRe}\nabla \cdot (\lambda^n \nabla T^{n+1}) \tag{2.8}$$

(this operation presents, in fact, the second stage of our scheme). Thus, momentum Equation (1.2) is approximated with first order in time and with second order in coordinate according to the equality

$$\frac{V^{n+1} - V^n}{\tau} = \mathbf{G} - \frac{\nabla P^{n+1}}{\rho^{n+1}} \tag{2.9}$$

where the operator $\mathbf{G}$ is determined by equations (2.2) and (2.4).

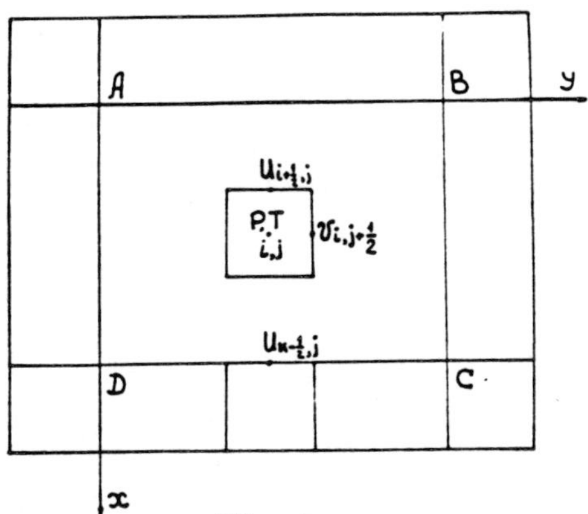

Fig. 1.

2. Let us write Equation (2.7) in a difference form for the uniform rectangular grid $\Omega = \Omega_x \times \Omega_y$ (Figure 1)

$$\Omega_x = \{X_i = i\Delta x,\ i = 0,\ k\},\ \Omega_y = \{y_j = j\Delta y,\ j = 0,\ l\}.$$

(The indices $i = 0,\ k,\ j = 0,\ l$ are related to fictitious cells.

$$\Lambda_x P^{n+1}_{i,j} + \Lambda_y P^{n+1}_{i,j} = \frac{(\nabla\cdot\tilde{\tilde{V}})_{i,j} - (\nabla\cdot V^{n+1})_{i,i}}{\tau} + \Lambda_x P^n_{i,j} + \Lambda_y P^n_{i,j} \qquad (2.10)$$

where

$$(\nabla\chi V)_{i,j} = \frac{U_{i+1/2,j} - U_{i-1/2,j}}{\Delta x} + \frac{V_{i,j+1/2} - V_{i,j-1/2}}{\Delta y},$$

$$\Lambda_x P_{i,j} = \frac{1}{\Delta X^2}\left[\frac{P_{i+1,j} - P_{i,j}}{\rho_{i+1/2,j}} - \frac{P_{i,j} - P_{i-1,j}}{\rho_{i-1/2,j}}\right],$$

$$\Lambda_y P_{i,j} = \frac{1}{\Delta y^2}\left[\frac{P_{i,j+1} - P_{i,j}}{\rho_{i,j+1/2}} - \frac{P_{i,j} - P_{i,j-1}}{\rho_{i,j-1/2}}\right],$$

Equation (2.10) comprises unknown values of $\tilde{\tilde{V}}$ at the boundary and of pressure in fictitious cells interconnected by (2.6). The technique suggested in Ref. [15] permits simultaneous exclusion of unknown values, with V^{n+1} at the boundary (or at a part of the boundary where this technical is applied) being assumed known. Let us consider, for example, the boundary CD: $i = K - 1$ and suppose $u^{n+1}_{k-1/2,j}$ to be given. Then we obtain:

$$\frac{P^{n+1}_{k-1,j} - P^{n+1}_{k-2,j}}{\rho^{n+1}_{k-3/2,j}\Delta x^2} + \Lambda_y P^{n+1}_{k-1,j} = \frac{U^{n+1}_{k-1/2,i} - \tilde{\tilde{U}}_{k-3/2,j}}{\tau\Delta x} + \frac{\tilde{\tilde{V}}_{k-1,j+1/2} - V_{k-1,j1/2}}{\tau\Delta y}$$

$$\qquad (2.11)$$

$$-\frac{1}{\tau}(\nabla\cdot V^{n+1})_{k-1,j} - \frac{P^n_{k-1,j} - P^n_{k-2,j}}{\tilde{\tilde{\rho}}_{k-3/2,j}\Delta x^2} + \Lambda y P^n_{k-1,j}.$$

The equation includes pressure values only for the interior cells and known values. If at some boundary part G_1 the value of V^{n+1} is not specified the above transformation cannot be performed and, therefore, it is necessary to set the boundary condition for the pressure: $P|_{G_1} = P_0$.

The method proposed in the paper deals with two modifications differing in solution of the pressure Equation (2.7).

3. The iterative procedure is suggested in Refs. [4, 11, 16] for the pressure equation solution with the simultaneous velocity field correction.

Initial approximations for iterative process are chosen as follows: $V^{n+1,0} = \tilde{\tilde{V}}$, $P^{n+1,0} = P^n$.

The first step is the pressure correction:

$$P_{i,j}^{n+1,k+1} - P_{i,j}^{n+1,k} = \frac{\omega}{\tau d_{i,j}} [\nabla \cdot V^{n+1,k} - \nabla \cdot V^{n+1}]_{i,j}, \tag{2.12}$$

where

$$d_{i,j} = \frac{1}{\Delta x^2}\left(\frac{1}{\rho_{i+1/2,j}^{n+1}} + \frac{1}{\rho_{i-1/2,j}^{n+1}}\right) + \frac{1}{\Delta y^2}\left(\frac{1}{\rho_{i,j+1/2}^{n+1}} + \frac{1}{\rho_{i,j-1/2}^{n+1}}\right) \tag{2.12a}$$

ω being the relaxation parameter.

The second step is the velocity correction:

$$\frac{U_{i-1/2,j}^{n+1,k+1} - U_{i-1/2,j}^{n+1,k+1}}{\tau} = -\frac{1}{\rho_{i-1/2,j}^{n+1}}\left(\frac{P_{i,j}^{n+1,k+1} - P_{i-1-1}^{n+1,k+1}}{\Delta x} - \frac{P_{i,j}^{n+1,k} - P_{i-1-1}^{n+1,k}}{\Delta x}\right), \tag{2.13a}$$

$$\frac{V_{i,j-1/2}^{n+1,k+1} - V_{i,j-1/2}^{n+1,k}}{\tau} = -\frac{1}{\rho_{i,j-1/2}^{n+1}}\left(\frac{P_{i,j}^{n+1,k+1} - P_{i,j,j}^{n+1,k+1}}{\Delta y} - \frac{P_{i,j}^{n+1,k} - P_{i,j,j}^{n+1,k}}{\Delta y}\right) \tag{2.13b}$$

In Ref. [17] it has been recommended to correct the velocity in grid-points $(i + \frac{1}{2}, j)$, $(i, j + \frac{1}{2})$ immediately upon the pressure changing in grid-point (i, j). It has also been pointed out that this method is equivalent to the successive overrelaxation method. Iterations are performed until either the convergence condition

$$\max_{i,j}|(\nabla \cdot V^{n+1} - \frac{1}{PrRe}\nabla \cdot (\lambda^n \nabla T^{n+1}))_{i,j}| < \varepsilon \tag{2.14}$$

is fulfilled or the given number of iterations is achieved. The velocity V^{n+1} is considered as the velocity of the $(n+1)^{\text{th}}$ time level.

4. Solution of Equation (2.7) for pressure occupies the major part of the computational process and, therefore, the search for the optimal algorithm is very real, especially as associated with the intensive development of parallel computations. In the present paper a numerical procedure based on

the conjugate gradient algorithm with preconditioning using the Fast Fourier Transform (FFT) for solving the Poisson equation is suggested. Note that similar algorithms are used to compute the flows of stratified fluids [19] and to solve a diffusion equation [20]. Following Ref. [20] we present an algorithm of a conjugate gradient method with preconditioning for solving the equation $Lp = f$.

Let some arbitrary initial approximation P_1 be given, then an initial residual $r_1 = f - Lp_1$ can be found. After that, equation $M\Phi_1 = r_1$ is solved, where Φ_1 is an auxiliary vector, M is the operator of preconditioning energetically equivalent to operator L. Then again, until the convergence condition

$$\max_{\Omega}|P_{i+1} - P_i| < \epsilon_1$$

is fulfilled, the following set of actions is executed ($q_1 = \varphi_1$):

$$\alpha_i = \frac{r_i,\, \varphi_i}{q_i,\, L_{qi}},$$

$$P_{i+1} = P_i + \alpha_i q_i,$$

$$r_{i+1} = r_i - \alpha_i L q_i,$$

$$\varphi_{i+1} = M^{-1} r_{i+1},$$

$$\beta_i = \frac{r_{i+1},\, \varphi_{i+1}}{r_i,\, \varphi_i},$$

$$q_{i+1} = \varphi_{i+1} + \beta_i q_i.$$

$(\bullet, \bullet)$ is the scalar product. Note the method should be applied only to a self-adjoint difference operator.

For using the method of conjugate gradients with preconditioning to solve Equation (2.7) the scalar product is defined according to Ref. [18] to meet the condition for the difference operator $L = \nabla\bullet(\nabla/\rho)^*$ to be self-adjoint. As for $\rho = $ const, $L = \nabla$ is the difference Laplace operator, it is natural to choose it as the operator of preconditioning $M = \Delta$, since it can be efficiently inverted by means of the Fourier method. (The equivalence of operators M and L can be trivially proved similar to Ref. [18].) In the pres-

---

* The finite-difference approximation of operator L has been written in number 2.

ent paper the method of solution based on the FFT implemented in the vector processor (VP) ES-1055M computer is used.

5. The Fourier method is explained in detail in Ref. [18]. Therefore we describe its use of equation $\Delta\Phi = r$ obtained from (2.10) for $\rho = 1$ and solve for preconditioning while exploiting the conjugate gradient method. It should be noted that the FFT is suitable only for a Laplace operator written as

$$A\varphi = \begin{cases} 2(\varphi_1 - \varphi_0)/\Delta x^2, & i = 0, \\ (\varphi_{i+1} - 2\varphi_i + \varphi_{i-1})/\Delta x^2 & i = 1, \overline{K - 1}, \\ -2(\varphi_k - \varphi_{k-1})/\Delta x^2, & i = K. \end{cases} \tag{2.15}$$

Expressions (2.15) and (2.10) for $\rho = 1$ differ by coefficients with indices $i = 0, k$. Therefore, the FFT cannot be used directly to solve System (2.10). The following method is suggested to overcome this difficulty. Let us write the equation with the number $i = k$ of system (2.10) in the following manner:

$$-2\frac{P^{n+1}_{k,j} - P^{n+1}_{k-1,j}}{\rho^{n+1}_{k-1/2,j}\,\Delta x^2} + \Lambda_y P^{n+1}_{k,j} = \Lambda_y P^n_{k,j} \tag{2.16}$$

While obtaining a stationary solution, from Equation (2.16) it follows

$$P^{n+1}_{k,j} = P^n_{k,j}. \tag{2.17}$$

(This condition is also used to compute unsteady flows with sufficiently small time steps.) Then, the equation $\Delta\Phi = r$ for $i = k, k - 1$ takes its common form.

$$-2\frac{\varphi_{k,j} - \varphi_{k-1,j}}{\Delta x^2} + A_y\varphi_{k,j} = r_{k,j}$$

$$\frac{\varphi_{k,j} - 2\varphi_{k-1,j} + \varphi_{k-2,j}}{\Delta x^2} + A_y\varphi_{k-1,j} = r_{k-1,j}$$

and the equation for pressure

$$\Lambda_x P^{n+1}_{k-1,j} + \Lambda_y P^{n+1}_{k-1,j} = \Lambda_x P^n_{k-1,j} + \Lambda_y P^n_{k-1,j} + f_{k-1,j} \tag{2.18}$$

in case where (2.17) is valid and identical to (2.11). FFT is not used with respect to the second variable and the resulting system is solved by the factorization method [18]. In this case as well as for boundary conditions of the first kind at the channel's exit there is no need to use the described method. The generalization of the suggested method is described in Appendix 1 to solve a three-dimensional equation of the elliptical type.

6. Let us check the efficiency of the conjugate gradient method with preconditioning by applying it to a well-known test [18].

An equation $\nabla\cdot(b\nabla y) = 0$ in the domain $[0, 1] \times [0, 1]$ is solved with the boundary conditions $y|_G = 0$. Initial approximation $y^0|_G = 0$, $y^0|_\omega = 1$ is chosen, where ω is the set of interior mesh-points. Coefficients of the equations take the form:

$$b_1 = 1 + c[(x_1 - 0.5)^2 + (x_2 - 0.5)^2],$$

$$b_2 = 1 + c[0.5 - (x_1 - 0.5)^2 - (x_2 - 0.5)^2].$$

Table 1 contains the results of a comparison of the suggested method (4th column) with the modified alternating triangle method [18] (1st column), the alternating-triangle method accelerated by conjugate gradient procedure [21] (2nd column), and with the "$\alpha - \beta$" iterations method [22] (3rd column).

Table 1

R	2				32				512			
Method N	1	2	3	4	1	2	3	4	1	2	3	4
16	-	-	-	5	-	-	-	9	-	-	-	10
32	18	11	16	6	23	12	18	11	24	13	27	11
64	26	15	59	6	34	16	63	11	36	18	94	12

The results demonstrate a high efficiency of the conjugate gradient method with preconditioning. Let us point out that the number of iterations is independent of N in this method, thus demonstrating its promise for computation on fine grids.

3. Numerical Results

This section deals with comparison of the efficiency of the described algorithms. We shall refer to the first as iterative and to the one based on FFT as direct. These algorithms are first applied to the isothermal flow $\rho =$

$T = 1$, i.e., to the incompressible flow. In this case the splitting scheme (2.2)–(2.7) reduces to the scheme proposed in Ref. [23], namely

$$\frac{\tilde{V} - V^n}{\tau} = -(V^n \cdot \nabla)\tilde{V} + \frac{1}{Re}\Delta\tilde{V} - \nabla P^n \tag{3.1}$$

$$\Delta P^{n+1} = \frac{\nabla\cdot\tilde{V}}{\tau} + \Delta P^n, \tag{3.2}$$

$$\frac{V^{n+1} - \tilde{V}}{\tau} = \nabla P^{n+1} + \nabla P^n \tag{3.3}$$

Then the pressure equation takes the form of the Poisson Equation (3.2) and no iterations are needed to use the direct algorithm.

The Poisson equation has been solved by means of the standard sequential scalar FFT algorithm [24] as well as of the vector modification implemented on the vector processor of the ES-1055M computer. The vector FFT algorithm provides speed-up equal to 11, 14.5, 19, and 25 for $N = 32$, 64, 128, and 1024, respectively. The overall execution time reduction for the pressure computation is 1.5 for a mesh size of 22 × 66.

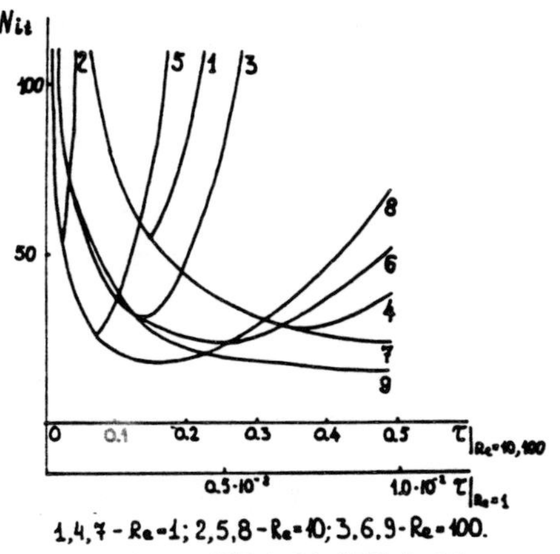

1,4,7 – Re=1; 2,5,8 – Re=10; 3,6,9 – Re=100.
1,2,3 – h=0.05; 4,5,6 – h=0.1; 7,8,9 – h=0.2.

Fig. 2.

Table 2

Re/h	0.2	0.1	0.05
1	2.0	0.6	0.13
10	3.0	0.8	0.25
100	5.0	2.35	0.68

Let us consider the development of the flow in a cylindrical channel with radius $R = 1$ and length $L = 2$ when the inlet velocity jumps instantly from zero to a given value. Due to the difficulties of the theoretical study of the stability of the two-dimensional scheme, numerical experiments have been performed. The largest time increments for different Reynolds numbers and the spatial increments of the uniform mesh ($h = \Delta r = \Delta Z$) are presented in Table 2. Note that stability is controlled by the first stage of the algorithm, hence an implicit approximation is essential: the largest time for the same Reynolds numbers and mesh increments. Note that this number is insensitive to h for a given Re when $\tau < \tau_{opt}$.

Comparison of the direct and iterative methods for a mesh size of 22 × 66 reveals that the execution time of the latter is 10 times greater ($\varepsilon = 10^{-3}$). It should be noted that the limitation of the number of iterations ($k \leqslant 20$) is justified when the steady flow is computed by the time-marching method. However, the limitation can result in large errors of the nonstationary computations. This issue will be discussed in detail later.

Let us consider the hyposonic nonisothermal flow in a plane channel with height $H = 1$ and length $L = 3.2, 6.4, 12.8$ (Figure 3). A parabolic velocity profile is assumed at the inlet as well as the equality $T_0 = 1$ for the inlet gas temperature. There are isothermal sections AB($T_w = 3$) and CD($T_A = 7/3$) on the lower and upper walls, respectively; the rest of the boundaries are considered adiabatic. A constant pressure is assumed at the exit.

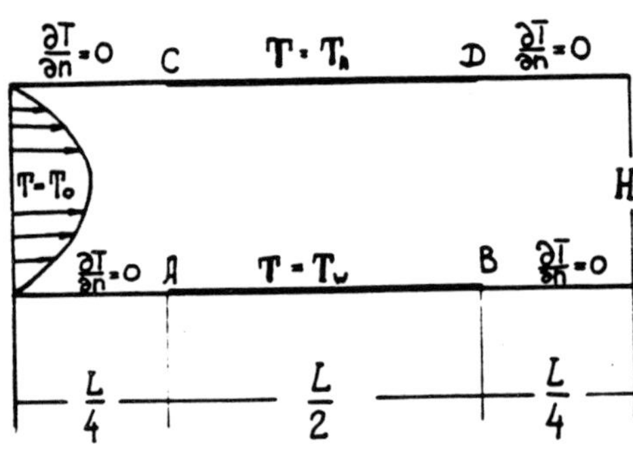

Fig. 3.

An advantage of the algorithm implementation on the vector processor in the case of the hyposonic flow is more pronounced than in the case of the incompressible flow because in addition to FFT algorithm the conjugate gradients method is readily vectorizable. For example, the vector speed-up for Re = 100, Fr = 100, $h = 0.1$, $\tau = 0.1$ equals 2.9 and 3.4, respectively; for the computation of the flow in the channel with length $L = 6.4$ (mesh size 12 × 66) and $L = 12.8$ (12 × 130).

A comparison of the direct and iterative algorithms has been carried out for both steady and unsteady computations. Table 3 contains the results of the steady flow computation (mesh size 12 × 34, $\Delta X = \Delta Y = 0.1$) for different Reynolds numbers. The dimensionless time needed for the mass-flow rate through the channel to reach the stationary value as well as the execution time) (in minutes) is presented for the direct and iterative algorithms. Note that due to the insufficient number of iterations ($K \leqslant 20$) in the latter method, the generalized incompressibility condition is not satisfied with the required accuracy and flow development is delayed.

The same mesh has been used to compute nonstationary flow in the channel with a cold wall at the initial time moment and subsequent gradual heating of wall sections AB and CD Re = 100, Fr = 100, $\tau = 0.025$. No limit on the number of iterations is set in this case. For an admissible error in the on the number of iterations is set in this case. For an admissible error in the generalized incompressibility condition 10^{-3} the number of iterations in the direct algorithm varies from 2 to 4 and from 10 to 200 in the iterative. The direct method requires 5-6 iterations to provide an accuracy of 10^{-6} whereas it is impossible to obtain the same accuracy level using the iterative method. The overall speed-up of the direct method equals 4-5 depending on the wall heating rate. The relative efficiency of this algorithm increases with the increase of the mesh size (especially of the number of cells along the channel); speed-up equals 12-15 for 22 × 130 mesh.

Table 3

Re	τ	t_{it}^{de}	t_{dir}^{de}	t_{it}	t_{dir}	$K = \dfrac{t_{it}}{t_{dir}}$
1	10^{-3}	0.38	0.38	26	18	1.43
10	10^{-2}	1.46	1.46	8.67	6	1.41
100	10^{-1}	10.6	3.0	5	1.5	3.32

Finally, we consider the computation of the nonisothermal flow in the long channel ($L = 12.8$). This flow is of considerable interest as it is associated with the study of the processes in vapor phase epitaxy reactors. The flow in a plane reactor is characterized by interplay of the forced motion of the inflowing gas and the upward natural convection over the hot lower surface (susceptor) and strongly influences the growth process [25]. Stream-function lines (a) and isotherms (b) of the forced convection dominated flow and the mixed convection flow are shown in Figures 4 and 5, respectively. The direct method was used, the mesh size was 22 × 30. The execution time for one computation was about 4 hours on the ES-1055M.

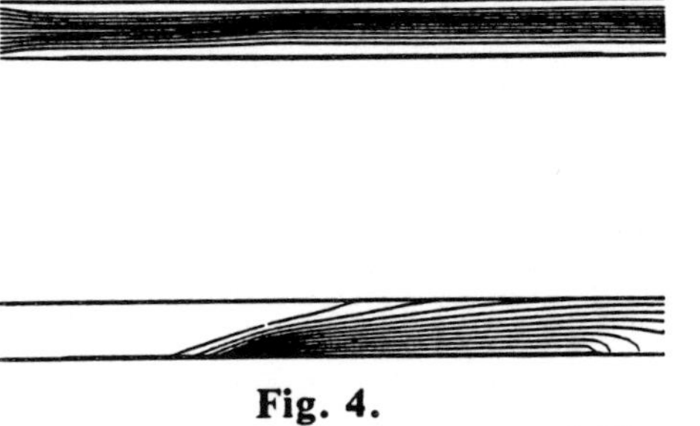

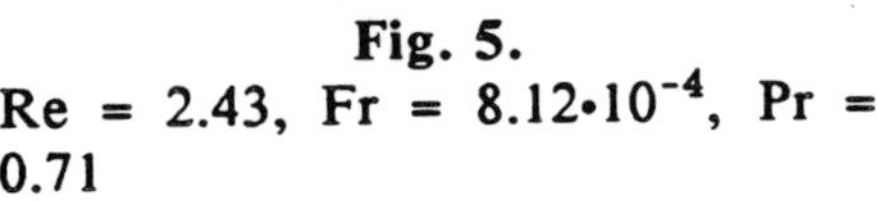

Fig. 4.	**Fig. 5.**
Re = 31.69, Fr = 13.72, Pr = 0.71	Re = 2.43, Fr = 8.12·10^{-4}, Pr = 0.71

Acknowledgment

The authors are grateful to A. A. Fursenko for helpful discussion.

Appendix

The method suggested in this paper can be easily generalized for computation of three-dimensional flows. As the first and third stages of the procedure are completely similar to those considered for the two-dimensional case, we dwell here on the stage dealing with the pressure equation solution. It is precisely the characteristic feature of three-dimensionality which manifests itself when solving the preconditioning equation. We denote the numbers of grid-points corresponding to each of three variables — i_1, i_2, i_3 as N_1, N_2, N_3, respectively, and assume that, for $i_1 = 0$ the equation is treated in the two-dimensional case. Then, the preconditioning equation takes the form:

$$(A_1 + A_2 + A_3)\varphi = r, \quad i_1 = \overline{1, N_1}; \quad i_2 = \overline{0, N_2}; \quad i_3 = \overline{0, N_3}. \tag{A.1}$$

Taking into account the boundary condition for pressure, we set

$$\varphi(N_1, i_2, i_3) = 0$$

and operator A_1 takes the form similar to (2.18)

$$A_1\varphi = \begin{cases} 2(\varphi_1 - \varphi_0)/h^2_1, & i_1 = 0 \\ (\varphi_{i_1+1} - 2\varphi_{i_1} + \varphi_{i_1-1})/h^2_1, & i = \overline{1, N_1 - 1}, \end{cases} \tag{A.2}$$

while operators $A_\alpha = 2, 3$, are written in the following manner

$$A_\alpha\varphi = \begin{cases} (\varphi_1 - \varphi_0)/h^2_\alpha, & i_\alpha = 0 \\ (\varphi_{i_\alpha+1} - 2\varphi_{i_\alpha} + \varphi_{i_\alpha-1})/h^2_\alpha, & i_\alpha = \overline{1, N_\alpha - 1}, \\ -(\varphi_{N_\alpha} - \varphi_{N_\alpha-1})/h^2_\alpha, & i_\alpha = N_\alpha \end{cases} \tag{A.3}$$

with h_α being the step of the grid by N_α coordinate. The solution equation (A.1) is sought by means of a representation

$$\varphi(N_1 - i_1, i_2, i_3) = \sum_{k_1=0}^{N_1} \sum_{k_2=0}^{N_2} \varphi_{k_1,k_2} (i_3)\mu_{k_1}^{(1)} (i_1)\mu_{k_2}^{(2)} (i_2) \tag{A.4}$$

where $\mu_{k_1}^{(1)} = \sin \dfrac{2k_1 - 1}{2N_1} \pi i_1$ and $\mu_{k_2}^{(2)}$ is the solution of the eigenvalue problem:

$$A_{2k_2}^{(2)} + \lambda_{k_2}^{(2)}\mu_{k_2}^{(2)} = 0 \tag{A.5}$$

Following [18], we write explicitly the eigenfunction $\mu_{k_2}^{(2)}$

$$\mu_{k_2}^{(2)} = T_i(Z_{k_2}) - (1 - Z_{k_2}) U_{i-1}(Z_{k_2}), \tag{A.6}$$

with $t_i(Z_k)$, $U_{i-1}(Z_k)$ being Chebyshev polynomials of the first and second kind, respectively, and Z_k found as the roots of a transcendental equation:

$$\frac{Z_k T_{N_2+1}(Z_k) - T_{N_2}(Z_k)}{Z_k + 1} = 0, \ k = \overline{0, N_2} \tag{A.7}$$

The eigenvalues of Problem (A.5) are expressed by Z_k by a formula:

$$\lambda_{k_2}^{(2)} = \frac{2}{h_2^2}(1 - Z_{k_2}), \ K_2 = \overline{0, N_2} \tag{A.8}$$

The unknown expansion coefficients $\tilde{\Phi}_{k_1,k_2} (i_3)$ are found from a system of equations:

$$A_3 \tilde{\Phi}_{k_1,k_3} (i_3) - (\lambda_{k_1}^{(1)} + \lambda_{k_2}^{(2)}) \tilde{\Phi}_{k_1,k_2} (i_3) = \tilde{r}_{k_1,k_2} (i_3) \tag{A.9}$$

with the right-hand $\tilde{r}$ defined from an expansion:

$$\tilde{r}_{k_1,k_2}(i_3) = \frac{2}{N_1} \cdot \frac{1}{(\mu_{k_2}^{(2)}, \mu_{k_2}^{(2)})} \sum_{i_1=0}^{N_1} \sum_{i_1=0}^{N_2} \rho_{i_1-1} r(N_1 - i_1, i_2, i_3)$$

$$\mu_{k_1}^{(1)}(i_1)\mu_{k_2}^{(2)}(i_2),$$

$$\mu_{k_2}^{(2)}\mu_{k_2}^{(2)} = \frac{N_2 + 1 + \frac{1}{2}\,T_{N_2+1}(Z_{k_2})\,U_{N_2}(Z_{k_2})}{1 + Z_{k_2}} \tag{A.10}$$

$$\rho_i = \begin{cases} 1/2, & i = 0 \\ 1, & i = \overline{1, N_1} \end{cases}$$

$$\lambda_{k_1}^{(1)} = \frac{4}{h_1^2}\,\sin^2\frac{2k_1 - 1}{4N_1}\,\Pi$$

System (A.9) is solved by the factorization method for $k_1 = \overline{1, N_1}$, $k_2 = \overline{0, N_2}$.

It is easy to estimate the number of operations required to solve Equation (A.1), assuming the FFT requires in the order of $2N \log_2 N$ operations, the factorization method $8N$, the eigenfunction problem expansion $2N^2$.

Consider two cases of the full number of operations estimation:

I. with the use of a vector-processor (VP).

II. with the use of CPU only.

1. The FFT is definite only for the complex value function. Therefore, the operations are executed with a vector of length $4N_1$, and the number of operations is estimated as $4N_1 \log_2 4N_1$.

2. The operation with a vector using VP is executed $K_{op}^{-1}(N)$ times faster than with the use of CPU. Therefore, the number of operations for VP is estimated with the help of coefficient $K_{op}(N)$ which reduces that number of operation. Noting this, it is easy to obtain the estimates of operation number:

$$\sum_{VP} = N_1 N_2 N_3 \{32\,K_{FFT}(4N_1)\log_2 \varphi N_1 + 4N_2\,K_{epe}(N_1) + 8K_{fact}(N_1)\}$$

$$\sum_{CPU} = N_1 N_2 N_3 \{4 \log_2 N_1 + 4N_2 + 8\}$$

which lead to the following conclusions.

1. It is recommended that the number of grid points without VP be chosen in the following order: $N_2 \leqslant N_1 \leqslant N_3$ (if it is possible to fulfill the FFT applicability condition: $N_1 = 2^{n_1}$.

2. With the use of VP the choice of the grid-points number depends on coefficients $K_{FFT}(4N_1)$, $K_{fact}(N_1)$, and $K_{epe}(N_1)$. The calculations have shown that even for $N_1 = 32$ it is all right to assume $N_2 \leqslant N_3 \leqslant N_1$.

3. Evidently, the described method of solution is efficient for not very large values of N_2. Otherwise we should use the procedure of number 5 along the N_2 coordinate as well. It is possible, if $N_2 = 2^{n_2}$ and the main storage capacity is suitable for corresponding objectives.

It is easy to perform similar estimations for the two-dimensional case.

References

1. P. Roach, *Computational Fluid Dynamics*, Albuquerque: Hermosa, 1976.
2. Yu. V. Lapin, O. A. Nekhamkina, and K. A. Pospelov, et al., *Itogi Nauki i Tekhn, Mekh. Zhedk. i Gaza*, vol. 19, 1985, pp. 80-185 (in Russian).
3. A. T. Redorchenko, *Zh. Vychis. Matem. i Matem. Fiz.*, vol. 21, no. 5, 1981, pp. 1225-1232 (in Russian).
4. J. D. Ramshaw and J. A. Trapp, *J. Comput. Phys.*, vol. 21, no. 4, 1976, pp. 438-453.
5. R. G. Rehm and H. R. Baum, *J. Res. NBS*, vol. 83, no. 3, 1978, pp. 297-308.
6. D. A. Nikulin and M. Kh. Strelets, *Proc. All-Union Conf. on Heat- and Mass-Transfer*, Minsk, vol. 1, pt. 1, 1980, pp. 111-115 (in Russian).
7. M. Kh. Strelets, In: *Dynamics of Nonuniform Compressible Medium*, Leningrad: Leningrad State University, 1984, pp. 70-83 (in Russian).
8. P. C. T. de Boor, *Int. J. Heat. Mass Transfer*, vol. 27, no. 12, 1984, pp. 2239-2251.
9. A. I. Zhmakin and Yu. N. Makarov, *Sov. Phys. - Dokl.* (USA), vol. 30, 1985, p. 120.
10. R. Peyret and T. D. Taylor, *Computational Methods for Fluid Flow*, New York: Springer-Verlag, 1983.
11. J. A. Young and C. W. Hirt, *J. Fluid Mech.*, vol. 56, no. 2, 1972, pp. 255-276.
12. A. E Kuznetsov, O. A. Nekhamkina, and M. Kh. Strelets, *High Temp.*, vol. 22, 1984, p. 862.
13. A. I. Chorin, *J. Comput. Phys.*, vol. 2, no. 21, 1967, pp. 12-26.
14. R. K. Agarwal, *A Third Order Accurate Upwind Scheme for Navier-Stokes Solutions at High Reynolds Numbers*, AIAA Paper, no. 81-0112, 1981.
15. C. R. Easton, *J. Comput. Phys.*, vol. 9, no. 2, 1972, pp. 375-379.
16. C. W. Hirt and J. L. Cook, *J. Comput. Phys.*, vol. 10, no. 2, 1972, pp. 324-340.

17.   J. A. Viecelli, *J. Comput. Phys.*, vol. 8, no. 1, 1971, pp. 119–143.

18.   A. A. Samarskii and E. S. Nikolaev, *Solution Methods for Difference Equations*, Moscow: Nauka, 1978 (in Russian).

19.   O. M. Belotserkovskii, V. A. Guschin, and V. N. Konshin, *Zh. Vychis. Matem. i Matem. Fiz.*, vol. 27, no. 4, 1987, pp. 594–609 (in Russian).

20.   J. Gary, S. McCormik, and R. Sweet, *Appl. Math. and Comput.*, vol. 13, 1983, pp. 285–309.

21.   M. P. Bogdanova, A. B. Kucherov, et al., *Implicit Iterative Methods. Analysis and Assessment*, Preprint no. 115, Moscow: Institute for Applied Mathematics, 1978 (in Russian).

22.   B. N. Chetverushkin, *Mathematical Simulation of Radiation Gas Dynamics Problems*, Moscow: Nauka, 1985 (in Russian).

23.   O. M. Belotserkovskii, *Numerical Simulation for Continuum Mechanics*, Moscow: Nauka, 1984.

24.   J. W. Coley, J. W. Tukey, *Meth. Comput.*, 1965, vol. 19, pp. 297–301.

25.   I. G. Vsesvetskii, A. I. Zhmakin, L. A. Kadinskii, and Yu. N. Makarov, *Computation of Gasdynamics in VPE Reactors on the Matrix Processor ES-1055M*, Preprint no. 1094, Leningrad: Physical-Technical Institute, 1987 (in Russian).

EXPERIMENTAL AND NUMERICAL STUDY OF SHOCKED FLOWS IN PLANE CURVILINEAR CHANNELS

P. A. Voinovich, V. A. Komissaruk, N. P. Mende,
A. N. Semenov, E. V. Timofeev, A. A. Fursenko

Abstract: This paper examines experimental and numerical study of shocked flows in plane curvilinear channels under the following headings; Introduction, Shock Wave Processes in Plane Channel Bends, Experimental Approach, Numerical Simulation, Discussion of Results.

1. Introduction

Studies of shock waves (discontinuous states of continuum) have been carried out for over a century, however, the number of publications continues to grow. To prove it we can compare two indices (Ref. [1, 2]): the first covers the period from 1950 through 1959 and contains a little less than 1500 entries; while the second shows the number of publications during the next decade as approaching six thousand. This fact is evidence that the final goal of the study, namely, to be able to predict its development for a wide range of conditions has not yet been reached. None of the existing approaches: experimental, analytical, or numerical, is able to provide sufficiently complete, accurate, and reliable information in many practically important cases. As a rule, only selected aspects of the processes encountered in practice are considered. Such an approach (discounting secondary phenomena) is necessary in order to classify shock wave interaction regimes, to determine boundaries and existence conditions of these regimes, i.e., to develop and refine process models [3]. However, synthesis of knowledge cannot be accomplished by means of collecting separate results, since the mutual effect of different factors must be taken into account. Thus experimental study of shock wave reflection and diffraction over obstacles and curvilinear walls demonstrates the complex shocked flow structure in curvilinear channels [5, 6], for example, but does not allow it to be predicted with sufficient accuracy *a priori*. At the present time it is practically impossible to determine the forces acting on each of several objects in their supersonic flight by means of an analysis of shock wave configurations. Experimental investigation also encounters difficulties: in the case of free flight the process is nonstationary due to the relative displacement of the objects and that prohibits the determination of the instant forces while the flow pattern in wind tunnel experiments is disturbed due to parasitic interference.

In general, numerical simulation of gasdynamic processes provides the best possibilities [8]. However, taking the capability of computers into account, the application of numerical methods should be based on clear understanding of the physical processes in question and experimental data as a benchmark for numerical schemes verification.

Three groups of authors specialize in different approaches to supersonic flow phenomena and have acquired the capability to judge the shortcomings of the methods used, each in his own field. In the present paper we intend to demonstrate the advantages of a combined approach to investigation of the same object.

The flow behind the shock wave in a plane curvilinear channel has been chosen as the object of study. An interferometer was used to obtain a general flow pattern and the density distribution in the concurrent flow. Accuracy was good enough to allow comparison with the numerical data.

Results of the study of different types of shock wave interactions and reflections enables us to identify discontinuities and vortex structures.

Finally, numerical flow simulation tested by comparison with experimental density distribution as well as with flow singularity positions provides complete information about the thermodynamic state of the gas and flow cinematic parameters.

2. Shock Wave Processes in Plane Channel Bends

An understanding of a rather complex gasdynamic process of shock wave propagation along the channel bend [6, 9] is best reached by separate analysis of two phenomena which occur simultaneously at the inner and outer walls of the curvilinear channel. Indeed, after the shock wavefront passes through a cross-section *AB* (Figure 1) and enters the curvilinear part of the channel, it will interact with outer wall *BE* at an acute incidence angle α, and with inner wall *AG* at an obtuse incidence angle. Traditionally the former process is referred to as shock wave reflection from the concave surface and the latter is called shock wave diffraction over the convex surface. The process of reflection from the outer wall and diffraction over the inner evolve autonomously without mutual influence for a certain time interval which can be easily estimated. Later the process interference results in the multiform pattern of shock wave interactions with each other and expansion waves, and also with slipstreams, boundary layers, and vortices. Knowledge of the independent development of the reflection and diffraction

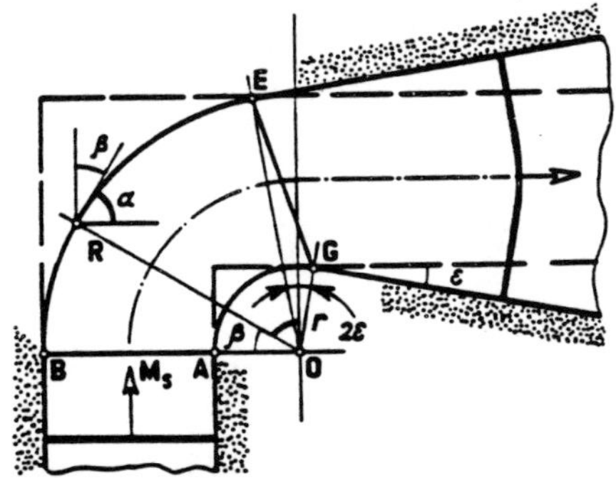

Fig. 1. Curved channel geometry.
ABEGA is curved part of channel; β - local wedge angle; α - local shock incident angle; $R = 45$ mm, $r = 15$ mm, $\varepsilon = 10°$.

processes allows us to trace the evolution of the main process features during interference. The shock wave diffraction over a convex surface belongs to gasdynamic phenomena which have been thoroughly studied and quite satisfactorily understood [10, 11]. Let us describe briefly the main features of the process. The schemes of weak and strong shock wave diffraction over a circular convex surface are presented in Figures 2 and 3.

The weak shock wave diffraction ($M_s < 2.068$; the concurrent flow behind the shock wave is subsonic) consists of, on the one hand, the formation of the diffracted (curved) shock wave *KF* which starts from the triple diffraction point *K* and reaches the wall at a right angle to point *F*; and on the other hand, arising from the disturbed flow region separated from the undisturbed flow by the head of expansion wave which is an arc of a sonic circle. This sonic signal originates from the shock wave propagation through the generatrix junction point *A* and at later times is convected at the concurrent velocity. The absence of flow separation is characteristic for weak shock wave diffraction.

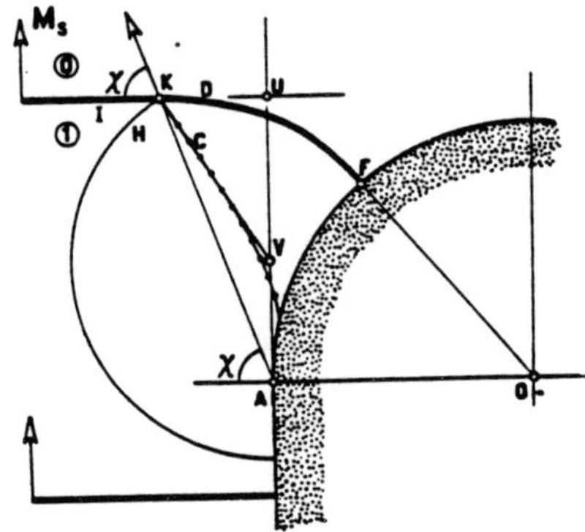

Fig. 2. Weak shock wave diffraction over a convex wedge.

A - a junction point of straight and curved surface parts; *K* - triple point; *I* - incident shock wave; D - diffracted shock wave; *H* - expansion wave head; *C* - contact discontinuity; *V* - sonic circle center. For the time moment equal to unity the interval lengths correspond to the following velocities: *AU* - absolute shock I velocity; *AV* - absolute concurrent flow velocity behind shock *I*; *UV* - relative flow velocity behind shock *I* in frame of reference attached to shock *I*; *KV* - relative flow velocity behind shock *I* in a frame of reference attached to the point *K* equal to sound speed in region 1.

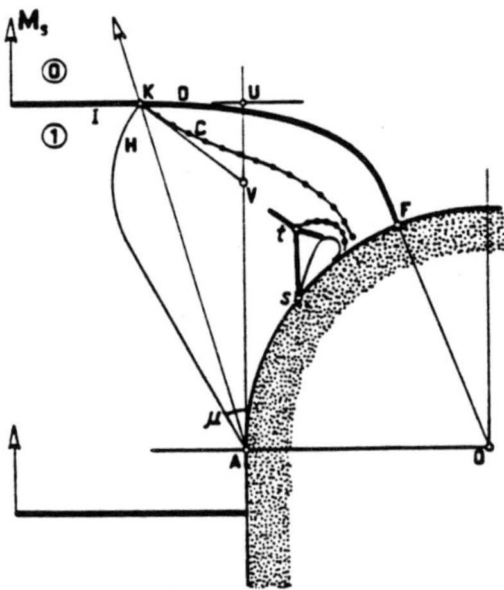

Fig. 3. Strong shock wave diffraction over a convex wedge.

Notation is the same as in Figure 2. μ - Mach angle; *s* - separation point; *t* - triple point of the λ-configuration arising due to separation.

The process of strong shock wave diffraction over a curvilinear surface ($M_s > 2.068$; the concurrent flow behind the shock wave is supersonic) has two distinctions compared to the preceding case (Figure 3). First the disturbed flow region is separated from the external flow by a composite boundary consisting of the Mach line disturbance coming from point *A* at Mach angle μ to the flow direction and the arc of the sonic circle originating from the same point *A* and convected by the concurrent flow. Second, the flow separation is observed at some point *s* for sufficiently strong shock waves, the recirculation zone structure depending on the shock wave Mach number [12]. At a certain shock wave Mach number a separation vortex and the λ-shock arise, the latter consisting of three shock waves and a slipstream coming from the triple point *t*. The gasdynamic system containing the separation vortex and the λ-shock provides the joint supersonic flow accelerated over the convex surface with flow immediately behind the diffracted shock front.

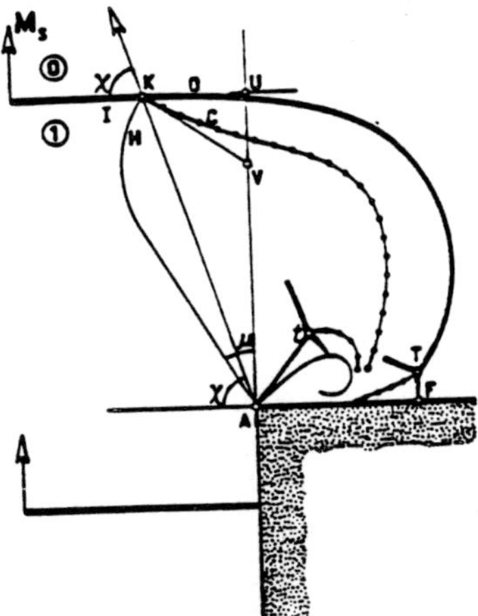

Fig. 4. Strong shock wave diffraction over step.

Notation is the same as in Figure 2. *T* - triple point at Mach reflection of diffracted shock wave.

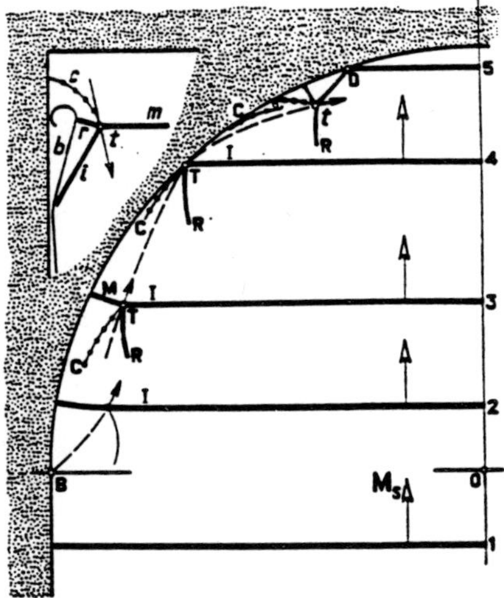

Fig. 5. Shock wave reflection from concave wedge.

Position 1 - Shock wave prior to interaction. Position 2 - Irregular reflection of shock wave I over initial part of curved surface. Mach reflection is not developed. Position 3: Mach reflection. *T* - triple point; *R* - reflected shock; *M* - Mach stem; *C* - contact surface. Position 4: Collision of Mach reflection with ramp. Position 5: Regular reflection of shock *I* and irregular collision of newly reflected shock *Dt* with the old shock *R*. In the left upper corner: bifurcation of the shock reflected from 90° ramp in the wall boundary layer b induced by incident shock *I*.

The scheme of strong shock wave diffraction over the step is presented in Figure 4. This type of diffraction differs from the one considered previously by an essential feature resulting from the presence of a corner point on the surface. Since in that case the separation occurs at practically arbitrary Mach numbers, for a sufficiently high Mach number the separation vortex turns out so strong as to considerably modify the velocity field in the vicinity of the diffracted shock wave. As a result the shock wave cannot be normal to the surface but reflects from it according to rules to be described later.

We have mentioned above that at the outer wall of the curvilinear channel nonstationary reflection of the shock wave from the solid concave surface takes place. It should be noted that at present this problem is actual and continues to be intensively studied experimentally, theoretically and numerically [10, 13-20]. The problem is of interest since during shock wave propagation along the curvilinear concave surface the angle of incidence varies continuously from 90° to zero and, as a consequence, all types of reflections from Mach reflection to a regular one can be observed.

Figure 5 presents the pattern of shock wave interaction with the circular concave surface at five different times.

Position 2 in this figure displays the initial stage of shock wave interaction with curved surface. First it is necessary to point out that the reflection process starts with zero wedge angle β and right local incidence angle α at junction point B. This type of reflection we call trivial Mach reflection since here the reflected wave is the arc of the sonic circle which arises when the wave passes point B and whose center is convected with the velocity of the concurrent flow behind the incident wave. The interaction pattern described is identical to the formation of the expansion wave head H and the triple diffraction point K in Figures 2-4.The interaction stage is shown in Figure 5, position 2 is remarkable in respect to a rather unusual reflection process: the reflected wave is very weak, it is practically impossible to identify the slipstream, the triple point is found with difficulty, the mach stem is characterized by smooth junction with both the incident shock and that normal to the wall. We note that the interaction in question is not Mach but its gasdynamic nature is close to the shock wave diffraction process. It has been suggested that such phenomenon be called von Neumann reflection [4].

As the shock wave propagates along the wall true Mach configuration with the triple point T is formed as seen in Figure 5, position 3. Two problems are of greatest interest at this interaction stage: at what local wedge (or local incidence angle) does the intrinsic mach reflection develop and what is triple point trajectory like? Elementary theoretical analysis cannot provide answers to these equations.

At some time the trajectory of the Mach reflection triple point runs into the wall (Figure 5, position 4) and the process of the collision of the Mach reflection with the solid wall starts. Understanding of the process development can be drawn from the shock wave pattern at the subsequent time instant shown in Figure 5, position 5.

The collision of triple Mach configuration with the wall can be considered a combination of two processes: the reflection of the incident shock wave from the wall and the interaction of new discontinuities formed in the course of the reflection with the existing elements of old triple configuration. For that matter it is important to know what type of incident shock wave reflection occurs in each specific case. If (as shown in the figure) the incident shock I reflects in a regular way, the arising reflected shock wave Dt must interact with the shock wave R of the old Mach configuration. A rough analysis of the conditions of the two shocks interaction pointed out shows that in this case one of the irregular types of shock-on-shock collisions [21, 22] takes place and the usual Mach configuration with triple point t is formed. We note that the new slipstream coming from point t couples at some distance from this point with the old slipstream C which exists up to the end of the observation.

Now we describe briefly the peculiarity of shock wave reflection in a channel with a 90° bend. The reflection is shown in the insert in the left upper corner in Figure 5. In an idealized case of inviscid gas the process is an elementary problem of the head-on reflection of the incident shock wave from a vertical wall. In the real process shock wave motion along the solid wall causes the growth of a boundary layer on the wall along which the reflected shock wave propagates. The reflected shock wave interacts with the boundary layer in such a way that the confined separation zone arises in the latter under the shock wave and the shock wave itself branches to form the well-known λ-configuration consisting of three shocks and a slipstream. The phenomenon just described is called a shock wave bifurcation in the boundary layer. An elementary calculation method for this phenomena is known [11].

A brief description of the nonstationary shock wave reflection from the curvilinear concave surface has been given above. For obvious reasons it is clear that the process develops in such a way that at any fixed time the interaction pattern depends not only on local reflection conditions but on the evolution history as well. Nevertheless, since in the course of the shock wave reflection from the curvilinear surface we encounter all the phenomena which we are familiar with in a "purified" form, it is justifiable to analyze the nonstationary process on the basis of the accumulated knowledge of such pure phenomena. The necessary information is presented in the Appendix.

3. Experimental Approach

3.1. Shock Tube

In order to stimulate supersonic discontinuous flow in a channel an explosive shock tube with a working cross-section allowing channel configuration variation was made. Its scheme is shown in Figure 6. The tube channel has a rectangular cross-section 30×30 mm^2 and a length of 1.5 m. The right (in figure) channel end enters receiver 5 and couples with working section 6. Assembling section 6 consists of two faces with optical windows of 200 mm in diameter. The required channel configuration is formed between the faces with the use of changeable details 7. The divergent "semi-nozzle" is shown in the figure: one of the details 7 continues the channel wall, the other is deflected.

The transverse shock tube channel dimensions have been selected on account of the peculiarities of the optical equipment used, namely, a grating interferometer with an image displacement equal to 100 mm. The flow in the study as a whole occupies the left half of the windows while the right one serves as a reference branch. With regard to the photographs, this interferometer does not differ in any respect from a Mach–Zehnder interferometer with a working field of 100 mm.

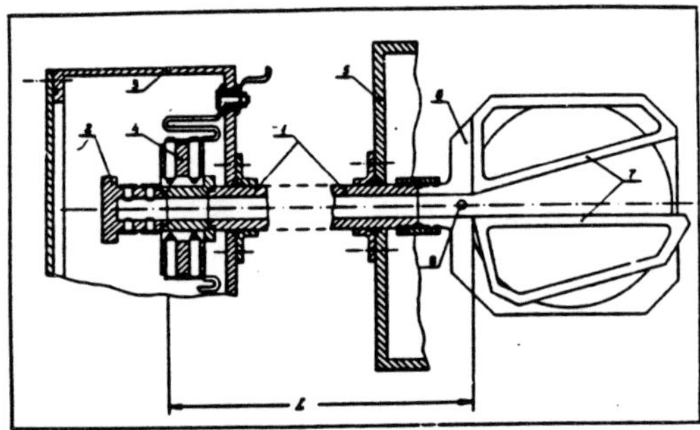

Fig. 6. Shock tube arrangement.
1 - shock tube channel; 2 - removable muffler containing electric percussion caps; 3 - hermetic jacket providing insulation of shock tube left end; 4 - device providing contact with electric percussion caps, cables to the movable contact device are installed through jacket 3; 5 - receiver containing the working section; 6, 7 - movable parts used to form the channel geometry in working section; 8 - piezoelectric gauge.

The required shock wave intensity has been provided by the choice of explosive charge value and the gas pressure in receiver 5.

3.2. Interferometer

The above-mentioned interferometer has been assembled on the basis of a Schlieren device IAB-451 [31]. The illumination and photographic systems of the interferometer have been modified for the purpose of obtaining two interferograms at close time intervals. Two ruby lasers were used with mutually perpendicular beam polarization planes. The birefringent prisms were used in the recording system for beamsplitting. The scale of photos were close to normal.

The interferometer has been adjusted to infinite fringe width, thus interference fringes coincide with isopycnics. Two-frame recording with a known time interval enables us to determine the shock wave velocity at the curvilinear channel entry as well as during the diffraction process. The measured velocity along with shock wave relations were used to calculate the gas parameters just behind the shock wave.

Laser initiation time has been related to the moment of explosion which initiates the shock wave. Thus the time elapsed since the shock wave passing the curvilinear channel entry can only be determined approximately. We have used the shock wavefront position in the channel for the purpose of comparison with numerical data.

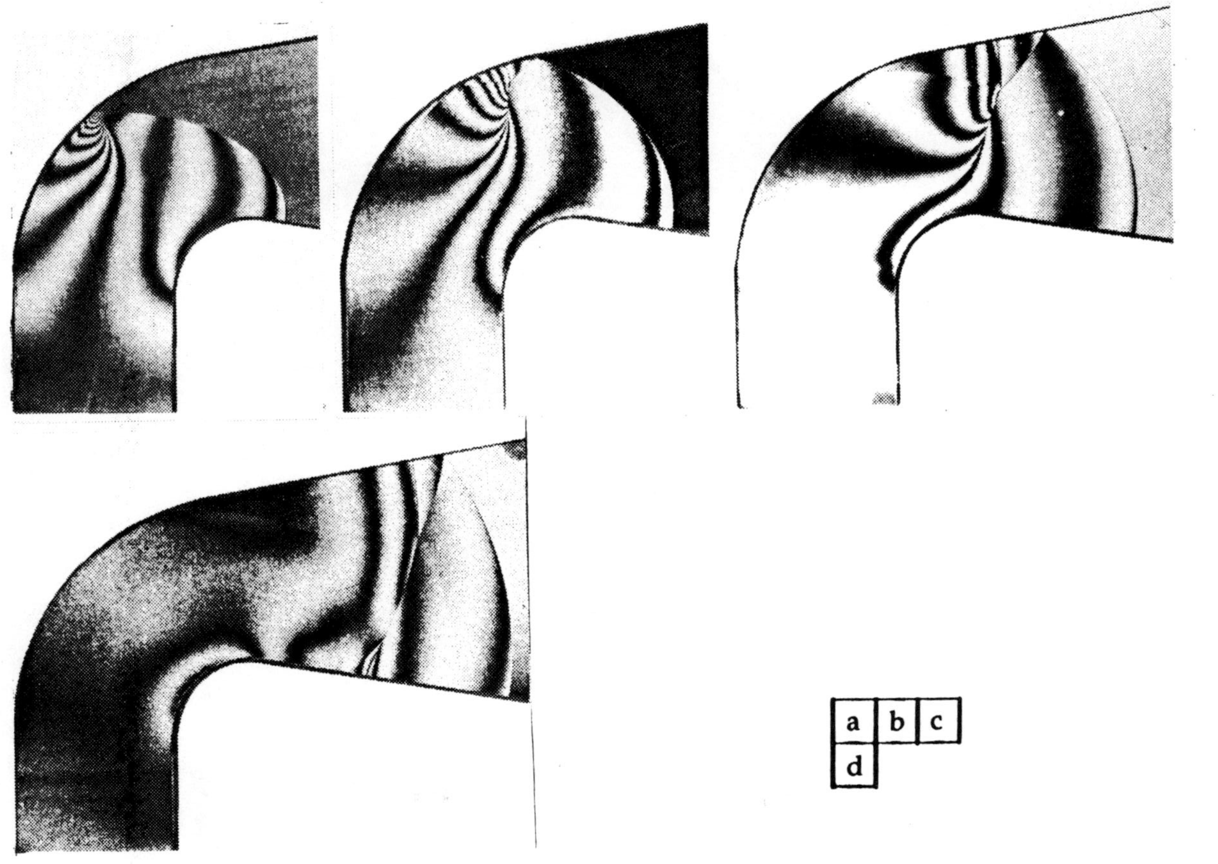

Fig. 7. Interferograms of flow field in the channel with a smooth bend and subsequent diffuser; $M_s = 1.2$.

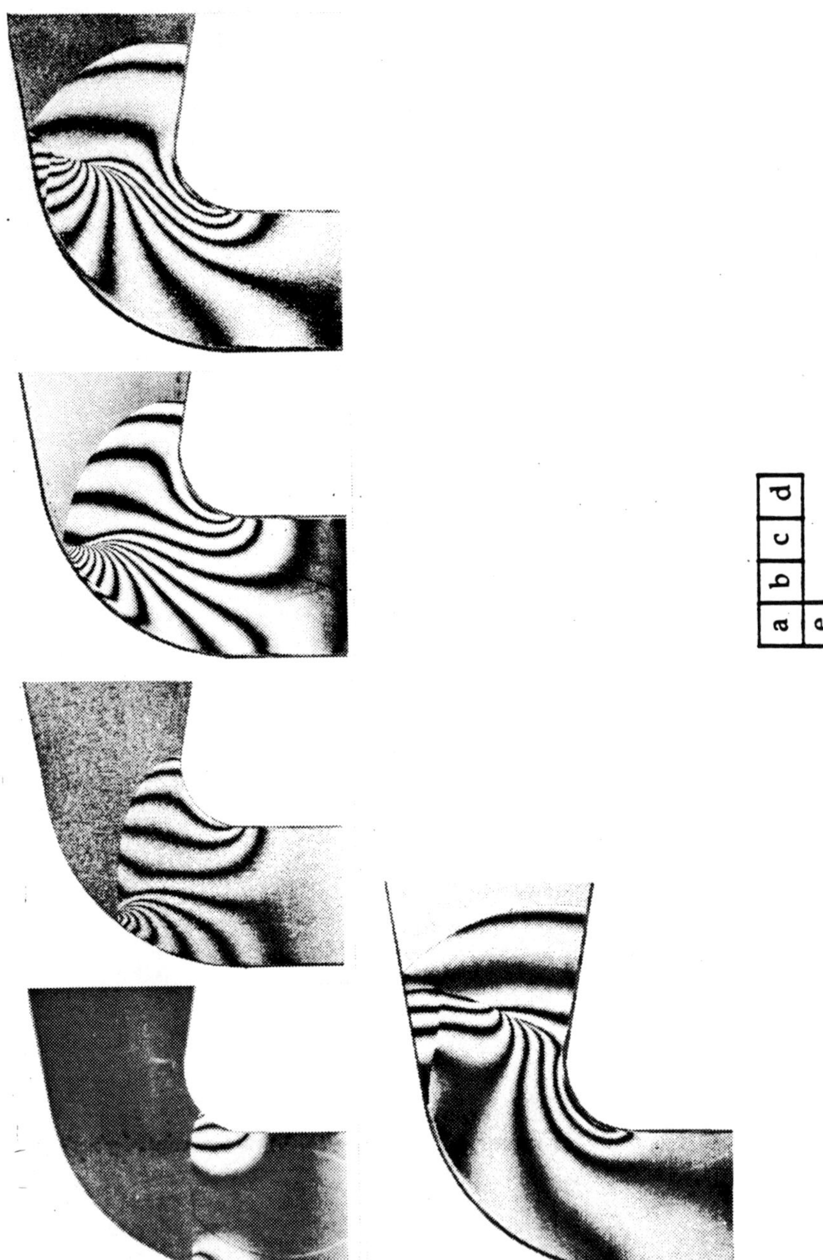

Fig. 8. Interferograms of flow field for $M_s = 1.35$.

2.3. Experimental Results

The interferograms of the flow behind the shock wave in the channel coupled smoothly to the plane divergent nozzle (the axes of the channel and the nozzle are mutually perpendicular) are shown in Figures 7–13. The shape of the junction as well as of the nozzle is shown in Figure 1. Five flow regimes corresponding to shock wave Mach numbers before the junction M_s = 1.2, 1.35, 1.7, 2.1, and 3.0 have been studied. The first three regimes are characterized by the subsonic air flow behind the shock wave. At M_s = 2.1 the flow Mach number slightly exceeds unity (M_f = 1.02); at M_s = 3.0 a supersonic flow with Mach number equal to 1.36 is observed behind the shock wave.

The photos of the flow behind the shock wave in the channel with a 90° angle are presented in Figures 14–18 for the same regimes. The shock wave is propagating upwards and to the right in all the photographs.

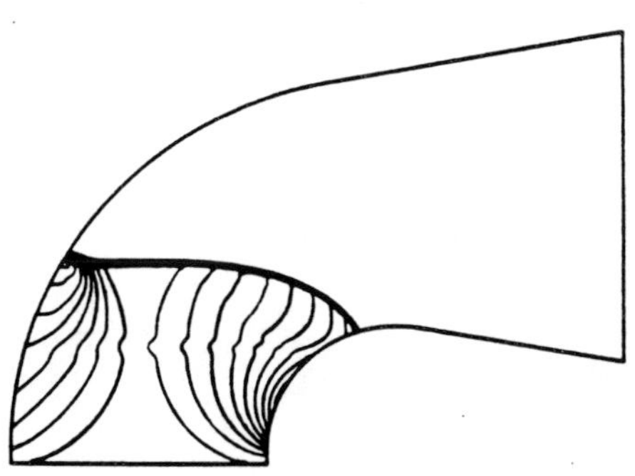

Fig. 9. Computed shock wave shape and isopycnics (increments 0.085) for M_s = 1.7. The Harten scheme [39] has been used for computations.

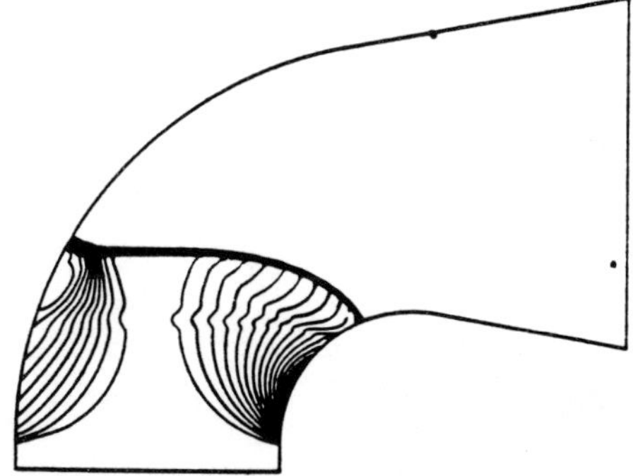

Fig. 11. Computed shock wave shape and isopycnics for (increment 0.085) for M_s = 2.1 (Harten scheme).

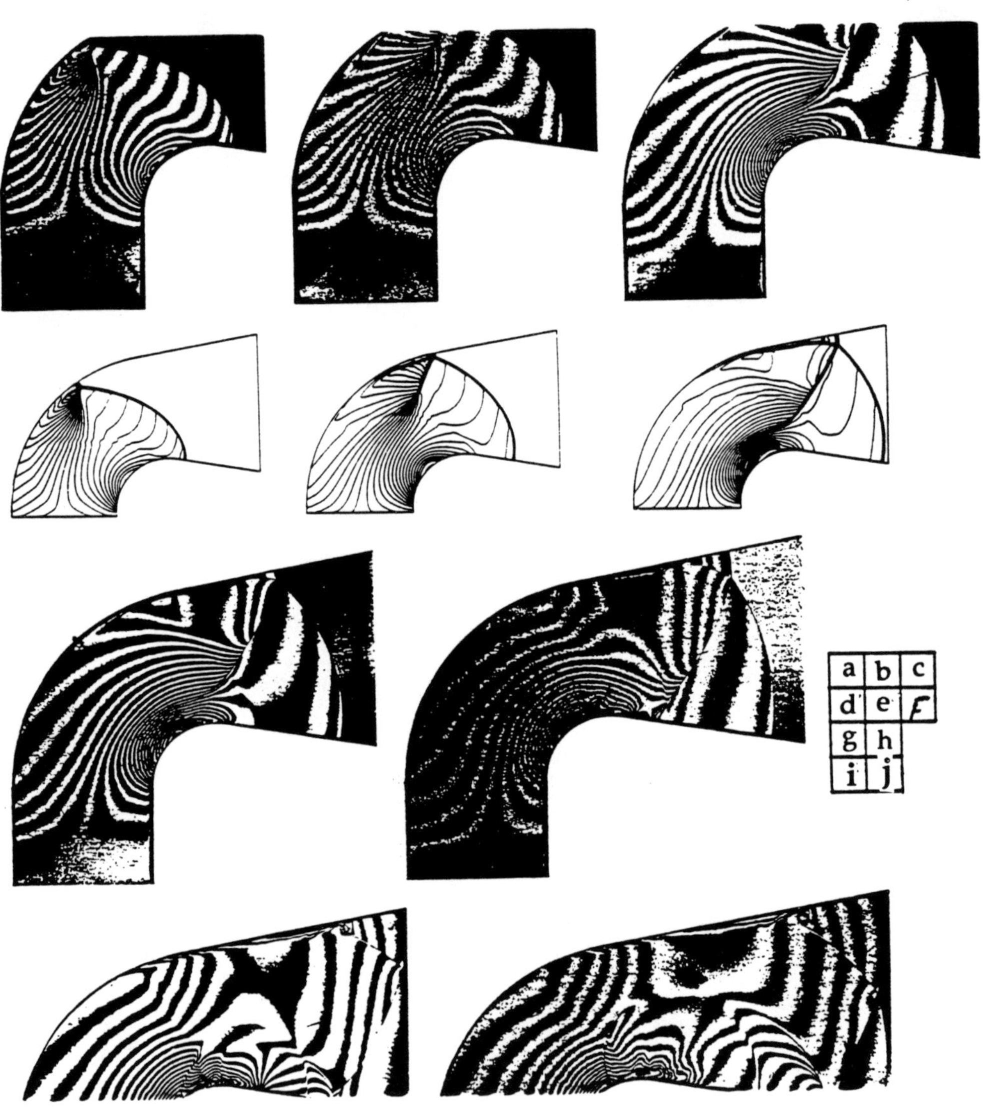

Fig. 10. Interferograms and numerical isopycnics for $M_s = 1.7$ (Harten scheme).

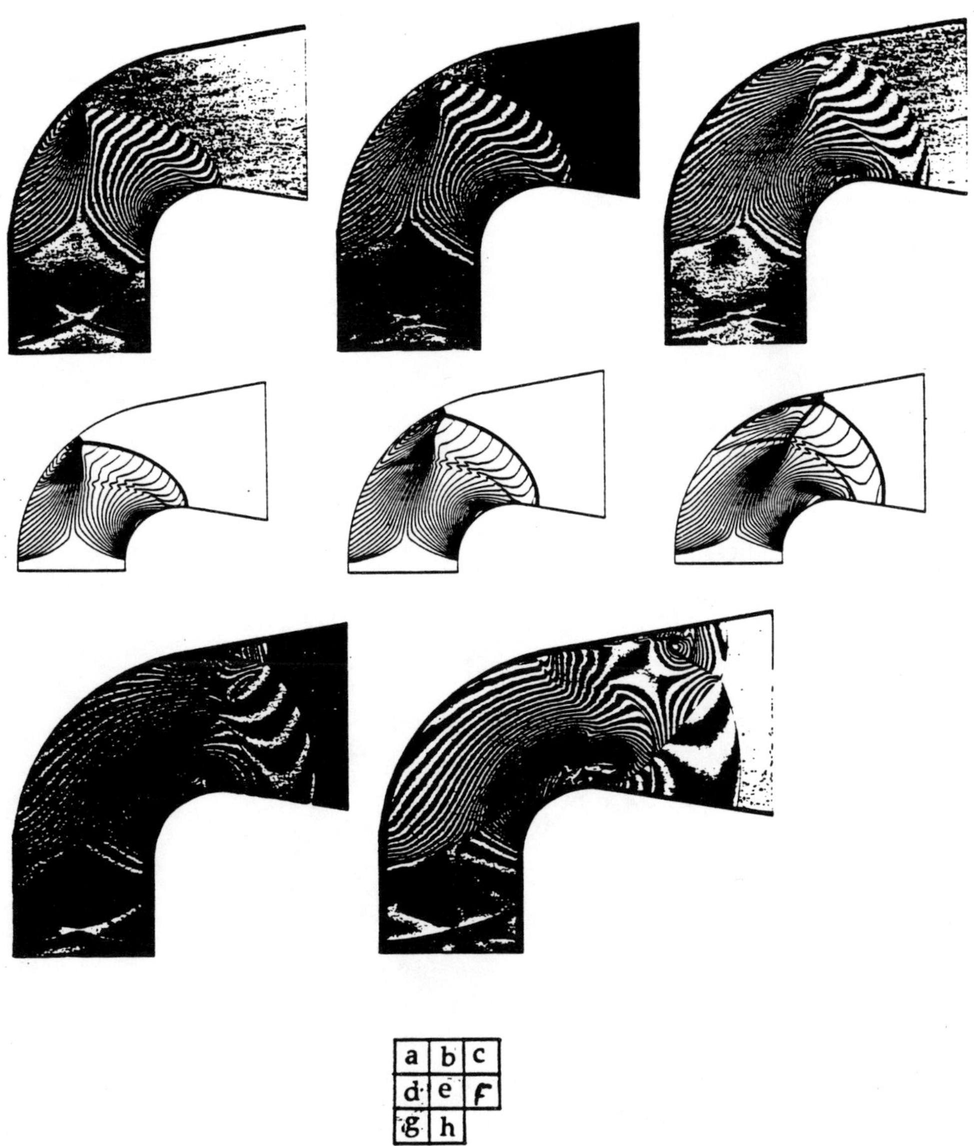

Fig. 12. Interferograms and numerical isopycnics for $M_s = 2.1$.

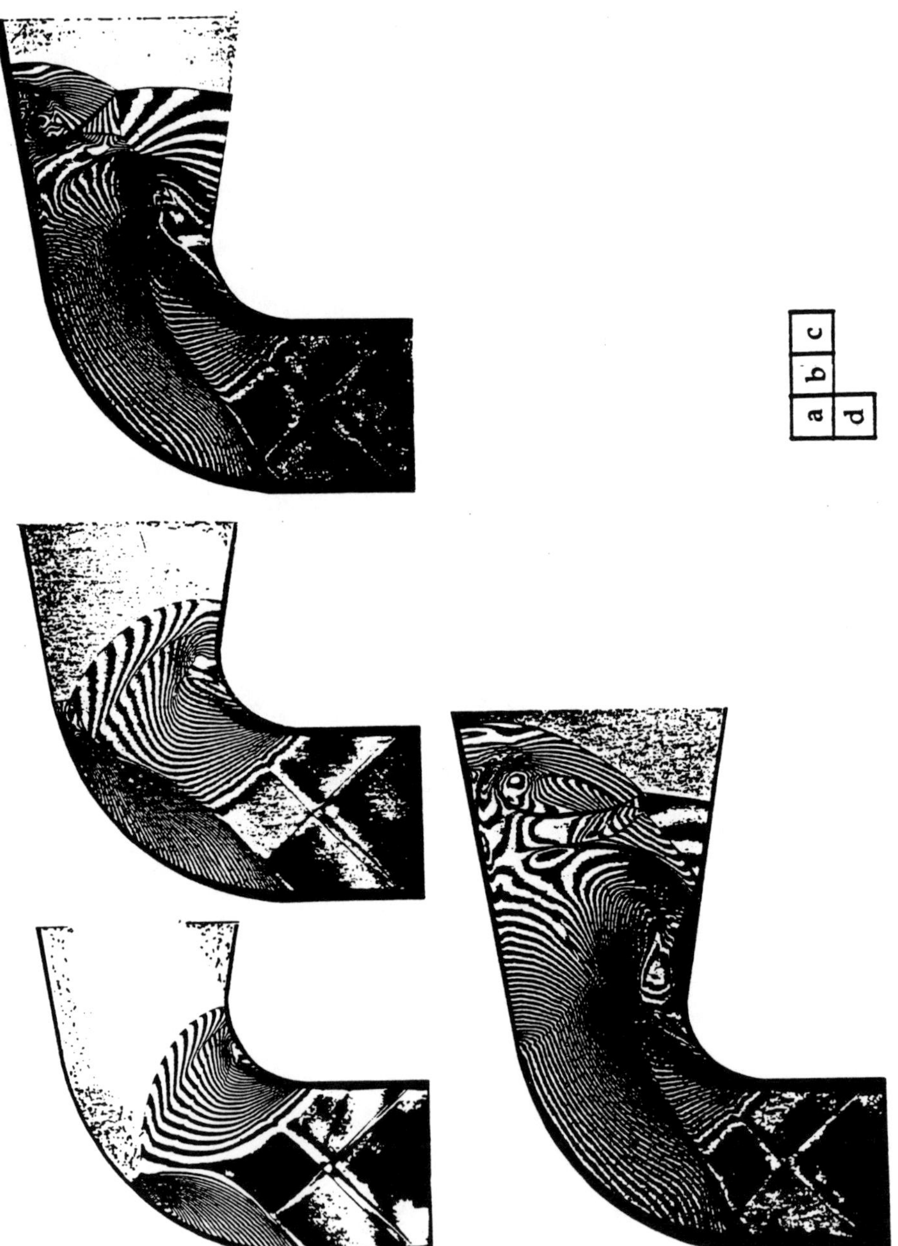

Fig. 13. Interferograms of flow field for $M_s = 3.0$.

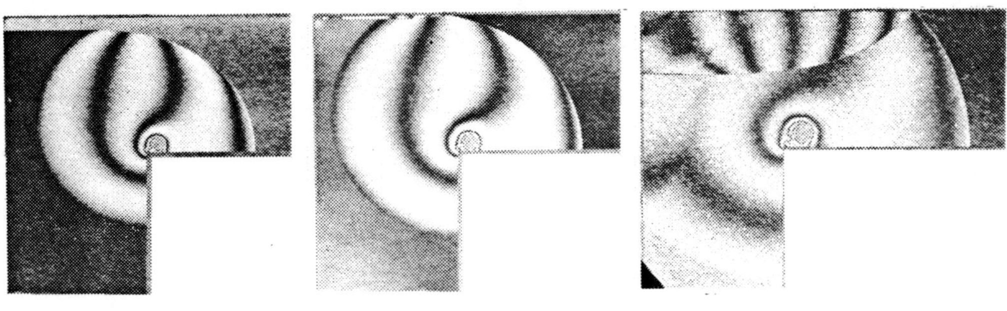

Fig. 14. Interferograms of flow field in the channel with an abrupt bend for $M_s = 1.2$.

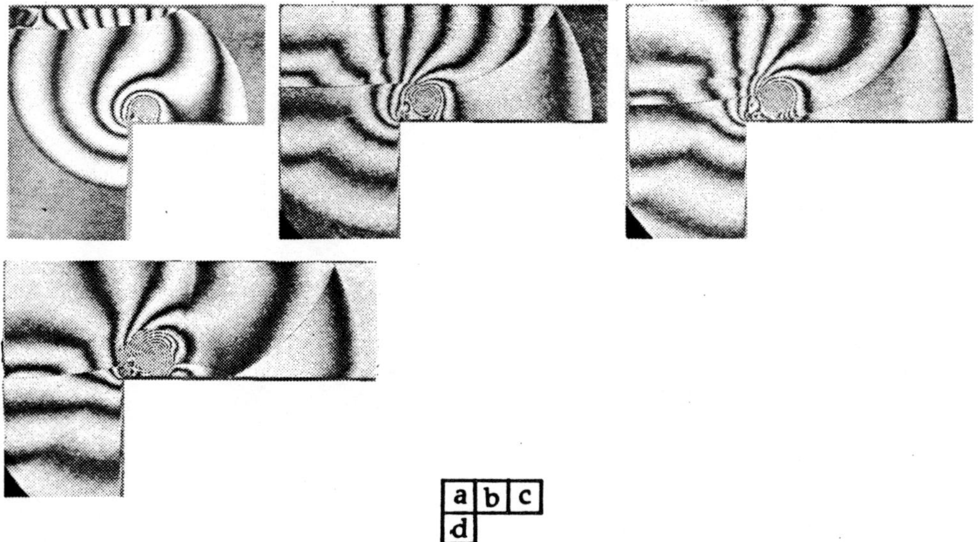

Fig. 15. Interferograms of flow field in the channel with an abrupt bend for $M_s = 1.35$.

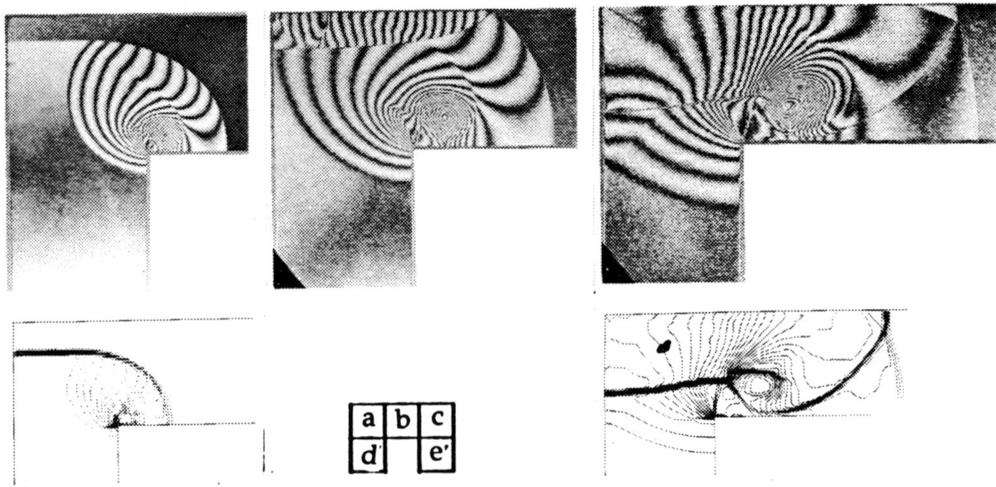

Fig. 16. Inteferograms and numerical isopycnics for $M_s = 1.7$. MacCormack scheme [36] has been used for computations.

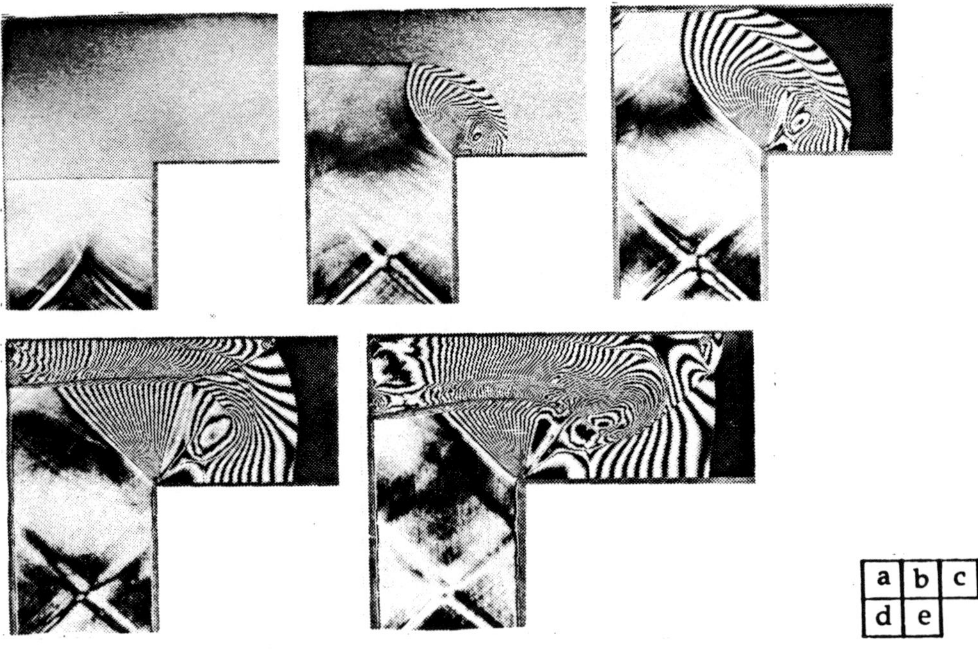

Fig. 17. Interferograms of flow field in the channel with an abrupt bend for $M_s = 2.1$.

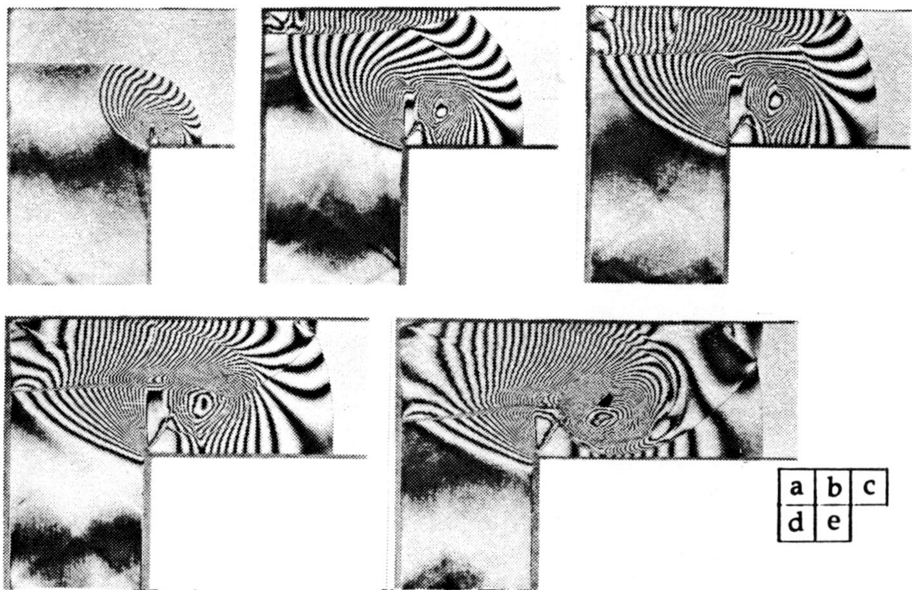

Fig. 18. Interferograms of flow field in the channel with an abrupt bend for $M_s = 3.0$.

It has already been noted that the shock wave in the channel is initiated by the explosion. As a consequence, the gas parameters behind the shock wave in the straight channel vary with the distance from the wavefront. We can judge the characteristic of density variation using the photos containing the flow region undisturbed by the bend (for example, Figures 13, 17, and others). In the photos taken at Mach numbers M_s not greater than 2.1 the relative density (referred to as the undisturbed gas density) variation between the adjacent fringes is equal to 0.085. At Mach number $M_s = 3.0$ the relative density increment is greater and equals 0.15. In the above-mentioned photos (such as Figure 13) it is seen that within the segment equal to the channel transverse dimension the optical path difference change is about one fringe that corresponds to the relative density increment presented. These variations have not been taken into account in the course of numerical simulation of the shock wave propagation. The parameters behind the shock wave have been assumed to be constant and equal to the initial parameter just before the bend.

In cases when numerical simulation have been carried out the computed isopycnics are placed along with the corresponding flow photos. The isopycnic increment is equal to the density variation between the adjacent interference fringes.

4. Numerical Simulation

The problem of shock wave propagation along the curved channel does not differ in its principal aspects from the problems of shock wave interactions with curvilinear surfaces which have been simulated numerically by many authors (see, for example, the review in [10]). However, the flow pattern is drastically complicated due to interference of the secondary waves caused by the incident shock wave interactions with the channel walls; computational difficulties may be encountered.

The approach used by the authors is based on the model of an inviscid non-heat conducting gas (the integral form of Euler equations in conservation form is used) and spatial splitting. Its computer implementation described in Ref. [34] has a modular structure and is in accordance with the rules adopted in OLYMPUS and SAFRA systems [5].

The software package includes several finite-difference schemes for shocked flows computation for arbitrary geometry and space dimension. The monotonous MacCormack schemes [36], various modifications of Godunov's method [37], Steger-Warming flux-splitting method [38], and TVD schemes by Harten [39] and Chakravarthy [40] are considered. Thus along with the solution of the main problem (the study of nonstationary shocked flow), we can compare the named finite-difference schemes and choose an optimal one for each concrete case. All the schemes are second order accurate in both time and space for smooth solutions.

Gasdynamic discontinuities in the computations presented have been captured. The grid had 250 cells along the channel axis and 100 cells in the transverse direction for a channel with a smooth wall junction. The cells were approximately square in the vicinity of the axis. In the bend region the cells were elongated near the outer wall and contracted near the inner one. In all cases the cell shape does not differ significantly from rectangular with the ratio of the long to short sides not greater than 2. The computation of the flow in the channel with smooth band has been carried out using the described grid by means of the Harten scheme with a minor modification. In the case of the channel with an abrupt bend a monotonous MacCormack scheme has been exploited, the grid size is 103×54 cells.

5. Discussion of Results

The initial stages of shock wave motion along the curved channel or in the vicinity of an abrupt bend are well described by the theory of Section 1. In the photos in Figures 7a, 8a, b, and 13a the processes of the diffraction over convex surface and the reflection from the concave surface do not influence each other. The photographs in Figures 14, 15, and 18 clearly demonstrate peculiarities of the flow behind both weak and strong shock waves which can be explained on the basis of the above stated theoretical concepts.

Nevertheless, numerical simulation is useful for these process stages if one is interested in the parameters in the whole flow field as well as in wall loadings important in practice. The computed results presented in Figures 9, 10a, 11 and in part, 12a and b correspond to these stages.

Numerical simulation becomes indispensable for later stages when an essentially nonlinear interference of diffraction and reflection processes occurs. Their mutual influence invalidates the application of the theoretical concepts developed for separated structures. The role of experiment is reduced to obtaining the general flow pattern and to measurement of available parameters (density in our case).

The complete flow description for late stages can be obtained by numerical methods only (see Figure 12c, c'). However, the inherent drawbacks of numerical methods associated with the model and the method itself as well as with its implementation should be kept in mind. For example, the isolines oscillations in Figure 11 which grow in time (see Figure 12c, c') are due to incorrect initial conditions. In addition, the model used does not allow study of viscous effects such as the boundary layer development and flow separation. Thus, in Figure 16c, c', it is seen that the vortex shape corresponds to that experimentally observed, but the flow inside the vortex is not described satisfactorily. Neither can the quantitative description be reached in cases when shock wave interaction with the boundary layer becomes essential (Figure 12d, e).

These considerations force us to stress once more the necessity for combining different approaches to shocked flow studies.

APPENDIX

Types of Shock Wave Reflections

The present knowledge of shock wave reflections is a reliable theoretical basis for analysis and interpretation of the complex nonstationary process of shock wave reflection from a curvilinear concave surface. In Figures 19 and 20 an attempt is made to present this knowledge in compact form.

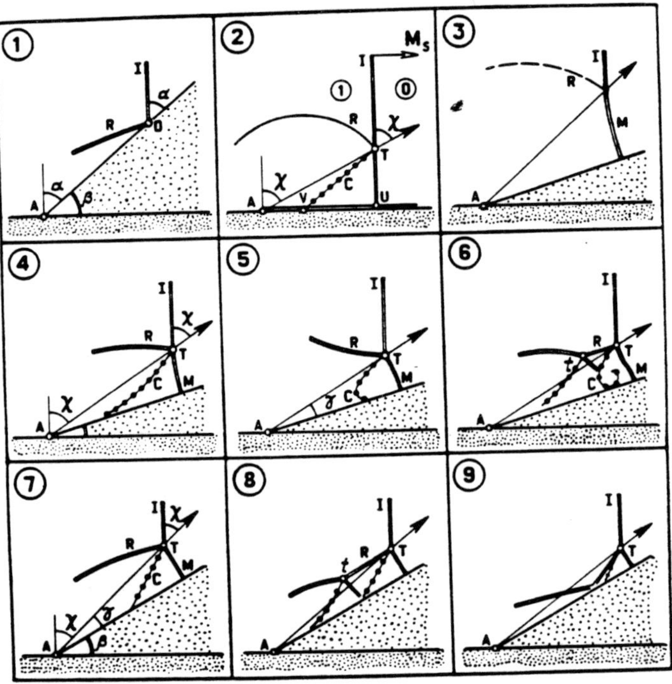

Fig. 19. Types of shock wave reflections from a wedge.

1 - Regular reflection; I - incident shock wave; R = reflected shock wave; α - shock wave incidence angle; β - wedge angle.

2 - Trivial mach reflection; T - triple point; R = reflected wave as circle's arc with the center in point V and the radius equal to the sound speed in region 1; C - slipstream; χ - angle between shock wave and the triple point trajectory.

3 - Von-Neumann reflection.

4 - Mach-Smith reflection; T - triple point; I, R - incident and reflected shock waves, respectively; M - Mach stem; C - slipstream.

5 - Single Mach-White reflection.

6 - Double Mach-White reflection; t - second triple point.

7 - Single Mach-Cabannes reflection.

8 - Double Mach-Cabannes reflection; t - second triple point.

9 - Terminal Mach-Cabannes reflection.

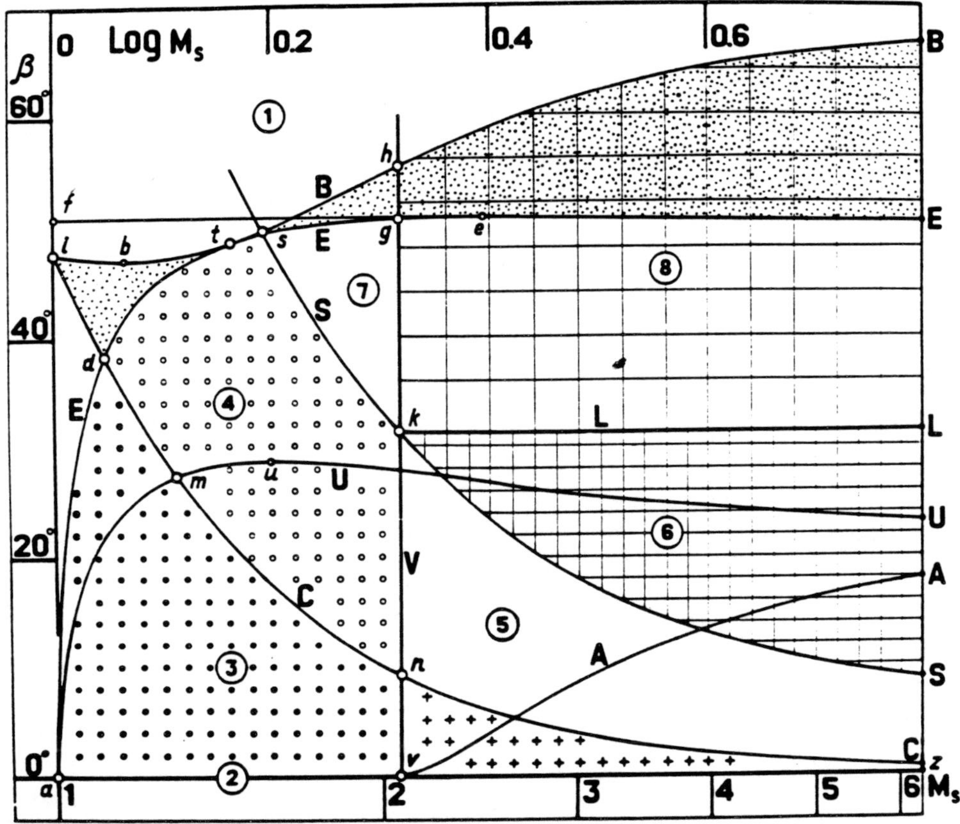

Fig. 20. Regions and transition lines of reflections of shock wave from a wedge in the (M_s, β)-plane.

M_s - incident shock wave Mach number; β - wedge angle. E - regular reflection boundary. Irregular reflection region is under the curve E; e - extremal curve point determining the region of anomalous regular reflection *afeta*. B and C - boundaries of intrinsic Mach reflection; B - transition between Mach reflection and collapse due to collision of two shock waves moving towards each other; b - extremal curve point; t - tangent point at the regular reflection boundary; C - transition boundary between Mach reflection and confluence due to the catching of one shock wave by the other; U - boundary of subsonic flow (relative to the wall reflection point) behind the incident wave, flow under curve U is subsonic; u - extremal point of the curve U; S - boundary of subsonic flow (relative to the triple point) behind the Mach reflected shock wave; flow under curve S is subsonic; V - boundary of subsonic flow (relative to the stationary frame of reference) behind the incident shock wave. A - boundary of bow shocks attached to the wedge vertex; attached shocks are under curve A. Both regular and Mach reflections are possible in regions *ibtdi* and *thBEet*. Axis *az* - trivial Mach reflection. Region *adska* - Mach-Smith reflection; subregion *amnva* - von Neumann reflection. Region *sBELks* - Mach-Cabannes reflection; subregion *shgks* - single subregion *kghBELk* - double Mach-Cabannes reflection. Region *vkLzv* - Mach-White reflection; subregion *vkSzv* - single, subregion *kLSk*, double Mach-White reflection.

Modern concepts of different types of pseudostationary reflection of the shock wave from a wedge based on experimental and theoretical data obtained for more than 40 years by researchers all over the world [4, 10, 23–29] are summed up in Figure 19.

Let us characterize briefly the known reflection types using Figure 20 in the determinant parameter plane for the pseudostationary shock wave reflection from the wedge. We stress that such a plot is necessary for analysis of nonstationary shock wave reflection from the concave wall of the curvilinear channel since the wave with a fixed mach number M_s propagation along this wall corresponds to the vertical motion on the graph in Figure 20 from the abscissa axis point M_s up to the point corresponding to geometrical parameters of the curvilinear channel boundary. It is possible to state that an interpretation of the nonstationary shock wave interaction with curvilinear concave wall can be considered, with some reservations, as scanning of the plot in Figure 20 along the vertical (or other) trajectories.

When designing a common plot, first of all the regular reflection boundary E has been found. It divides the entire region of the reflection existence ($1 \leqslant M_s < \infty$; $0° \leqslant \beta \leqslant 90°$) into the subregions of regular (above curve E) and irregular (below curve E) reflection. Note that the conception of "Mach reflection" was not used.

The regular reflection theory provides one more boundary U containing some information about irregular reflection, namely: irregular reflection below the curve U cannot generate a shock wave, while above the curve a reflected shock wave can exist.

The theory of pseudostationary mach reflection by Cabannes [3, 4] gives two (B and C) boundaries and presents a surprise: it turns out that the regions of irregular and Mach reflections (*atEza, itBcmi*) do not coincide in an astonishing way. The first unexpected fact is that the true Mach reflection region occupies two characteristic subregions of regular reflection (*ibtdi* and *thBEet*). Thus, the problem of the reflection duality at corresponding determinant parameters arises. In addition, the Mach reflection region does not cover the spacious irregular reflection region *admnCza*. The question of the type of reflection in the latter has not yet been solved. In our opinion, this region of determinant parameters presents a field for testing different heuristic solutions. The simplest one is a trivial Mach reflection (Figure 19.2) and axis *az* in Figure 20) since it is not a limiting case of the existing theory but has been constructed using common sense on the basis of the simplest gasdynamic considerations. It should be noted that the trivial Mach reflection model accurately describes the real process of weak shock wave diffraction from a thin wedge [30].

The regions *ibtdi* and *thBEet* where both regular and Mach reflections are possible theoretically are also of interest. The first one which is entirely within the low Mach number range has probably never been thoroughly studied and rather simply ignored. This fact may be associated with the dubious property found theoretically for the mach reflection in the region *ibtdi*, namely, in this case the reflected shock waves are the shocks of a strong family. Nevertheless, this situation does not differ essentially from the one existing in the true Mach reflection region adjacent to boundary *dt* from below. The remarkable property of the region in question is the fact that as a whole it belongs to the anomalous regular reflection region; i.e., to the region of the process with a reflection pressure greater than the shock deceleration pressure (the pressure of head-on shock wave reflection).

The nature of the second mixed reflection region *thBEet* is more explicit. It is quite unambiguously established for this region [15, 16] that pseudostationary, and also truly nonstationary process of Mach reflection developing in the direction $RR \rightarrow MR$ (regular-reflection $\rightarrow$ Mach reflection) cannot enter the regular reflection region. Another situation occurs for stationary and truly nonstationary reflection with the process direction $MR \rightarrow RR$. In this case the Mach configuration is observed in the mixed region of reflection.

In order to demonstrate the properties of the pseudostationary Mach reflection from the wedge a classification of this phenomenon has been developed based on the consistent use of the concept of the interference of two simultaneous processes: the flow around the wedge vertex and the shock wave reflection from the inclined wedge surface. To this end it is necessary to determine, first, what type of perturbation impulse (compression or expansion) will result from the vicinity of the wedge vertex A to the reflection point, and secondly, if it could catch up with the triple point. The required analysis was carried out with the use of the boundaries of subsonic flow behind the reflected shock wave (boundary S) and of the concurrent flow behind the incident one (boundary V). Exploiting the boundaries mentioned above allows us to distinguish three main types of pseudostationary Mach reflection: Mach-Smith reflection (region *adskva*); Mach-Cabannes reflection (region *shBLks*), and Mach-White reflection (region *vkLzv*). The principle of Mach reflection division into types permits us to easily describe their ancestral signs.

Mach-Smith reflection is characterized by the fact that a perturbation impulse in the form of an expansion wave is sent from the flow region in the vicinity of the wedge vertex A to the reflection region and it reaches the triple point without hindrance. As a result the reflection pattern is distorted: all the discontinuities of Mach configuration are curved. In Figure 19.4 we should pay attention to the orientation of each discontinuity convexity or concavity relative to its motion direction.

The characterization of Mach-Smith reflection presented is extremely general. A more thorough analysis of the Mach-Smith reflection region in the determinant parameters plane (Figure 20) with the use of characteristic boundaries dividing this region reveals the variety of the process peculiarities. Indeed, in the course of the analysis of the von Neumann paradox of Mach reflection [4] attention has been drawn to the subregion *amnva* which is part of the non-Mach reflection region *aimnCza*. The process of shock wave interaction with the wedge in this region is remarkable in two respects: first, the reflected wave is to be formed as a result of subsonic flow deflection, and second, the flow around the wedge is subsonic near its vertex. These facts lead to the idea that under given conditions a simple compression wave instead of the shock wave will arise during the reflection process. It is possible to construct the simplest model of this phenomenon named von Neumann reflection (Figure 19.3).

The main distinctive feature of Mach-White reflection (Figure 19.5 and 19.6) is the existence of a spiral vortex which is formed as a result of the slipstream rolling up under the action of the compression impulse coming from the wedge vertex vicinity.

The Mach-White reflection region *vkLzv* in Figure 20 is divided by the sonic line *S* into two characteristic subregions. The shock wave interaction with the wedge in one of them (*vkSzv*) is such that the reflection process is distorted essentially by the compression impulse coming from the wedge vertex vicinity and reaching the triple point without hindrance. In this case the character of the discontinuity bending is such as shown in Figure 19.5. In the Mach-White subregion above the sonic curve *S* as well as in the one just considered, the compression impulse comes from the wedge vertex vicinity to the reflection place, but in contrast to the preceding case it cannot reach the triple point. This fact results in a new shock wave and a second triple point *t* (from which another slipstream comes) between the reflected shock wave and the slipstream. In order to distinguish types of the Mach-White reflection in subregions 5 and 6 in Figure 20 we can employ an international nomenclature suggested in [17]: then it is reasonable to call the processes in subregions 5 and 6 single and double Mach-White reflection respectively.

We note that the nature of the single Mach-White reflection is, probably, by far richer than appears at first sight since the existence region of this reflection type *vkSzv* contains the part *vnCzv* of intrinsic irregular reflection which can manifest itself in an unusual form.

Of all the types of Mach reflection the Mach-Cabannes reflection to the most extent corresponds to the idealized concept of the process. Owing to this consideration it has been isolated theoretically. The Mach-Cabannes reflection can be observed in the region *shBLks* of the determinant parameters and in this entire region the flow behind the reflected shock wave is supersonic relative to the triple point. As a consequence, the determinant property of the Mach-Cabannes reflection is the following: the process is not disturbed by external effects. However, two types of the Mach-

Cabannes reflection can be distinguished depending on the character of the wedge vertex flow influence on the reflection process. In the subregion *vshgk* this influence is determined by the expansion impulse which cannot catch the triple point. The reflection pattern for this case is shown in Figure 19.7. In the Mach-Cabannes reflection subregion *khBLk* (to the right of the vertical *kh*) the necessary conditions for shock wave development from the compression impulse coming from the wedge vertex vicinity are met. This shock wave does not reach the triple point but generates a new three-shock Mach configuration and a second triple point *t* (Figure 19.8). As in the preceding case in order to identify two types considered the following terms can be used: single and double Mach-Cabannes reflection.

We note that recently [23] the limiting form of the Mach-Cabannes reflection has been obtained experimentally: the reflected shock wave practically rests on the wall and a new wave governed by the regular reflection laws comes from the meeting point (Figure 19.9). This reflection type can be obtained, probably, in air at very high Mach numbers, but it is easily observed in gases of a complex molecular structure, i.e., in media with a low adiabatic index [29].

In conclusion, the authors hope that this brief review of shock wave reflection problems will be useful for the interpretation of the complex gasdynamic flow pattern in curvilinear channels, and both experimental and numerical studies of such flows will enrich and make more precise the theoretical concepts of reflection processes.

References

1. *Shock Waves: Annotated Index of Home and Foreign Literature (1950–1959)*, Moscow: Nauka, 1964, 187 pp. (in Russian.

2. *Shock Waves: Annotated Index of Home and Foreign Literature (1960–1969)*, Moscow: Nauka, 1979 (in Russian).

3. A. N. Semenov, *Analytical Solutions in the Elementary Theory of Mach Reflection. I. The Models of Mach Reflection; II. The von Neumann Theory; III. The Theory of H. Cabannes*, Preprints nos. 1005, 1006, 1007, Leningrad: Phys. Tech. Inst., 1986, 49, 50 35 pp (in Russian).

4. A. N. Semenov, *The Properties of Solutions from Elementary Theory of Mach Reflection*, Preprint no. 1159, Leningrad: Phys. Tech. Inst., 1987, 49 pp (in Russian).

5. V. P. Goloviznin, et al., *Shock Wave Propagation Along Plane and Axisymmetric Channels*, Preprint no. 709, Leningrad, Phys. Tech. Inst., 1981, 48 pp. (in Russian).

6.   A. I. Zhmakin, et al., *Shock Wave Propagation Along Plane Channel Bends*, Preprint no. 950, Leningrad: Phys. Tech. Inst., 1985, 18 pp. (in Russian).

7.   V. A. Komissaruk and N. P. Mende, *Pis'ma v Zh. Tekh. Fiz.*, vol. 6, no. 17, 1980, pp. 1025–1030 (in Russian).

8.   A. A. Samarskii, *Vestnik AN SSSR*, no. 5, 1979, pp. 39–49 (in Russian).

9.   Fernley, M. A. Nettleton, *Central Electricity Research Laboratories*, Laboratory Note, 1980, 15 pp.

10.   T. V. Bazhenova, et al., *Nonstationary Interactions of Shock and Detonation Waves in Gases*, Moscow: Nauka, 1986, 206 pp. (in Russian).

11.   T. V. Bazhenova and L. G. Gvozdeva, *Nonstationary Interactions of Shock Waves*, Moscow: Nauka, 1977, 274 pp. (in Russian).

12.   M. P. Syshchikova and M. K. Berezkina, In: *Aerophysical Studies of Supersonic Flows*, Moscow-Leningrad: Nauka, 1967, pp. 7–13 (in Russian).

13.   M. P. Syshchikova, A. N. Semenov, and M. K. Berezkina, *Pis'ma v Zh. Tekhn. Fiz.*, vol. 2, no. 2, 1976, pp. 61–66 (in Russian).

14.   P. A. Voinovich, F. D. Popov, and A. A. Fursenko, *Ibid.*, vol. 4, no. 6, 1978, pp. 313–316 (in Russian).

15.   L. F. Henderson and A. Lozzi, *J. Fluid Mech.*, vol. 96, pt. 3, 1979, pp. 541–559.

16.   K. Takayama, G. Ben-Dor, and J. Gotoh, *AIAA J.*, vol. 19, no. 9, 1981, pp. 1238–1240.

17.   G. Ben-Dor and J. M. Dewey, *AIAA J.*, vol. 23, no. 10, 1985, pp. 1650–1652.

18.   K. Takayama and G. Ben-Dor, *AIAA J.*, vol. 23, no. 12, 1985, pp. 1853–1859.

19.   G. Ben-Dor, K. Takayama, *AIAA J.*, vol. 24, no. 4, 1986, pp. 682–684.

20.   K. Matsuo, et al. In: *Shock Waves and Shock Tubes*, Stanford: Stanford Univ. Press, 1986, p. 235.

21.   A. L. Starykh, *Chis. Metody Mekh. Splosh. Sredy*, vol. 17, no. 6, 1986, pp. 119–124 (in Russian).

22.   V. G. Dulov, G. A. Lukjanov, and V. N. Uskov, *Model. v. Mekh.*, vol. 1 (18), no. 3, 1987, pp. 38–61 (in Russian).

23.   T. Ikui, et al., *Memoirs Fac. Eng. Kyushu Univ.*, vol. 41, no. 4, 1981, pp. 361–380.

24.   T. Ikui, et al., *Bulletin JSME*, vol. 25, no. 208, 1982, pp. 1513–1520.

25.   K. Matsuo, et al., *Bulletin JSME*, vol. 29, no. 248, 1986, pp. 422–438.

26.   T. V. Bazhenova, L. G. Gvozdeva, and M. A. Nettleton, *Prog. Aerospace Sci.*, vol. 21, no. 4, 1984, pp. 249–331.

27.   I. I. Glass, *Some Aspects of Shock-Wave Research*. UTIAS Review no. 48, Toronto, 1986, 40 pp.

28.   I. I. Glass, *AIAA J.*, vol. 25, no. 2, 1987, pp. 214-229.
29.   J. T. Urbanovicz, *Pseudo-Stationary Oblique Shock-Wave Reflections in Low Gamma Gases - Isobutane and Sulphur Hexafluoride*, UTIAS Technical Note no. 267, Toronto, 1988, 40 pp.
30.   J. M. Dewey and D. J. McMillan, *J. Fluid Mech.*, vol. 152, 1985, pp. 49-81.
31.   A. N. Semenov and M. P. Syshchikova, *Fiz. Goreniya i Vzryva*, vol. 11, no. 4, 1975, pp. 596-608 (in Russian).
32.   P. I. Kovalev, et al., *Optics and Laser Tech.*, vol. 11, no. 3, 1979, pp. 137-142.
33.   I. A. Dukhovsky, et al., *Ibid.*, vol. 17, no. 3, 1985, pp. 148-150.
34.   P. A. Voinovich, A. I. Zhmakin, and A. A. Fursenko, *Program for Computation of Multidimensional Nonsteady Inviscid Shocked Flows*, Preprint no. 1268, Leningrad: Phys. Tech. Inst., 1988, 41 pp. (in Russian).
35.   S. A. Gaifulin, V. Ya. Karpov, and T. V. Mishchenko, *SAFRA. Functional Software, OLYMPUS System*, Preprint no. 27, Moscow: Inst. Appl. Math., 1980, 70 pp. (in Russian).
36.   A. I. Zhmakin and A. A. Fursenko, *Zh. Vichisl. Matem. i Matem. Fiz.*, vol. 20, no. 4, 1980, pp. 1021-1031 (in Russian).
37.   A. N. Kraiko, H. I. Tilljaeva, and S. A. Shcherbakov, *Zh. Vichisl. Matem. i Matem. Fiz.*, vol. 26, no. 11, 1986, pp. 1679-1694 (in Russian).
38.   W. K. Anderson, J. L. thomas, and B. van Leer, *AIAA J.*, vol. 24, no. 9, 1986, pp. 1453-1460.
39.   A. Harten, *J. Comp. Phys.*, vol. 49, 1983, pp. 357-393.
40.   S. R. Chakravarthy and S. Osher, In: *Lect. Appl. Math.*, vol. 22, pt. 1, 1985, pp. 57-86.

NUMERICAL SIMULATION OF SUPERSONIC NONUNIFORM FLOWS ABOUT BLUNT BODIES

Yu. P. Golovachov, N. V. Leontieva

Abstract: The study of supersonic flows about the bodies immersed in non-uniform streams with coordinate dependent parameters is important for a number of aerodynamic problems. For instance, such flows occur when a body is placed in a supersonic jet, when it moves in an aerodynamic wake or in a nonuniform medium of another type.

This paper presents some results of numerical simulation of such flows. The results correspond to perfect diatomic gas flows about the nose of a spherically blunted body. The gas specific heat ratio c is equal to 1.4. Depending on the problem of inviscid gas, boundary layer, reduced, or full Navier-Stokes equations are used. The validity of the above-referred mathematical models for numerical simulation of the flow considered is investigated. Calculation results demonstrate the oncoming stream nonuniformity effects on flow field structure, drag, and heat transfer parameters on the body surface.

1. Mathematical Models and Numerical Methods

Under the conditions considered supersonic flow past a blunt body is accompanied by formation of a detached shock wave. For nonseparated flows with high Reynolds numbers flow field and wall pressure distribution may be calculated within the framework of an inviscid gas model, described by Euler equations. These calculations were performed with the use of the implicit finite-difference constant direction scheme by Lubimov and Rusanov [1] along with a bow shock fitting procedure. Inviscid gas flows with recirculation zones were calculated by means of Godunov's method [2] and the flux-corrected MacCormack scheme [3]. Both shock fitting and shock capturing procedures were applied in these calculations to treat the bow shock wave. The method from Ref. [1] is of second order accuracy in spatial coordinates and of first order accuracy in time; the method in Ref. [2] is of first order accuracy in all arguments; and the method in Ref. [3] is of second order accuracy for smooth solutions.

Inviscid gas flow calculations provide the outer boundary conditions for the boundary layer problem. Solution for this problem allows us to obtain heat flux and shear stress values at the body surface. A compact higher-order accurate finite-difference scheme [4] was used to solve the boundary layer equations.

In nonstationary problems the validity of the above approach for calculating heat transfer and shear drag is limited by flows with not only high Reynolds numbers but sufficiently small Strouhal numbers as well. During the body flight in a significantly nonuniform medium at Strouhal numbers Sh $\gtrsim$ 1 intensive compression and expansion waves arise and propagate through the shock layer. It does not allow the boundary layer theory to be applied to describe gas flow near the body surface.

Therefore, the viscous shock layer model is applied to investigate nonstationary flows. Within the framework of this model the whole flow area between the bow shock wave and the body surface is described with the viscous gas dynamics equations. The same formulation of the problem is used to calculate stationary flows with recirculation zones and rarefied gas flows.

Within the framework of a viscous shock layer model gas flow is described by full or reduced Navier-Stokes equations. The last set of equations includes all the terms of Euler equations and only the most important terms accounting for molecular momentum and energy transport phenomena [5]. For axisymmetric flows using spherical coordinates (R, θ) and conventional notation the reduced Navier-Stokes equations are written as follows:

$$\frac{\partial \rho}{\partial t} + u\frac{\partial \rho}{\partial R} + \frac{v}{R}\frac{\partial \rho}{\partial \theta} + \frac{\rho}{R}(R\frac{\partial u}{\partial R} + \frac{\partial v}{\partial \theta} + 2u + v\,\mathrm{ctg}\,\theta) = 0$$

$$\rho\left[\frac{\partial u}{\partial t} + u\frac{\partial u}{\partial R} - \frac{v^2}{R} + \frac{v}{R}\frac{\partial u}{\partial \theta}\right] + \frac{\partial P}{\partial R} = 0$$

$$\rho\left[\frac{\partial v}{\partial t} + \frac{v}{R}\frac{\partial v}{\partial \theta} + u\frac{\partial v}{\partial R} + \frac{uv}{R}\right] + \frac{1}{R}\frac{\partial P}{\partial \theta} = \frac{\partial}{\partial R}\left(\mu\frac{\partial v}{\partial R}\right)$$

$$\rho\left[\frac{\partial h}{\partial t} + u\frac{\partial h}{\partial R} + \frac{v}{R}\frac{\partial h}{\partial \theta}\right] = \frac{\partial P}{\partial t} + u\frac{\partial P}{\partial R} + \frac{v}{R}\frac{\partial P}{\partial \theta} + \mu\left(\frac{\partial v}{\partial R}\right)^2 + \frac{\partial}{\partial R}\left(\lambda\frac{\partial T}{\partial R}\right)$$

$$(1)$$

Reduced and full Navier-Stokes equations are integrated with the use of conventional or modified Rankine-Hugoniot relations to treat the bow shock wave. The boundary condition for temperature and no-slip conditions for gas velocity are prescribed at the body surface. Symmetry relations and approximate boundary conditions $\partial^2 f/\partial\theta^2 = 0$, where f is any unknown function, are used at the other two boundaries of the computational domain.

Viscous shock layer equations are integrated by means of the implicit finite-difference constant direction scheme [5]. Stationary solutions are obtained as the result of the time-asymptotic procedure. Iterations are applied to calculate the solution for a new time step. At each iteration the sets of one-dimensional difference equations are successively solved on the normals to the body surface with the use of the "progonka" algorithm. The derivatives on other spatial coordinates are calculated using previous iteration data. This method is of second order accuracy in the spatial coordinates and of first order accuracy in time. For the most part viscous shock layer calculations were performed on moving computational grids. The grid points were clustered in the region of steep gradients.

2. Flow from a Supersonic Source Past Spherical Bluntness

The flow from a supersonic source may be considered as the simplest mathematical model approximately describing the flow field in a supersonic underexpanded jet.

The flow pattern is sketched in Figure 1. Pressure p_*, density ρ_*, and velocity $V_* = \sqrt{\gamma p_* / \rho_*}$ are prescribed on the spherical source with radius r_*. Gas parameters in front of the bow shock wave that has been formed ahead of the body nose are obtained from the analytical solution for equations of stationary supersonic radial inviscid gas flow. The integration of these equations under the above boundary conditions on the source surface yields the following formulas for gas parameters at the point located at distance r from the source center

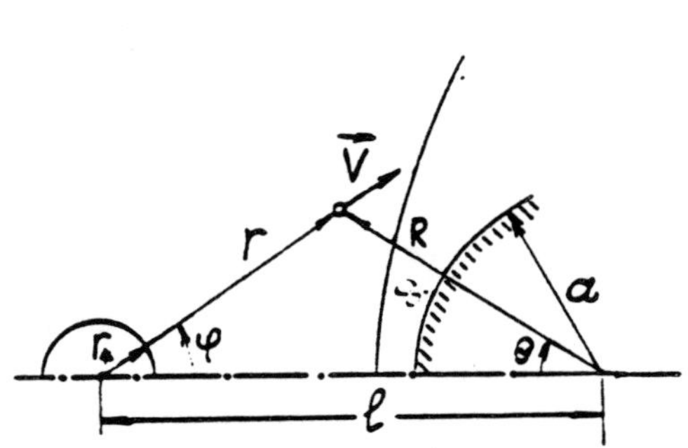

Fig. 1.

$$W = \left(\frac{r_*}{r}\right)^2 \left(\frac{2}{\gamma+1}\right)^{1/(\gamma-1)} \left(1 - \frac{\gamma-1}{\gamma+1}W^2\right)^{-1/(\gamma-1)}$$

$$\frac{P}{P_*} = \left(\frac{r_*}{r}\right)^2 \frac{\gamma + 1}{2W}\left(1 - \frac{\gamma - 1}{\gamma + 1}W^2\right) \qquad (2)$$

$$\frac{\rho}{\rho_*} = \left(\frac{r_*}{r}\right)^2 \frac{1}{W}.$$

Here $W = V/V_*$, V is the absolute value of the base velocity vector. Distance r, and the angle of velocity vector inclination to the symmetry axis, φ are easily expressed through the coordinates (R, θ) by the following formulas

$$r = [(l - R\cos\theta)^2 + R^2\sin^2\theta]^{1/2} \qquad (3)$$

$$\varphi = \operatorname{arctg}\frac{R\sin\theta}{l - R\cos\theta}$$

Pressure p_* and density ρ_* on the source surface and distance l are prescribed so that the required values of Mach number M_0 and Reynolds number Re_0 would be ensured on the symmetry axis ahead of the bow shock wave. In this case the oncoming stream nonuniformity is governed only by the body nose-to-source radii ratio $d = a/r_*$, and $d = 0$ corresponds to the uniform free-stream.

Figures 2 through 4 present some results of viscous shock layer calculations for Mach number $M_0 = 6$ illustrating the free-stream nonuniformity and rarefaction effects. These results refer to the flow about a sphere with wall temperature $T_w = 0.15\ T^*$ where T^* is a stagnation temperature. Gas viscosity is assumed to be $\mu \sim \sqrt{T}$, Prandtl number $Pr = 0.7$. Solid and dashed curves in Figure 2 show the positions of bow shock waves and sonic lines obtained for $Re_0 = 35{,}500$ and different free-stream nonuniformity. Figure 3 shows the longitudinal gas velocity component profiles in a shock layer on the ray $\theta = 30°$. Solid lines 1, 2 show the results of the solution of

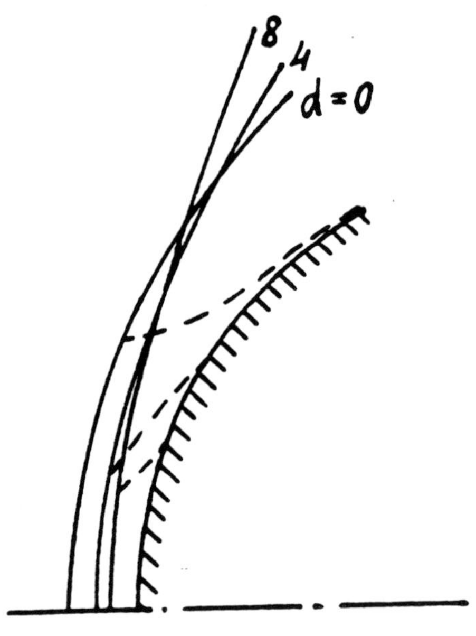

Fig. 2.

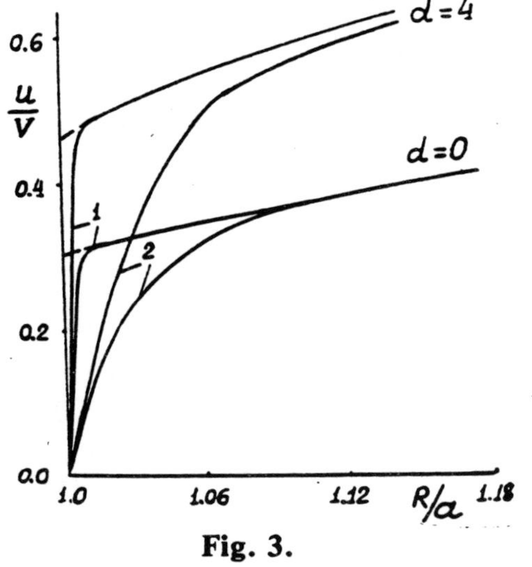

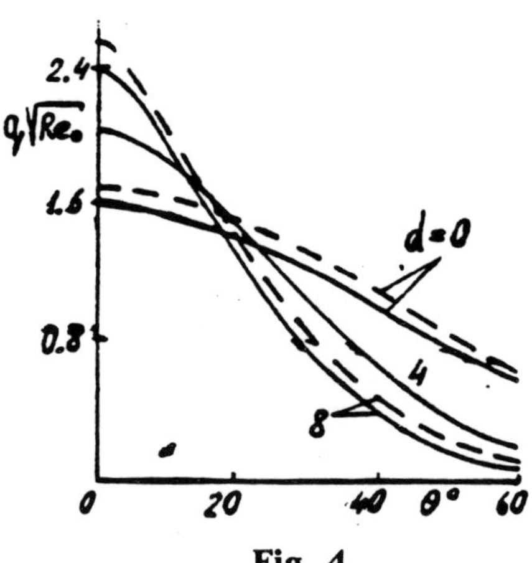

Fig. 3.   Fig. 4.

Navier–Stokes equations at Re_0 = 35000 and 355. Dashed lines depict the results of inviscid flow calculations. Heat transfer parameter variations along the body surface are shown in Figure 4. Here solid and dashed lines correspond to the same Reynolds number values Re_0 = 35,500 and 355, heat flux values are referred to as $\rho_0 V_0^3$. The figures show that oncoming stream non-uniformity significantly effects the flow structure and heat flux on the sphere surface. For the Reynolds number range considered, heat transfer behavior closely follows the boundary layer theory law $q \sim \sqrt{Re}$.

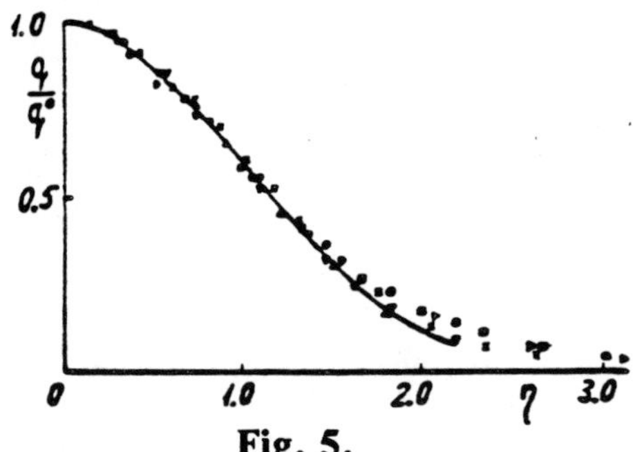

Fig. 5.

Based on the results of inviscid flow and boundary layer calculations, approximate similarity properties have been established for the flow studied. The variation of shock layer thickness, pressure, heat flux, and skin friction coefficient distributions along the body surface become universal when the angular coordinate is scaled by its value at the sonic point on the body surface in an inviscid gas flow θ^s. To illustrate the similarity, heat flux distributions along the sphere surface are plotted in Figure 5 as functions of the coordinate $\eta = \theta/\theta^s$. Here the solid line depicts the result of the boundary layer calculation for uniform oncoming stream at Mach number M_0 = 6 and wall temperature factor $k = T_w/T_* = 0.15$. Other symbols correspond to the following flow parameters:

$\bullet - d = 1, \triangle - d = 2, \triangledown - d = 4, \triangleright - d = 8, M_0 = 6, k = 0.15;$

$\blacksquare - d = 4, M_0 = 3, k = 0.22;$

$x - d = 4, M_0 = 10, k = 0.06.$

Open circles represent the solution for Navier-Stokes equations at $d = 4$, $M_0 = 6$, $Re_0 = 177$, $k = 0.15$. Similar properties to those mentioned above have been used by the author of Ref. [6] to construct relations that may be used to approximately determine shock layer thickness, pressure, heat flux, and shear stress on the body surface provided that only the data on inviscid gas flow in uniform free stream are disposed.

3. Supersonic Wake Flow Past a Spherical Bluntness

Supersonic far wake behind an axisymmetric body at zero angle of attack is the parallel axisymmetric stream with constant static pressure and axal velocity minimum and temperature maximum. The variations of the above functions are described with the following formulas [7]:

$$V(r) = V(\infty)[1 - d \, \exp(-br^2)], \quad T(r) = T(0)\left\{1 + c\left[1 - \frac{V^2(r)}{V^2(0)}\right]\right\}, \quad (4)$$

where r is a distance from the wake axis referred to the body nose radius. Parameters b, c, d are determined by comparison with the results of wake flow calculation or experimental data.

The sphere center is supposed to be located on the wake axis. The main feature of the flow considered is a formation of the recirculation regions in the shock layer. Full Navier-Stokes equations are used here to describe the flow field. Molecular transport coefficients are calculated as in the previous problem.

Figures 6 and 7 present the calculation results for flow past a sphere immersed in a wake behind the spherically blunted cylinder at $M_\infty = 3$, $Re_\infty = 200$, $k = 0.257$, $L = 3.6 \, R_0$. Here L, R_0 are the length and radius of the cylinder, respectively; M_∞ and Re_∞ are Mach and Reynolds numbers of the uniform free-stream. Sphere surface is assumed to be adiabatic. Calculation results published in [8] for the wake flow behind the cylinder were used to determine the parameters of formulas (4). Figure 6 presents the positions of bow shock waves and stream lines in the shock layer for some values of sphere radius a and distance between the bodies x. The secondary vortex is seen in Figure 6c arising when the oncoming stream becomes more nonuni-

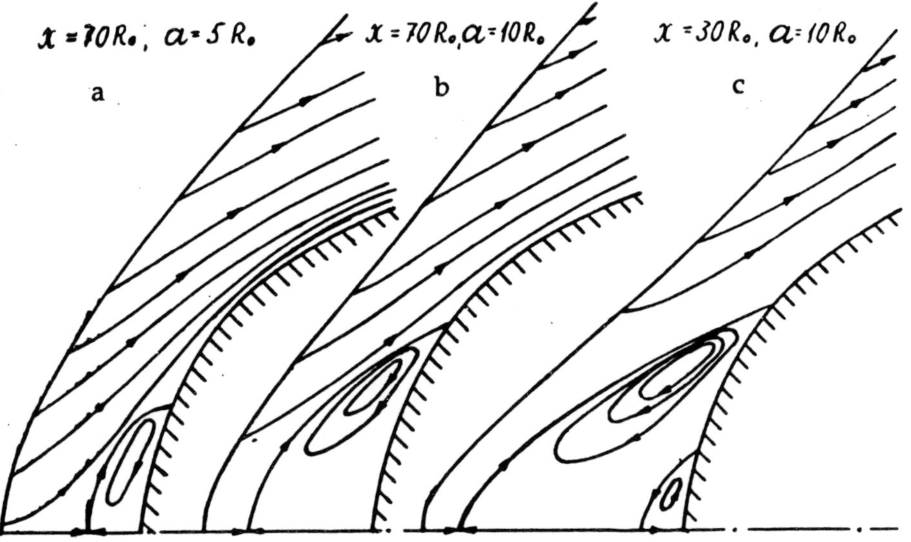

Fig. 6.

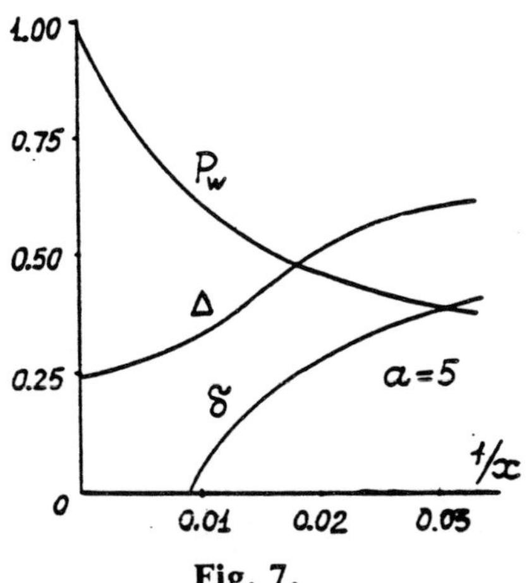

Fig. 7.

form. Gas density (Reynolds number) increase and sphere surface cooling result in the same variation of shock layer flowfield. Figure 7 demonstrates the dependences of some flow characteristics on the distance between the bodies. Here p_w is wall pressure at the stagnation point on the sphere surface, Δ is shock layer thickness and δ is recirculation zone thickness at the symmetry axis. All distances refer to cylinder radius R_0, pressure to $\rho_\infty V_\infty^2$. In this case the recirculation zone arises at $x = 110$, the secondary vortex does not appear.

Figure 8 shows the dependence of heat transfer parameter variation along the sphere surface on Reynolds number. Here $M_0 = 6$, $T_w = 0.15\, T_0^*$, $b = 7.2$, $c = 3.0$, solid lines correspond to $Re_0 = 3550$, dashed lines to $Re_0 = 177$. Reynolds number Re_0 is calculated with the sphere radius and oncoming stream parameters on the wake axis. Its value is changed by means of gas density variation. Essential reduction of the heat transfer parameter in a vicinity of the stagnation point in nonuniform flow when Reynolds number increase is a result of recirculating flow formation.

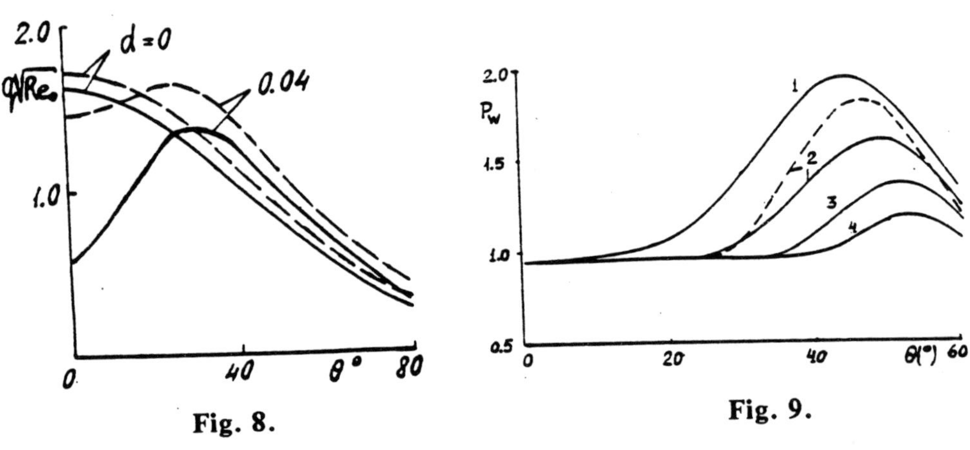

Fig. 8.   **Fig. 9.**

Figure 9 demonstrates Reynolds number and body surface cooling effects on wall pressure distribution. Here $M_0 = 6$, $b = 7.2$, $c = 3.0$, d $= 0.1$; curves 1 through 4 depict the results for Reynolds numbers $Re_0 = 177$, 533, 1770, and 5000. Solid lines correspond to the adiabatic wall case, dashed line to the cold wall case with $T_w = 0.3\ T_0^*$. Pressure values refer to $\rho_0 V_0^2$. Oncoming stream nonuniformity results in formation of the wall pressure maximum behind the reattachment point of recirculating flow. As the Reynolds number increases the wall pressure maximum decreases and moves downstream. Surface cooling results in an increase of pressure maximum and its motion in the reverse direction.

Calculation results for the shape of the bow shock wave and pressure distribution qualitatively agree with wind tunnel experimental data [10] obtained for the flows past a sphere in the wakes of some axisymmetric bodies at $Re \approx 10^6$.

The problem considered here was used as an example to investigate the validity of some mathematical models in calculating the flows with extensive recirculation zones. With this end in view the numerical solutions obtained for full Navier-Stokes equations, reduced equations (1) and Euler equations were compared [11, 12].

It was found that for large enough Reynolds number flows reduced Navier-Stokes equations provided quite accurate results for the shock layer structure, wall pressure variation along the body surface, heat flux and shear stress distributions downstream of the reattachment point of the recirculating flow. Nevertheless, within the recirculation zone for the Reynolds number range considered $Re_0 \leq 10^4$ the results obtained with the use of full and

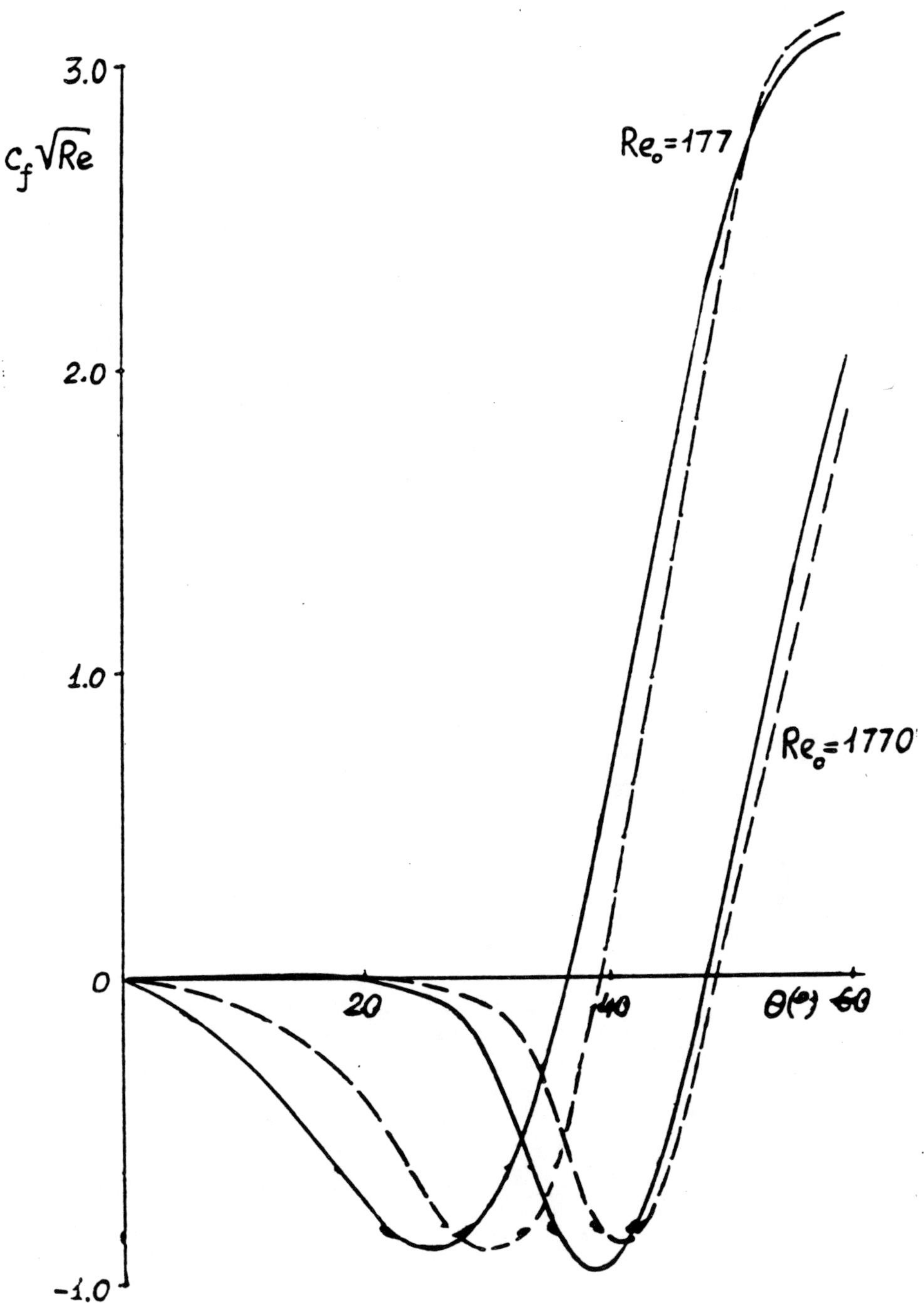

Fig. 10.

reduced Navier–Stokes equations differ significantly. As an example, the skin friction coefficient variation along the sphere surface is presented in Figure 10 for the adiabatic wall case at $M_0 = 6$, $b = 7.2$, $c = 3.0$, $d = 0.1$. Skin friction coefficient C_J is defined as the shear stress to $\rho_0 V_0^2$ ratio. The solutions of full Navier–Stokes equations are depicted by solid lines, the solutions of reduced equations by dashed lines.

Today many authors discuss the very important question of the possibility of numerical simulation of the separated flows on the basis of Euler equations and the adequacy of these results to real gas dynamic phenomena [11, 13–15]. It is generally supposed that high Reynolds number flows are simulated when Euler equations are used. Consider this question using, as an example, wake-type flow past a sphere that is calculated within the framework of inviscid gas model by means of Gudonov's method of first order accuracy [2]. Discretization error of finite-difference methods of this kind looks like numerical viscosity with a coefficient that is proportional to the mesh cell size and local velocity value. Inviscid flow calculations are usually performed using the computational grids that contain no more than some tens of equally spaced nodes in the spatial coordinates. Gas velocity is large enough in the zones where the molecular transport processes are important. So Reynolds numbers Re_a calculated with the use of free stream parameters and local value of the numerical viscosity coefficient lie in the range $10^2 - 10^3$. Numerical solutions of viscous gas equations for these Reynolds numbers demonstrate strong dependence of the flow characteristics on this parameter. It shows an inconsistency of numerical simulation of the flows considered on the basis of Euler equations with the use of first order accuracy methods and the above computational grids even if the difference between the structures of molecular and numerical viscosities is ignored.

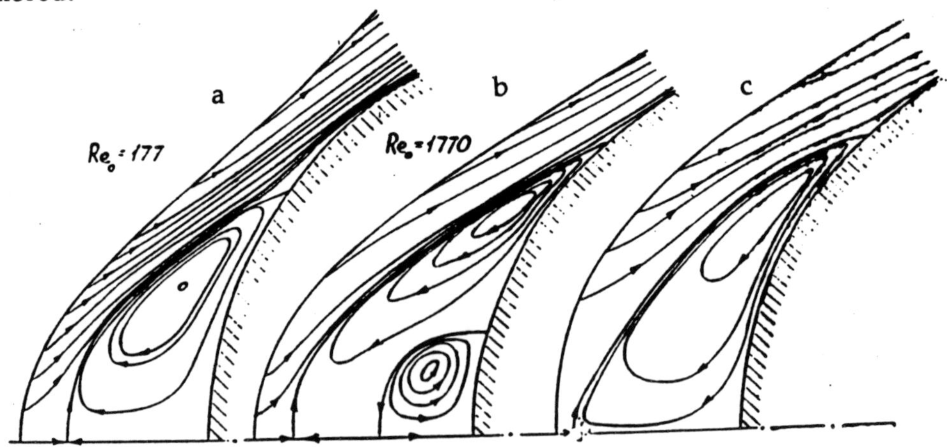

Fig. 11.

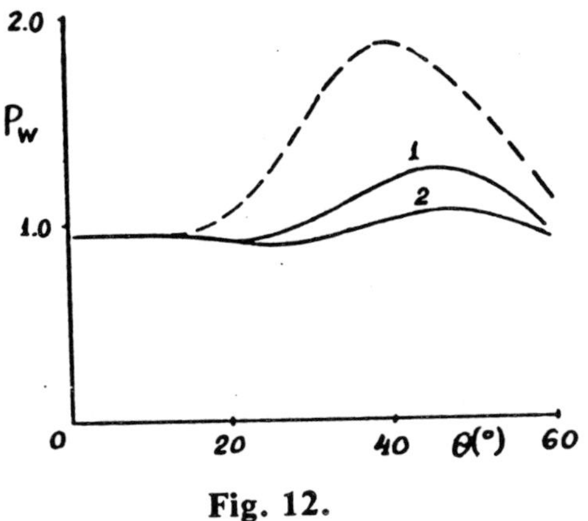

Fig. 12.

This conclusion is confirmed by the results that are presented in the next two figures for $M_0 = 6$, $b = 7.2$, $c = 3.0$, and $d = 0.1$. Figures 11a and b show the positions of bow shock wave and stream lines ahead of a sphere obtained by solving Navier-Stokes equations for the wall adiabatic case at two Reynolds number values. Figure 11c depicts the results of numerical solution of Euler equations. All the above calculations were performed on the same computational grid with mesh spacing $\Delta\theta = 2.5°$ and $\Delta\xi = 0.02$, where ξ is a distance from the sphere surface referred to the shock layer thickness. Wall pressure distributions on the sphere surface obtained within the framework of an inviscid gas model are plotted in Figure 12. Curves 1 and 2 correspond to calculations on the grids with mesh spacing: 1 - $\Delta\theta = 5°$, $\Delta\xi = 0.05$; 2 - $\Delta\theta = 2.5°$, $\Delta\xi = 0.02$. We may see some resemblance of results obtained with the use of viscous and inviscid gas models. Thus, the solutions of Euler equations show the existence of a recirculating flow in the shock layer, results obtained on grids 1 and 2 demonstrate a trend to wall pressure maximum reduction and its removal from the front stagnation point of a sphere when viscosity decreases. However, an essential difference between the solutions of Navier-Stokes and Euler equations is seen which is not only quantitative, but also qualitative. The solutions of Euler equations do not show the secondary vortex. It was noted in [11] as well that consistent numerical solutions of Euler equations were not obtained because the dependence of the results on mesh spacing did not vanish with mesh refinement. The dashed curve in Figure 12 depicts wall pressure distribution calculated on the grid with grid points located on the lines parallel to the symmetry axis. There were about 15 × 15 mesh cells in the shock layer region. A bow shock capturing version of Godunov's method was used. The difference between this pressure distribution and the results obtained with the use of spherical coordinates is mainly explained by numerical viscosity dependence on the mesh cell shape.

Numerical solutions for the problem considered within the framework of an inviscid gas model were also obtained by means of method in Ref. [3] of second order accuracy in the regions of smoothness functions variation. In qualitative aspects these results are the same as the results obtained by Godunov's method. They show only one vortex in the shock layer and are strongly dependent on size and shape of the mesh cells. The resemblance of the numerical simulation results obtained by two methods is easily explained because the use of flux-correction operator in method [3] results in local accuracy reduction to the first order. This reduction occurs in the regions of steep gradients where the numerical dissipation effect on the solution is most important.

4. Supersonic Flow Past a Sphere Moving Through Thermal Nonuniformities

The gas nonuniformities considered here are plane infinite layers with constant pressure and variable temperature. Unsteady axisymmetric flows past a sphere under its flight across a layer with constant speed are investigated. Calculations are performed with the use of reduced Navier–Stokes equations (1). Under the flight conditions considered this mathematical model provides nearly the same result as full Navier–Stokes equations. Gas viscosity is determined according to the Satherland formula for air, Prandtl number Pr = 0.7, sphere surface temperature $T_w = T_0 = 300$ K.

Three variants for temperature variation within the nonuniformity are considered. They are shown in Figure 13 and are expressed by formula

$$T(x) = T_0\{1 - d \exp[-(x/b)^2]\}^n \tag{5}$$

Power $n = \pm 1$ corresponds to variants 1 and 2. In these cases Formula (5) is applied for $x \leqslant 0$. In variant 3 this formula with $n = -1$ is applied for $-\infty < x < \infty$.

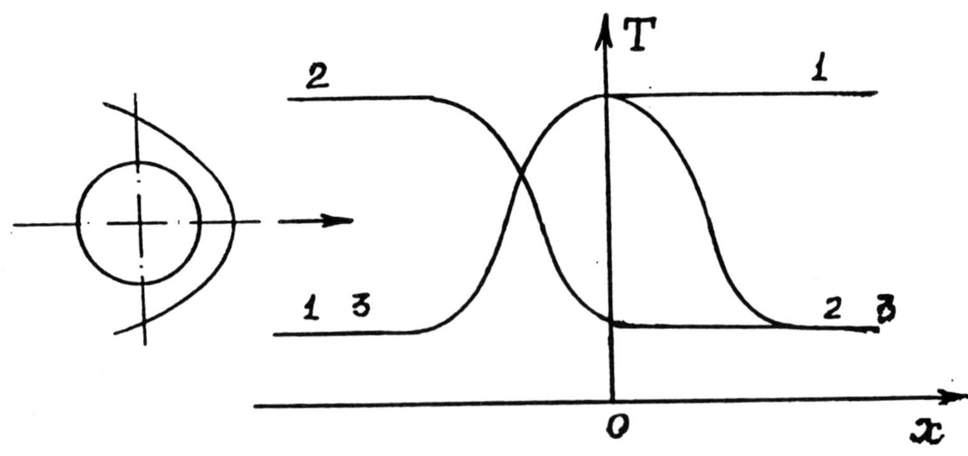

Fig. 13.

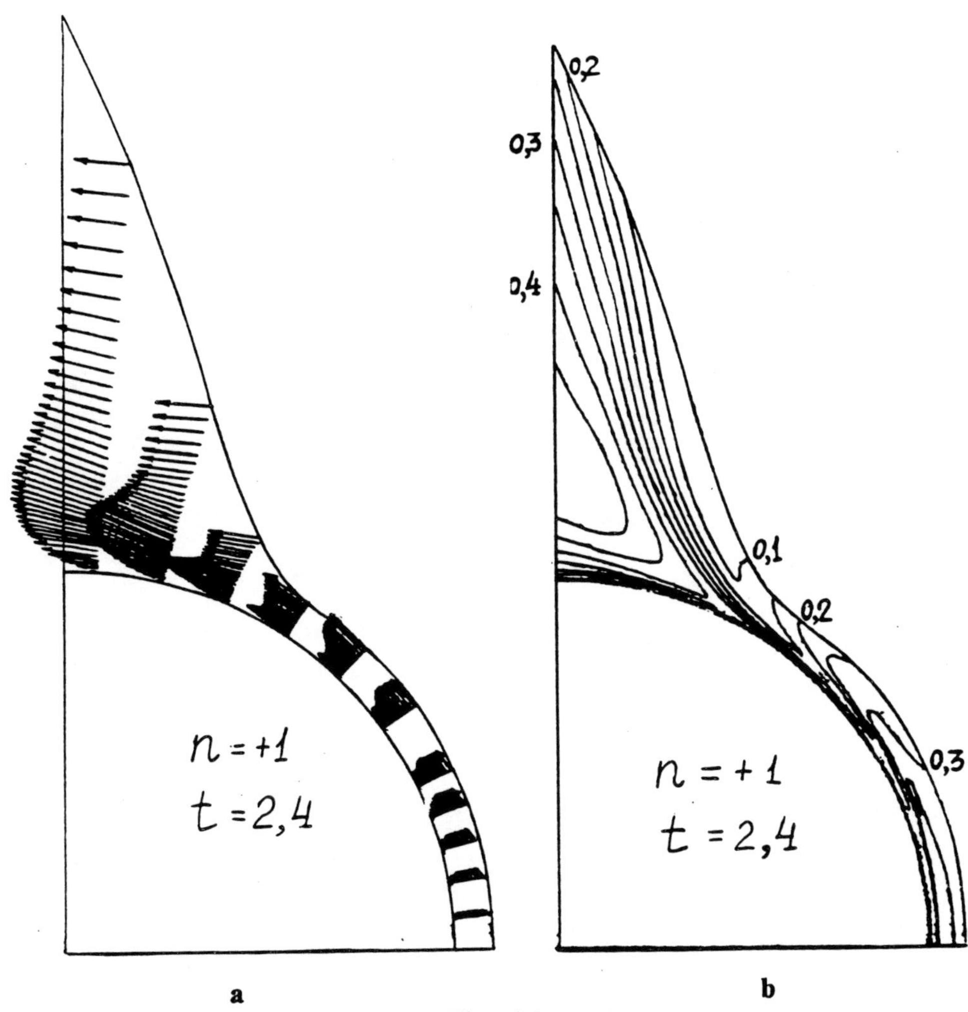

Fig. 14.

Figures 14 through 17 present some calculation results for variants 1 and 2 at $d = 0.9$ and $b = 0.45\,a$, where a is the sphere radius. At these values of parameters b and d gas temperature is changed ten times on the length equal by an order of magnitude to the sphere radius, Strouhal number Sh is approximately equal to unity. A stationary flowfield obtained at $t \to \infty$ for $n = -1$ is used as the initial data for $n = +1$. Speed of sphere motion is assumed to be the same for both variants. Mach and Reynolds numbers calculated with the use of flight speed, sphere radius, and gas parameters in the region of higher density are $M = 6$, $Re = 10^4$.

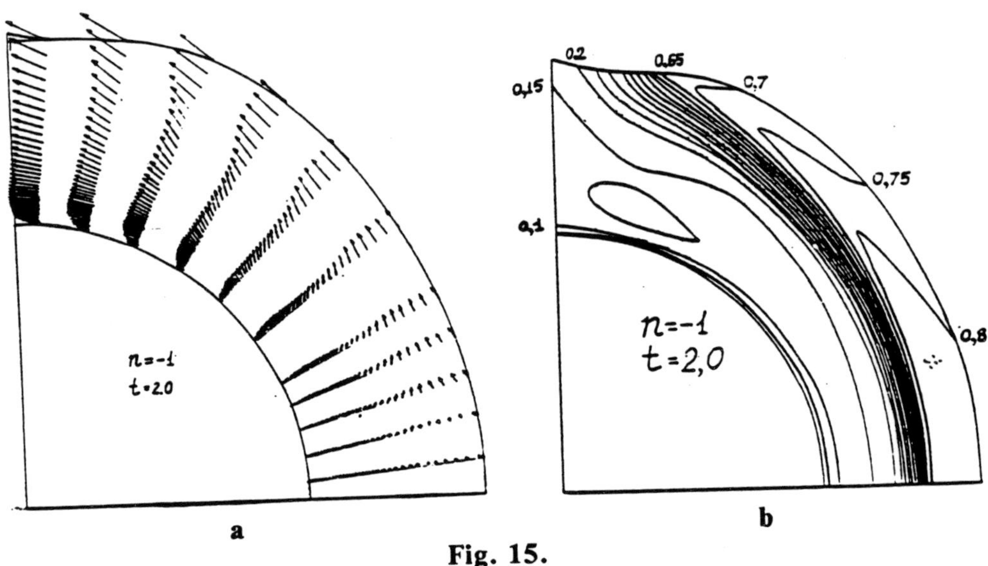

Fig. 15.

Calculation results are presented in nondimensional form. Time t is referred to as a/V, temperature T to $V^2/2R^*$, heat flux q to $\rho_* V^3$, where V is the speed of sphere motion, ρ_* is maximum gas density value at $|x| \to \infty$, R^* is specific gas constant.

Figures 14 and 15 show the variation of the bow shock shape and some details of nonstationary flow structure in the shock layer during the sphere flight through areas with decreasing ($n = +1$) and increasing ($n = -1$) temperature. Velocity maps and isotherms are plotted for the moments when the middle plane of a sphere approaches the plane of extreme temperature gradient. The rise of jet-type flows inside the shock layer that is the result of abrupt diminution of the speed of bow shock wave motion from the body surface in a vicinity of the stagnation line at $n = -1$ and diminution of the shock layer thickness in this region at $n = +1$ are clearly seen. Isotherms are plotted with spacing $\Delta T = 0.05$ ($T_w = 0.04$). Figure 15b shows the formation of significant temperature gradient in the shock layer as the body enters the area with increasing gas temperature. At moment $t = 2$ the region of steep temperature variation is approximately in the middle of the shock layer and moves from the bow shock to the body surface. In the case $n = +1$ non-monotonous temperature variation along the bow shock is seen in Figure 14b that is explained with an increase of bow shock intensity during the body motion in the area with decreasing gas temperature.

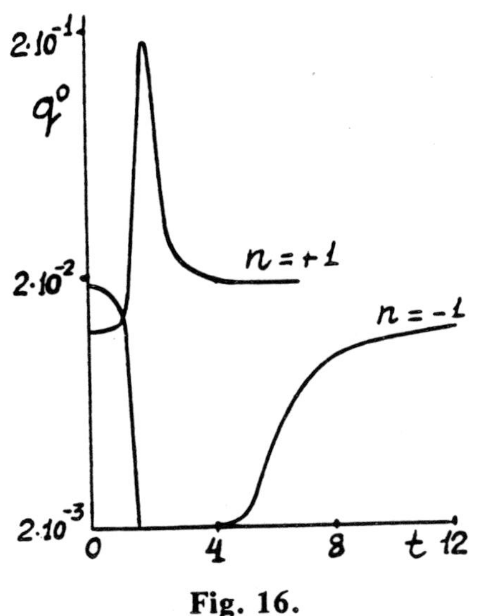

Fig. 16.

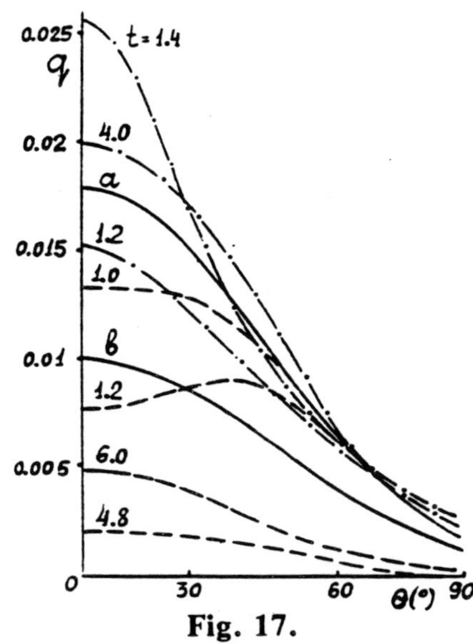

Fig. 17.

Figure 16 shows heat flux versus time at the stagnation point of a sphere. A substantial nonstationarity of heat transfer process is evident. The stationary flow regime is reached through less time as the body enters the area with increasing gas density because of lower values of the shock layer thickness. In this case the heat flux overshoot exceeds the value corespondent to the stationary flow by approximately an order of magnitude.

Heat flux distributions along the sphere surface are plotted in Figure 17. Solid curves depict the results for stationary flows at $x \to \pm\infty$. Curve $a(b)$ corresponds to the initial moment for $n = -1(+1)$ and the final moment of time integrations for $n = +1(-1)$. Heat flux distributions for $n = -1$ and $n = +1$ are shown by the dashed and broken lines. Numbers above the curves denote the nondimensional time. It is seen in this figure that at some time maximum heat flux values occur outside the vicinity of the front stagnation point of the sphere.

It is interesting to study the nonstationary effects dependence on the length scale of nonuniformity. Calculation results obtained for the values of parameter b in the range $0.2 \le b/a \le 4.5$ show that the characteristic time for the flow parameters variation is determined by the ratio b/V for $b \ge a$ and by the ratio a/V for $b \le a$. The first of these ratios characterizes a body stay within the nonuniformity while the second characterizes the time required for the small disturbances to propagate through the shock layer.

Figures 18 and 19 present some calculation results for the third variant of gas temperature variation within the nonuniformity (see Figure 13). Figure 18 demonstrates the Reynolds number effect on nonstationary heat transfer at the stagnation point of a sphere $M = 2$, $b = 0.45\,a$, and $d = 0.5$. Separation of the curves correspondent to different Reynolds number values witnesses for an invalidity of the boundary layer theory.

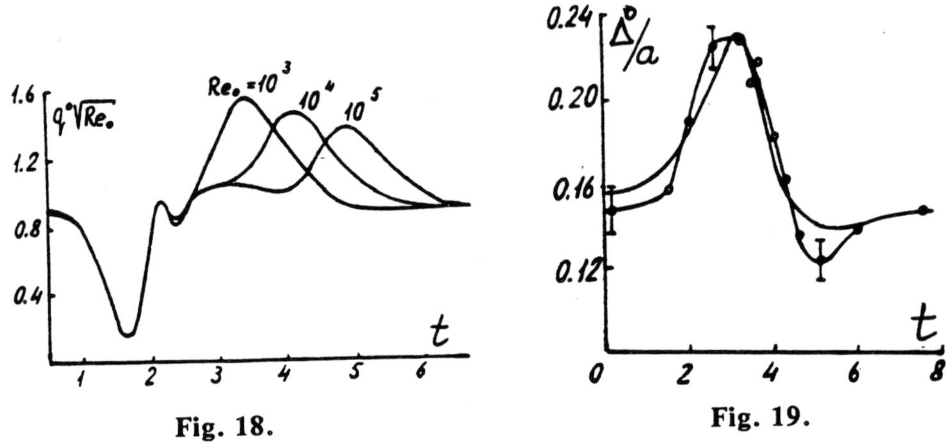

Fig. 18.   Fig. 19.

Figure 19 presents a comparison calculated variation of the bow shock wave detachment on the symmetry axis with experimental data [16]. Here $V = 1.98$ km/sec. $a = 1$ cm; static pressure and gas temperature out of the nonuniformity and maximum temperature in the nonuniformity are: $p_0 = 0.131$ atm, $T_0 = 343$ K, $T_{max} = 700$ K. Length scale of the nonuniformity is equal to sphere diameter by an order of magnitude. The solid curve corresponds to calculation results, the curve with solid circles depicts experimental data.

5. Conclusion

Numerical results demonstrate various oncoming stream nonuniformity effects on supersonic flow past a blunt body; namely variation of shape of the bow shock wave, rise of recirculating, jet-type flows, and intensive compression waves inside the shock layer. These phenomena essentially alter wall pressure, heat flux, and shear stress distributions along the body surface. Calculation results show that a viscous shock layer model must be used for most of the flows considered.

References

1. A. N. Lyubimov and V. V. Rusanov, *Gas Flows Near Blunt Bodies*, vol. 1, Moscow: Izd. Nauka, 1970 (in Russian).
2. S. K. Godunov, A. V. Zabrodn, M. Ya. Ivanov, A. N. Krako, and G. P. Prokopov, *Numerical Solving Multidimensional Gas Dynamics Problems*, Moscow: Izd. Nauka, 1976 (in Russian).
3. A. I. Zhmakin and A. A. Fursenko, *Zh. Vychisl. Mat. i Mat. Fiz.*, vol. 20, no. 4, 1980, pp. 1021-1031 (in Russian).
4. V. V. Shchennikov, *Zh. Vychisl. Mat. i Mat. Fiz.*, vol. 5, no. 1, 1965, pp. 142-144 (in Russian).
5. Yu. P. Golovachev and F. D. Popov, *Zh. Vychisl. Mat. i Mat. Fiz.*, vol. 12, no. 5, 1972, pp. 1292-1303 (in Russian).
6. Yu. P. Golovachev, *Int. J. Heat Mass Tr.*, vol. 28, no. 6, 1985, pp. 1165-1171.
7. T. C. Lin, B. L. Reeves, and D. Siegelman, *AIAA J.*, vol. 15, no. 8, 1977, pp. 1130-1137.
8. N. S. Kokoshnskaya, B. M. Pavlov, and V. M. Paskonov, *Numerical Investigation Supersonic Viscous Flows About Bodies*, Moscow: Izd. Moscow Univ., 1980 (in Russian).
9. Yu. P. Golovachev and N. V. Lenntieva, *Izv. Akad. Nauk SSSR. Mekh. Zhid. i Gaza*, no. 3, 1985, pp. 143-148 (in Russian).
10. V. S. Khlebnikov, *Uch. Zap. TsAGI*, vol. 9, no. 6, 1978, pp. 108-114 (in Russian).
11. Yu. P. Golovachev, N. V. Leontieva, and V. V. Tsimbalov, *Dif. Uravneniya*, vol. 21, no. 7, 1985, pp. 1267-1269 (in Russian).
12. Yu. P. Golovachev, N. V. Leontieva, *Zh. Vychisl. Mat. I Mat. Fiz.*, vol. 27, no. 11, 1987, pp. 1739-1744 (in Russian).
13. A. Kumar and M. D. Salas, *AIAA Paper*, 84-0339.
14. R. W. Newsome, *AIAA J.*, vol. 24, no. 4, 1986, pp. 552-561.
15. S. R. Chakravarthy and D. K. Ota, *AIAA Paper*, 86-0440.
16. V. P. Goloviznin, G. I. Mishin, Yu. L. Serov, and I. P. Yavor, *Zh. Tekh. Fiz.*, vol. 57, no. 7, 1987, pp. 1433-1435.

NONLINEAR ESTIMATION OF AERODYNAMIC CHARACTERISTICS FROM DISCRETE FREE-FLIGHT DATA

N. P. Mende

Abstract: **Plane motion of an axisymmetric object in free flight is considered. The longitudinal center of the mass coordinate in an earthbound coordinate system is used as an independent variable in momentum equations.**

To identify the parameters of aerodynamic coefficient expansion in angle-of-attack and Mach numbers, a method for separate treatment of equations of a system (decomposition) in the course of successive approximations is utilized. Parameter estimation is carried out by means of an interaction process of residual squares sum minimization by use of the Gauss-Newton method.

Attention is paid to the questions of the adequacy of the obtained mathematical model, the importance of the parameters to be estimated and an estimate of confidence intervals of the determined aerodynamic coefficient dependences. The role of the metrological verification of the ballistic facility measuring systems is described.

The results of joint processing of several trajectories of a sharp cone at an angle of attack up to 40° and Mach number equal to 2 are presented.

Introduction

The problem of the determination of aerodynamic forces and moments acting on a body in free flight is the traditional one for investigations carried out using a ballistic facility [1]. These studies span several decades. However, only recently have methods using the theory of nonlinear model construction been applied to this problem [2-4] following their intensive development and use in other fields (see, for example, references in [5]). Unfortunately, mathematical statistics methods are not applied to the estimation of aerodynamic coefficient errors in the cited works.

Mathematical models of object motion in free flight are dynamic models [6]. They represent dynamic Euler equations containing both first and second derivatives of the object coordinates with respect to time. The experiment linear and angular coordinates are measured at fixed times. Hence, the model equations cannot be used immediately to estimate the unknown parameters (aerodynamic coefficients) entering into these equations.

Authors of previously published works have overcome this difficulty in different ways.

The simplest approach consists of substituting finite difference for derivatives [7]. Clearly, this method is invalid if the measurement point spacings are large; its errors are very difficult to estimate even in favorable cases.

In order to diminish the effect of random measurement errors and to ease the requirements for point spacing, different trajectory approximations are used along with subsequent differentiation of analytical dependences [7]. The approximate expression form is chosen to minimize, for example, the Euclidean norm of the residual down to the level of the mean-square error of the measurements. This approach allows us to obtain smoothly varying derivatives, but nevertheless numerical differentiation of the data is still necessary. The problem of estimation of parameter errors is still difficult.

The use of splines to approximate the trajectory has been attempted [8], but in our opinion, it does not advance the case since the requirements and limitations of the previous method remain. The authors of Ref. [8] use spline interpolation to obtain preliminary parameter estimates needed for parametric identification.

The problem is considerably simplified in those rare cases when the model equations can be integrated analytically [9]. Then we can obtain the reduced equations for model [6]:

$$\eta = \mathbf{f}(x, \mathbf{C}),$$

where η is the vector of the observations, $\mathbf{f}$ is the vector of functions, x is the independent variable, $\mathbf{C}$ is the vector of aerodynamic coefficients (vectors are shown in bold). In this case proper arrangement of the experiment allows us to solve the question of model adequacy by relating the measured quantities. The problem of determining the values of the parameters entering into the equations remains. However, such situations are very rare and attempting to approach them by means of a special experimental arrangement can result in a very laborious experiment [9].

Instead of the above-mentioned trajectory approximation prior to its differentiation, we can approximate the final results, integrate the equations, and obtain algebraic expressions relating the results of the measurements to the unknown parameters. This approach has been described in [2] as an example of the method of initial parameter approximations determination in the nonlinear estimation procedure employing the Gauss–Newton method. The expressions approximating aerodynamic coefficients are substituted into the differential equations, then the latter are formally integrated. The expressions obtained contain single and double integrals with variable upper limits to be determined numerically using discrete trajectory data. This method as well as numerical differentiation requires close point placing and generates errors which are impossible to estimate.

Later we will consider the case in which motion equations are to be integrated numerically, thus functions $f(x, C)$ are determined implicitly. Such an approach has been used, for example, in [2], and also in [3, 4]. However, the authors of the latter two works, developing the iteration method of analytical parameter continuation, paid primary attention to consideration of the general motion characteristic (i.e., three-dimensional) with relatively simple aerodynamic coefficient dependences on the motion conditions. As a consequence the problem of the iteration process convergences for "poor" initial approximations, with unknown parameters. At the same time, as experience with trajectory data processing reveals, determination of the appropriate initial parameter approximations is not difficult. The problems of model construction with essentially nonlinear aerodynamic coefficients and estimation of model adequacy and errors of parameters determination appear to be more important.

In carrying out the present work the authors take [2] as the fundamental study where, apparently for the first time, the Gauss-Newton method was applied to estimation of the nonlinear aerodynamic characteristics with the use of the trajectory data on the "short" part of the trajectory.

This approach is based on the independent treatment of the plane oscillation equation. The equation for the center of mass motion turned out to be very efficient and to provide good stability [2]. We extend the method of [2] to the equations for the center of mass plane motion taking into account weak coupling of all three equations except, of course, the "strong" angle of attack influence of aerodynamic coefficients.

In addition, in the present work, in contrast to [2] attention is paid to the error estimation, and in particular, to the importance of parameters to be estimated, since we must constantly solve the latter problem as the model becomes more complicated in the course of the goal function minimization.

1. Equations of Plane Free Motion of Axisymmetric Object in a Gaseous Medium

In what follows we use an informal approach to the solution of the equation system. Thus we are interested in the concrete form of the equations and, in particular, the form of aerodynamic force components and moment representation.

We write the scalar equations of motion [9]:

$$\frac{d^2x}{dt^2} = -(C_x \cdot \cos\theta + C_y \cdot \sin\theta) \cdot \frac{\rho S}{2m} \cdot V^2 \tag{1}$$

$$\frac{d^2y}{dt^2} = -(C_y \cdot \cos\theta + C_x \cdot \sin\theta) \cdot \frac{\rho S}{2m} \cdot V^2 - g, \tag{2}$$

$$\frac{d^2\vartheta}{dt^2} = C_m \cdot \frac{\rho S}{2m} \cdot r^{-2} \cdot l \cdot V^2, \tag{3}$$

$$\vartheta = \alpha + \theta = \alpha + \operatorname{arctg}\,(dy/dx). \tag{4}$$

The coordinate system, acting forces, and cinematic relations are shown in Figure 1; the nomenclature is given at the end of the paper.

Introducing the derivatives with respect to the longitudinal coordinate x we rewrite Equations (1)–(3) as

$$t'' = kt'(C_x + C_y \cdot y') \cdot [1 + (y')^2]^{1/2} \tag{1'}$$

$$y'' = kC_y[1 + (y')^2]^{3/2} - (t') \cdot g \tag{2'}$$

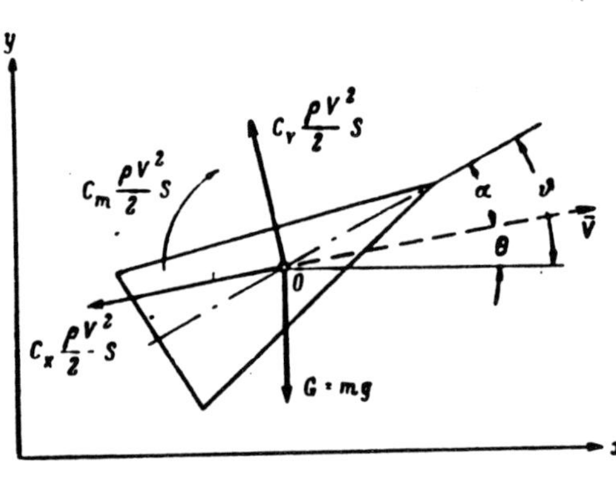

Fig. 1. Coordinate axes, aerodynamic forces and moment.

$$\vartheta'' = k(C_x + C_y \cdot y')[1 + (y')^2]^{1/2} \cdot \vartheta' - kr^{-2}lC_m[1 + (y')^2]. \tag{3'}$$

Here $k = \rho S/2m$; the trait denotes the derivatives with respect to the longitudinal coordinate x.

We note that in the equations no assumptions have been made regarding the dependences of the aerodynamic coefficients C_x, C_y and C_m on the conditions of motion. In the following, however, it is essential that the body moves almost horizontally and the angle $\operatorname{arctg}(y')$ between the velocity vector and x axis is small (hence the term "longitudinal coordinate", i.e., almost along the direction of motion).

Aerodynamic coefficients C_x, C_y, and C_m are assumed to depend on the angle of attack α, angular velocity $\omega = d\vartheta/dx$, and Mach number M. The dependence on the Reynolds number, which is proportionate to the velocity, can be neglected on the short part of the trajectory. We represent the aerodynamic forces and the moment as the sum of static and dynamic components:

$$C_x = C_x(\alpha, M) + C_x^\omega(\alpha, M) \cdot \alpha \frac{l}{V} \cdot \frac{d\vartheta}{dt} + C_x^{\omega^2}(\alpha, M) \cdot (\frac{l}{V} \frac{d\vartheta}{dt})^2 =$$

$$= C_x(\alpha, M) + C_x^\omega(\alpha, M) \cdot [1 + y')^2]^{-1/2} \cdot l \cdot \alpha \cdot \dot{\vartheta}' + \qquad (5)$$

$$+ C_x^{\omega^2}(\alpha, M) \cdot [1 + y')^2]^{-1} \cdot l^2 \cdot (\vartheta')^2,$$

$$C_y = C_y(\alpha, M) + C_y(\alpha, M) \cdot \frac{l}{V} \cdot \frac{d\vartheta'}{dt} =$$

$$= C_y^\alpha(\alpha, M) \cdot \alpha + C_y^\omega(\alpha, M) \cdot [1 + (y')^2]^{-1/2} \cdot l \cdot \vartheta', \qquad (6)$$

$$C_m = C_m(\alpha, M) + C_m(\alpha, M) \cdot \frac{l}{V} \cdot \frac{dv}{dt} =$$

$$= C_m^\alpha(\alpha, M) \cdot \alpha + C_m^\omega(\alpha, M) \cdot [1 + y')^2]^{-1/2} \cdot l \cdot \vartheta', \qquad (7)$$

$$\omega = \frac{\omega \cdot l}{V}.$$

The inclusion of two terms containing the angular velocity and its square into Expression (5) for C_x is a matter of opinion. These terms are introduced for constructing a model which takes into account the delay in the flow pattern reconstruction caused by the change in the angle of attack in real motion (hysteresis phenomenon). Such delay is observed, for example, for bodies with a front separation zone [8]: the local extremum C_x may not correspond to the zero angle of attack and the value of C_x at $\alpha = 0$ may differ from the static coefficient. The displacement of the local extremum along the α axis depending on the sign of the angular velocity is provided by the term containing $\alpha \cdot \vartheta'$ while its magnitude is corrected by the even function $(\vartheta')^2$. Inclusion of the terms mentioned above allows us to verify this assumption.

We do not take into account the damping term proportional to the rate of change for the angle of attack in the expression for the moment coefficient (7), but consider only the damping dependence on angular velocity. The point is that the angles of attack and pitch of motion in the ballistic range differ by a small quantity, hence it is impossible to separate the effects of $d\alpha/dt$ and $d\vartheta/dt = \omega$ due to model degeneration: the corresponding terms in conditional and normal equations will be practically linear dependent. Therefore, coefficient C_m^{ω} in Expression (7) takes into account both rotational derivatives (with respect to the angular velocity and to the rate of change of the angle of attack).

Now we consider the choice of approximative dependences taking moment coefficients as an example. We represent the damping and static coefficients by a truncated multiple Taylor series in the vicinity of zero angle of attack and the Mach number averaged along the trajectory, grouping the terms in the following way

$$C_m^{\omega}(\alpha, M) = (M - M_{\text{av}}) \sum_{i=1}^{k} a_i \cdot \alpha^{2i-2} + \sum_{i=1}^{L} b_i \cdot \alpha^{2i-2}, \tag{8}$$

$$C_m^{\alpha} \cdot \alpha = (M - M_{\text{av}}) \sum_{i=1}^{R} d_i \cdot \alpha^{2i-1} + \sum_{i=1}^{T} e_i \cdot \alpha^{2i-1}. \tag{9}$$

The first items in (8) and (9) take into account the linear dependence of the moment coefficients (expressing the dependence on the angle of attack given by the second terms) on the Mach number. As a rule, experiments in regimes with close velocities are chosen for joint processing. As regards the Mach number variation in each experiment, it is small on the short part of the trajectory. Therefore, even the linear dependence in the Mach number terms in (8) and (9) are often unimportant; introduction of a more complex dependence is unnecessary.

Similar expansions are used for the drag and lift coefficients.

Having more accurately defined the form of the dependences of aerodynamic coefficients of the conditions of motion, we write, as an example, the moment equation in the form of a structural model

$$\vartheta'' + f(\alpha, C_m, C_x, C_y) \cdot \vartheta' + h(\alpha, C_m, C_x, C_r) = 0, \tag{10}$$

and add to it by cinematic relation

$$\vartheta = \alpha + \psi(\alpha, C_{\bar{x}}, C_y). \tag{11}$$

Here C_m, C_x, and C_y are the vectors of expansion parameters for corresponding coefficients; $\psi(\alpha, C_{\bar{x}}, C_y) = \text{arctg}(y')$ is the value determined by the center of mass motion, hence, the dependence on C_m is implicitly expressed here by means of α.

Prior to the discussion of the methods for parameter estimation we will formulate the general approach which takes into account the peculiarities of the interrelation of Equations (1')-(3').

2. Estimation Algorithm for Aerodynamic Coefficients

Analyzing the interrelation of Equations (1′)–(3′) we show that the terms in Equation (3′) caused by the center of mass motion weakly influence the object oscillations as compared with the aerodynamic moment action. Indeed, the quantity y' is of the order 0.01, the factor k, as a rule, does not exceed 0.01 m^{-1}, while the quantity $k \cdot r^{-2} \cdot 1$ is of the order 0.1 m^{-2}.

It is this fact that allowed the authors of [2] to consider the moment equation alone. Equations (1′) and (2′) can be treated independently as the first approximation if prior to it the parameters of Equation (3′) have been estimated. These parameters allow us to describe the variation of the angle of attack along the trajectory, and thus to take account of dependences $C_x(\alpha)$, $C_y(\alpha)$.

The considerations presented permit us to propose the following algorithm for the estimation of aerodynamic coefficients.

1. The approximate averaged values $C_x = $ const, $C_y^\alpha = $ const, and the initial conditions given, we estimate the coefficients of the damping and static moment expansions in arguments α and M. It is sufficient to choose quantities from the range of real possible values of C_x and C_y as their approximate values (such *a priori* information is always available).[1]

The quantity $y'(x)$ in all Equations (1′)–(3′) as well as the quantity $t'(x)$ in Equation (2′) is found from the solution of Equations (1′) and 2′) for given coefficient values.

To choose the initial approximation $Cm = $ const it is sufficient to determine the wavelength of the oscillation using the plot $\vartheta(x)$. The value of ϑ measured at the first recording point of the trajectory is accepted as $\vartheta(0)$; the quantity $\vartheta'(0)$ is found graphically or by means of approximating the derivative by the finite differences if the adjacent measurement point spacing is small enough to correctly determine at least the sign of the derivative. The initial conditions from Equations (1′) and (2′) are found similarly.

The technique of model construction will be discussed at length later.

2. Using the coarse pitch motion model we estimate the first terms in the left approximation for Equation (2′) and use the solutions of Equations (1′) and (3′) as auxiliaries. The angle of attack entering into the lift expression is found by integration of Equation (3′) with the use of the relation $\alpha = \vartheta - \mathrm{arctg}(y')$.

---

[1] Acceptable approximation for the first terms in the C_m expansion in the angle of attack may be found even if we assume $C_x = 0$, $C_y^\alpha = 0$.

3. Using the results of items 1 and 2 we estimate the drag coefficients by Equation (1′); the other two equations are used for calculation of α and y'.

4. The estimates of the lift and drag coefficients being obtained, we return to item 1 and determine the next approximation of the moment coefficient.

Acting in accordance with the stated algorithm, we gradually improve the model until it becomes adequate. this will be discussed in detail later.

The approach described allows us to apply the estimation procedure at each step to only one of the three equations, diminishing the problem dimension by neglecting the derivative of the function in question (t, y, or ϑ) with respect to parameters of other equations. This fact favorably influences both the stability and the convergence rate of the solutions.

3. Goal Function

As has been shown in Section 1, the structural equation of the dynamical model in the question reads as

$$\mathbf{g}(\eta(x), \frac{d\eta}{dx}, \frac{d^2\eta}{dx^2}, \mathbf{C}_m, \mathbf{C}_x, \mathbf{C}_y) = 0, \tag{12}$$

$$\eta(0) = \eta_0, \quad \eta'(0) = \eta'_0, \tag{13}$$

where the following notation is used: $\mathbf{g} = (g_1, g_2, g_3)^T$ is the vector of functions (dynamic Euler equations), T denotes transposition; x is an independent variable (a longitudinal center of mass coordinate in the earthbound coordinate system); $\eta = (t, y, \vartheta)^T$ is a vector of observations (response functions) of the time of motion, the transverse center of mass coordinate and the angle of pitch; $\mathbf{C}_m = (C_1, C_2, ..., C_m)^T$, $\mathbf{C}_x = (C_{x1}, C_{x2}, ..., C_{xn})^T$, $\mathbf{C}_y = (C_{y1}, C_{y2}, ..., C_{yl})^T$ are vectors of unknown parameters of aerodynamic coefficient expansions.

Due to the nonlinearity of Equation (12) it is impossible to obtain a reduced model, i.e., an explicit expression $\eta = \eta\,(x, \mathbf{C}_m, \mathbf{C}_x, \mathbf{C}_y)$.

In accordance with the algorithm described in the preceding section we will treat one response function, namely, that of the angle of pitch, since the parameter estimation starts from the moment equation. The necessary equations for the other two functions (time and the y coordinate) are obtained similarly).

Following [2] parameter estimation is performed by minimization of the goal function, i.e., the sum of squares of the measured response function value deviations from the chosen model (in the sequel "residuals"). The least-mean-square method (LMS) is used without weighting since for the algorithm accepted the goal function contains the measurement results of the quantities for the same dimension with equal accuracy:

$$\Phi(C) = \sum_{i=1}^{N} (\vartheta_{exp}(x_i) - \vartheta_{mod}(x_i))^2 \tag{14}$$

Here $\Phi(C)$ is the goal function; C is the vector of the parameters to be estimated containing (see Section 2) the parameters of a single equation, in this case the aerodynamic moment expansion coefficients; $V_{exp}(x_i)$ are the results of the angle of pitch measurements at points x_i; $V_{mod}(x_i)$ are the angle of pitch values obtained by the accepted model at the same points.

The function $V_{mod}(x)$ cannot be obtained explicitly, but the values $V_{mod}(x_i)$ can be determined by numerical integration of Equation (10) with the use of (11) for given parameter approximations.

In connection with this point we introduce the following vector notation: $\tilde{\vartheta} = (\vartheta_1, \vartheta_2, ..., \vartheta_N)^T$ is the vector of measurements in one experiment; $x = (x_1, x_2, ..., x_N)^T$ is the vector of coordinate x values; $\vartheta(x,, C) = \vartheta_1(x_1, C), \vartheta_2(x_2, C),..., \vartheta_N(x_N, C))^T$ is the vector of the calculated response function values at points x_i for a given approximation C.

Then we can write

$$\tilde{\vartheta} = \vartheta(x, C) + \xi \tag{15}$$

where $\xi = (\xi_1, \xi_2, ..., \xi_N)^T$ is the vector of residuals. For the adequacy model $\xi \sim N (0, \sigma_\vartheta^2, 7)$, where σ_ϑ^2 is the uniform variance of the measurements at points x_i, 7 is the unit matrix.

Now the goal function reads as:

$$\Phi(C) = \xi^T \cdot \xi \tag{16}$$

and the parameter estimates are determined from the condition

$$\hat{c} = argmin(\xi^T \cdot \xi) = argmin[(\tilde{\vartheta} - \vartheta(x, c))^T \cdot (\tilde{\vartheta} - \vartheta(x, c))]$$

We shall place the multiplication sign, namely a point between the brackets containing the factors in contrast to ones expressing the dependence.

We note that the independent variable values are assumed to be known exactly. For relatively slow varying functions ϑ and y this assumption is quite acceptable. As regards the time of flight, it leads to variance of the time measurements and variance of the coordinate x will influence the estimate of the residual variance. We shall return to this question during the discussion of model adequacy and the importance of the parameters.

As in [2], in order to minimize the goal function nonlinear in parameters, we apply the Gauss–Newton method which uses the response function linearization in the form

$$\phi(\mathbf{x},\ \mathbf{C}) = \phi(\mathbf{x},\ \mathbf{C}) + \bar{\bar{P}} \cdot \Delta\mathbf{C} \tag{17}$$

Here $\Delta\mathbf{C} = \Delta C_1,\ \Delta C_2,\ ...,\ \Delta C_m)^T,\ \Delta C_j = C_j - C_{0j},\ j = 1, 2, ..., m,$

$$\bar{\bar{P}} = \left\| \begin{array}{cccc} \left.\dfrac{\partial\vartheta(x_1,\ \mathbf{c})}{\partial C_2}\right|_{C_0} & \left.\dfrac{\partial\vartheta(x_1,\ \mathbf{c})}{\partial C_2}\right|_{C_0}, & \cdots & \left.\dfrac{\partial\vartheta(x_1,\ \mathbf{c})}{\partial C_m}\right|_{C_0} \\ \cdot & & \cdot \\ \left.\dfrac{\partial\vartheta(x_N,\ \mathbf{c})}{\partial C_1}\right|_{C_0} & \left.\dfrac{\partial\vartheta(x_N,\ \mathbf{c})}{\partial C_1}\right|_{C_0}, & \cdots & \left.\dfrac{\partial\vartheta(x_N,\ \mathbf{c})}{\partial C_m}\right|_{C_0} \end{array} \right\| \tag{18}$$

(bold symbols denote a matrix). Then,

$$\Phi(\mathbf{C}) = \tilde{\phi} - \phi(\mathbf{x},\ \mathbf{C}_0) - \mathcal{P} \cdot \Delta\mathbf{C})^T \cdot (\tilde{\phi} - \phi(\mathbf{x} - \mathbf{C}_0) - \mathcal{P} \cdot \Delta\mathbf{C}) \tag{19}$$

Expression (19) is similar to the residual sum of squares for the, linear regression: the matrix of partial derivatives (still unknown) P replaces the matrix of independent variable values (plan matrix) and the residual values $\tilde{\phi} - \phi(\mathbf{x},\ \mathbf{C}_0)$ calculated at the point $\mathbf{C}_0$ replace the measured response values [10].

By Expansion the problem is reduced to a linear one. However, due to the approximate character of Equation (17) the determination of the estimates

$$\hat{\mathbf{C}} = \mathbf{C}_0 + \Delta\mathbf{C} \tag{20}$$

does not allow us to find the minimum (19) in one step as in estimating the parameters of the linear regression. Nevertheless, the fact that the iteration Gauss–Newton algorithm is a sequence of linear regression problems [6] will be used later in the course of the confidence intervals estimation of unknown parameters. However, the unknown matrix $\mathcal{P}$ must be determined first.

4. Sensitivity Equations

The elements of partial derivative matrix (18) can be determined from the sensitivity equations obtained by differentiation of the initial equation of motion (10) with respect to unknown parameters.

Let us introduce additional notation. The columns of matrix P containing the derivatives with respect to the jth unknown parameter at all the points x_i we denote as

$$\mathbf{P}_j(\mathbf{x},\ \mathbf{C}) = (P_j(x_1,\ \mathbf{C}),\ P_j(x_2,\ \mathbf{C}),\ ...,\ P_j(x_N,\ \mathbf{C}))^T \equiv$$

$$\equiv \left[\frac{\partial \vartheta(x_1,\ \mathbf{C})}{\partial C_j},\ \frac{\partial \vartheta(x_2,\ \mathbf{C})}{\partial C_j},\ ...,\ \frac{\partial \vartheta(x_N,\ \mathbf{C})}{\partial C_j}\right]^T \equiv \frac{\partial \phi(\mathbf{x},\ \mathbf{C})}{\partial C_j}. \tag{21}$$

Among the parameters to be estimated are also the initial conditions (13) for which by analogy with (21) we write

$$\mathbf{P}_{\vartheta_0}(\mathbf{x},\ \mathbf{C}) = \frac{\partial \phi(\mathbf{x},\ \mathbf{C})}{\partial \vartheta_0},\ \mathbf{P}_{\phi_0}'(\mathbf{x},\ \mathbf{C}) = \frac{\partial \vartheta(\mathbf{x},\ \mathbf{C})}{\partial \vartheta_0}. \tag{22}$$

In addition, according to the algorithm described in Section 2, we assume

$$\frac{\partial \phi(\mathbf{x},\ \mathbf{C})}{\partial C_{xj}} = \frac{\partial \phi(\mathbf{x},\ \mathbf{C})}{\partial C_{yj}} \equiv 0 \tag{23}$$

and, finally, we accept the assumption

$$\frac{\partial \phi(\mathbf{x},\ \mathbf{C})}{\partial C_j} \cong \frac{\partial \alpha(\mathbf{x},\ \mathbf{C})}{\partial C_j}. \tag{24}$$

Expression (24) is obtained from (4) by neglecting the derivatives with respect to C_j of the quantity y' which is itself small as compared with α (generally y' depends on C_j through the angle of attack, see (11)).

Now differentiating Equation (10) with respect to all the parameters to be estimated and taking into account that

$$\frac{\partial \vartheta}{\partial C_j} = P_j,\ \frac{\partial \vartheta'}{\partial C_j} = P_j',\ \frac{\partial \vartheta''}{\partial C_j} = P_j'' \tag{25}$$

we can write the sensitivity equations

$$P_j'' + f \cdot P_j' + \left[\vartheta \cdot \frac{\partial f}{\partial \alpha} + \frac{\partial h}{\partial \alpha}\right] \cdot P_j = \vartheta' \frac{\partial f}{\partial C_j} + \frac{\partial h}{\partial C_j}, \tag{26}$$

$$j = 1,\ 2,\ ...,\ m;$$

$$P''_{\vartheta_0} + f \cdot P'_{\vartheta_0} + \left[\vartheta' \frac{\partial f}{\partial \alpha} + \frac{\partial h}{\partial \alpha} \right] \cdot P_{\vartheta_0} = 0, \tag{27}$$

$$P_{0'} + f P'_{\vartheta_{0'}} + \left[\vartheta' \frac{\partial f}{\partial \alpha} + \frac{\partial h}{\partial \alpha} \right] \cdot P_{\vartheta_{0'}} = 0, \tag{28}$$

Functions f and h are defined in (10). Initial conditions for the sensitivity equations read as

$$P_j(0) = 0, \quad P'_j(0) = 0, \quad j = 1, 2, \dots, m,$$

$$P_{\vartheta_0}(0) = 1, \quad P'_{\vartheta_0}(0) = 0, \tag{29}$$

$$P\vartheta_0(0) = 0, \quad P'_{\vartheta_0}(0) = 1.$$

The components of the vectors (21) and (22) which are the columns of the matrix P are determined by the solution of the sensitivity equations along with Equation (10).

5. System of Normal Luminescence Equations and Estimate Parameters

The normal equations of the least-mean-square method (LMS) are derived by differentiating the goal function (14) with respect to the unknown parameters and equating the obtained equations to zero (the conditions of the local minimum of the function). The system of the linear algebraic equations has the following form:

$$P^T \cdot P \cdot \Delta C = P^T (\tilde{\phi} - \phi(x, \ C_0)). \tag{30}$$

For the nonsingular matrix P System (30) has a unique solution

$$\Delta C = (P^T \cdot P)^{-1} \cdot P^T \cdot (\tilde{\phi} - \phi(x, \ C_0)), \tag{31}$$

which defines unambiguously the vector of corrections ΔC, the latter providing better correspondence of the model to the observation results as compared with the initial approximation of the unknown parameter C_0.

In order to minimize the goal function the iteration process is constructed as follows:

$$C^{(t+1)} = C^{(t)} + [\mathcal{P}(C^{(t)})^T \cdot \mathcal{P}(C^{(t)})]^{-1} \times$$

$$\times \mathcal{P}(C^{(t)})^T \cdot (\tilde{\phi} - \phi(x, C^{(t)})),$$

(32)

where t is the iteration number, C with superscript in brackets is the next approximation of the unknown parameters vector.

Algorithm (32) has a maximum convergence rate. It usually works well when the nonlinearities of the aerodynamic characteristics are relatively weak and the information about the initial approximation of parameters is reliable. However, the large convergence rate in the case of strong nonlinearity (for example, when static stability of the object is lost) sometimes results in the divergence of the iteration process, since the calculated corrections $\Delta C^{(t)}$ may give the new values $C^{(t+1)}$ which are far from the region where linear approximation (17) is valid. In this case in accordance with the recommendations in [10] we used Hartley algorithm

$$C^{(t+1)} = C^{(t)} + \gamma \cdot \Delta C^{(t)},$$

(33)

where $\Delta C^{(t)}$ is the solution of Equation (30) and γ is a parameter controlling the step length ($0 < \gamma \leqslant I$). The acceptable value of γ is chosen by an operator in the interactive regime by calculation control.

6. Importance Estimation of Aerodynamic Coefficient Expansions Parameters

The initial number of parameters of the aerodynamic coefficient expansions (the components of the vector C) is chosen on the basis of *a priori* information about the object and, perhaps, intuitional considerations. There are two questions to be answered once the estimates of parameters have been obtained.

1) Are all the parameters which have been determined important?

2) Is the accepted model adequate to the measurement results?

Since an adequate model can contain unimportant coefficients, the first question should be answered before the second.

In order to examine the hypothesis concerning the importance of the determined parameters it is necessary to estimate the confidence interval of the parameter in question and to compare it with the parameter value: if the confidence interval includes the zero parameter value, the parameter must be considered unimportant.

The most general method of confidence interval estimation is the Monte Carlo method (its application to the problem in question will be discussed later). However, it requires large execution time and is inconvenient for an operative estimation of the importance in the course of the model construction. To this effect we have used the fact that the Gauss-Newton method is the sequence of the linear regression problems. It follows that the methods of regression parameter estimation can provide acceptable estimates for the coincidence intervals of the unknown parameters during the iteration process as well as, in particular, at its final stages when the coefficient corrections become negligibly small and linear approximation (17) becomes valid in the small solution vicinity considered. However, in contrast to the linear regression in this case parameter corrections rather than the parameters itself are sought. But the error of the negligibly small correction of the parameter is nothing other than the error of the parameter itself.

The considerations presented are by no means a proof of the procedure to be described, but its practical usefulness has been confirmed repeatedly by comparison with the estimates obtained using the Monte Carlo method.

Thus retaining the terminology used for estimation of the variance and importance of the linear regression coefficients, we shall use for determination of the statistical characteristics of the model parameter estimates, the covariance matrix constructed on the basis of the partial derivatives

$$D = (P^T \cdot P)^{-1}. \tag{34}$$

Then the variances of the parameters to be estimated are determined by the diagonal elements of the matrix D and the estimate of the response function variance:

$$S_{C_j}^2 = S_{\Delta}^2 C_j = D_{jj} \cdot S_{\vartheta}^2 = \frac{(\tilde{\phi} - \phi(\mathbf{x},\ \mathbf{C}))^T \cdot (\tilde{\phi} - \phi(\mathbf{x},\ \mathbf{C}))}{N - m} \cdot D_{jj} \tag{35}$$

where S^2 are the estimates of the variances of the quantities marked by a subscript, the angle of pitch variance being estimated by the residual sum of squares; N is the total number of the angle of pitch measurements; m is the number of parameters to be estimated.

Let us consider the estimates of the variances of the moment coefficient dependences. We write the expression for the coefficient of the total aerodynamic model in the vector form using Equations (8) and (9) as

$$C_M = [\varphi(M,\ \alpha) + f(M,\ \alpha)]^T \cdot \mathbf{C}, \tag{36}$$

where

$$\varphi(M, \alpha) = [(M - M_{av}), (M - M_{av}) \cdot \alpha^2,...(M - M_{av})\alpha^R, 1, \alpha^2,...\alpha^T]^T;$$

$$f(M, \alpha) = [(M - M_{av}) \cdot \alpha, (M - M_{av}) \cdot \alpha^3,...(M - M_{av}) \times$$
$$\times \alpha^R, \alpha, \alpha^3,...\alpha^T]^T;$$

$$C = (a_1, a_2, ..., a_k, b_1, b_2, ..., b_L, d_1, d_2, ..., d_R, e_1, e_2, ..., e_T)^T.$$

Then for the known covariance matrix of vector **C** from the linear transform property it follows

$$S^2_{C_M} = [\varphi(M, \alpha) + f(M, \alpha)]^T \cdot D \cdot S^2_{\vartheta} \cdot [\varphi(M, \alpha) + f(M, \alpha)]$$

Separate estimates for the variances of Functions (8) and (9) are given by expressions

$$S^2_{C_M^\omega} = \sum_i \sum_j D_{ij} \cdot \alpha^{2i-2} \cdot \alpha^{2j-2} \cdot S^2_{\vartheta}, \tag{37}$$

$$S^2_{C_M^\alpha} = \sum_k \sum_l D_{kl} \cdot \alpha^{2k-1} \cdot \alpha^{2l-1} \cdot S^2_{\vartheta} \tag{38}$$

where the summation is performed on the superscripts of the covariance matrix elements corresponding to the coefficients of expansions (8) and (9).

The validity for the application of linear estimation methods to our case can be verified by means of the direct variance and covariance calculations on the base sample obtained with the use of statistical simulation. We have performed this verification at the final stage of the construction of the motion model.

Assuming that not only the direct measurement results but the components of the parameter vector in question also have a Gaussian distribution (the results obtained under this assumption can also be verified by statistical simulation), we estimate the confidence intervals for the parameters C_j with the use of the Student criterium.

The quantity

$$t = \frac{\hat{C}_j - \tilde{C}_j}{}$$

(where $\tilde{C}_j$ is the true coefficient value; $\hat{C}_j$ is the estimate obtained and S_{C_j} is the estimate of the least-square deviation (see (35)) is governed by the Student distribution with $(N - m)$ degrees of freedom. For a given probability $P = P\{|t| < t_{crit}\}$ we find, with the use of the Student distribution table, $t_{crit}(P)$. Then with probability P we have

$$|\hat{C}_j - \tilde{C}_j| < S_{C_j} \cdot t_{\text{crit}} = t_{\text{crit}} \sqrt{\overline{D_{jj}}} \cdot S_\vartheta. \tag{39}$$

Expression (39) defines the half-width of the confidence interval for the true value of the coefficient C_j. If the confidence interval contains the zero coefficient value, i.e., the inequality

$$t_{\text{crit}} \sqrt{\overline{D_{jj}}} \cdot S_\vartheta > |\tilde{C}|, \tag{40}$$

is fulfilled, then the zero-hypothesis is accepted $H_0 : \tilde{C}_j = 0$ and the coefficient C_j is recognized as unimportant.

Condition (40) is checked for each parameter during model improvement as we seek the best approximation.

The coefficient which turns out to be unimportant is excluded and then the estimates of the other coefficients and their variances are recalculated.

As is seen from (39) the width of the confidence interval is determined for a given level of importance by the plan of the experiment (on which the elements of matrix D depend) and the magnitude of the residual variances S_ϑ^2. We can diminish the residual variance by choosing the more complex model (up to interpolation). The question of the limitations of this approach arises in respect to this point. This is very important in the case of dynamic models described by differential equations. The model should only be complicated up to the adequacy level.

7. Adequacy of the Mathematical Model

In the cited works [1-4, 7, 9] insufficient attention is paid to the estimation of the model adequacy. In [9] this question does not arise because of *a priori* adequacy. In [7] the adequacy is checked without the use of statistical criteria. In [2] the question was not discussed and in [3, 4] it does not arise, apparently, due to the fact that trajectories are found by solving a direct problem of motion.

Strictly speaking, in the case of nonlinear aerodynamic characteristics, the adequacy of their description cannot be established on the basis of the trajectory data analysis alone without using *a priori* physical information. Indeed, the model being chosen, we can control its fitness by deviation of the calculated trajectory from the measured coordinates only using some criterium. Since the trajectory is obtained by twice-repeated integration, it is evident that a multitude of functions in the right-hand side of the motion equations exists which satisfy the conditions of the chosen criteria. Therefore, we can consider the adequacy of the chosen model in respect to the trajectory data only. The good correspondence between the trajectory obtained by integration and the measured coordinate values only establishes the fact that there is no reason to discard the model on the basis of the available data. The model can be rejected either by physical considerations or by using new data obtained, for example, for different initial conditions. However, no additional data can prove the validity of the model.

The adequacy of the model in this sense is checked by the residuals, i.e., by the deviation of the measurement results from the values calculated by using the model. The measurement errors are to be estimated.

Hypothetical variances of trajectory data can be estimated by summing the variances caused by successive stages of the measurements. To accomplish this we need to know the device errors at the stage of the optical system adjustment and the bonding of their benchmark devices to the unique coordinate system as well as the errors of the coordinate measurements in the photos.

However, the total error depends on a number of factors, some of which are practically impossible to take into account. Among the latter are temperature and season variations, errors caused by light refraction in the steep density gradient regimes in the flow around the object and others; for example, errors during model manufacturing. In addition, flaws in tuning equipment are always possible. In connection with this point it is necessary to estimate real variances in addition to their hypothetical values, or, to put it in another way, to attest the ballistic facility as a whole as a measuring instrument. One of the ways to accomplish this consists of launching heavy spheres. However, the variances of angular coordinates, naturally, cannot be estimated. The method advised in [12] consists of numerous repeated measurements at the same points and cannot be applied to ballistic experiments. It is impossible to reproduce the initial conditions (velocity, angle of attack, and angular velocity) with the required accuracy in a series of experiments. As a consequence, different quantities will be measured at the same points and averaging appears to be useless.

In order to obtain estimates of the angular coordinate variances we must choose a well-studied object with a known characteristic moment.

In experiments with a well-studied object as well as in the case of launching spheres into a vacuum, the adequacy question is solved *a priori*: in both cases the form of the mathematical model is known with an accuracy of constant parameters and the number of parameters is also known. In this case the residuals are the random errors and a distinct trend of deviation from the random character unambiguously shows the presence of a systematic error.

In practice we are forced to use both methods, i.e., launching spheres and, for example, sharp cones oscillating around a transverse axis.

The first and second cases provide reliable control of linear and angular coordinate errors, respectively.

We note, however, that although the moment characteristics of a cone have been thoroughly studied, they can differ from the assumed one in real motion due to, say, uncontrollable defects of the real cone. In this connection it is reasonable to bear in mind the hypothetical coordinate variances in the course of measuring system verification. It was our experience that these hypothetical variances allowed us to detect the presence of a small nonzero balancing angle of attack in the series of experiments carried out.

In accordance with these considerations, we act in the following way.

1. The hypothetical variances of the linear coordinates and time are calculated with the use of error estimates for separate stages of the tuning of the equipment and the measurements.

2. The trajectory data for spheres are processed and the estimates of item 1 are corrected, the longitudinal coordinate x being assumed to be known exactly. The last assumption results in a corresponding increase of the time variance, but, practically, it has an effect on the variance of the slow varying transversal coordinate of the center of mass.

As a result the estimates of the variances of both the transverse coordinate and the flight time are obtained with the number of degrees of freedom equal to $N - m$; where N is the total number of the coordinate measurements in all launches and m is the number of the parameters to be determined. In particular, for launching the spheres into vacuum, m is twice the number of experiments to be processed since only two initial conditions are unknown in each launch.

When the dependences $y(x)$ and $t(x)$ are approximated in the case of launching of spheres, the hypothesis concerning the equality of the residual variances of y and t to their hypothetical values has been checked. The zero hypothesis $H_0 : S_y^2 = \sigma_y^2$ and $S_t^2 = \sigma_{t\Sigma}^2$ can be verified with the use of the criterium χ^2:

$$\chi_y^2 = (N - m)\frac{S_y^2}{\sigma_y^2} \text{ and } \chi_t^2 = (N - m)\frac{S_t^2}{\sigma_{t\Sigma}^2} \tag{41}$$

with an alternative hypothesis $H_1 : S_y^2 > \sigma_y^2$ and $S_t^2 > \sigma_{t\Sigma}^2$. Here σ_y^2 and $\sigma_{t\Sigma}^2$ are the hypothetical variances; S_y^2, S_t^2 are the variance estimates obtained by the residual sum of squares. The value of $\sigma_{t\Sigma}^2$ in (41) is calculated under the assumption that it includes the effect of the error of the coordinate x (the latter has been assumed to be known exactly when estimating S_t^2):

$$\sigma^2_{t\Sigma} = \sigma^2_t + \frac{1}{V^2_{av}} \cdot \sigma^2_x \qquad (42)$$

Here V_{av} is the average value of the object velocity on the part of the trajectory in question. From now on the subscript Σ will be omitted and the sum (42) will be denoted by σ^2_t.

If there is no reason to reject the zero hypothesis, we can accept the residual variances S^2_y and S^2_t with the number of degrees of freedom equal to $N - m$ as the estimates of the variances of y and t. If the zero hypothesis is rejected, we should analyze the deviations which constitute the residual sum of squares: if the deviations are of a random nature, the same quantities S^2_y and S^2_t are to be taken as the estimates of the corresponding variances. If systematic errors are present, the excessively large variances may be caused either by errors during tuning or by inadequacy of the mathematical model used in the course of the measuring system verification.

3. At the final stage of the measuring system verification several trajectories of the oscillating cones have been processed. The amplitudes of the angle of attack oscillations were chosen in order, on the one hand, to average the variances for the different deviations of the flow from the symmetric one (the optical distortions due to the refraction will manifest themselves in different ways) and, on the other hand, to remain within the range of the almost linear variation of the static moment; i.e., to provide *a priori* the adequacy of the quasilinear model. The range has been chosen on the basis of the known data for the aerodynamic moment of the sharp cone [14].

The adequacy of the aerodynamic moment to the trajectory data is judged, as in the preceding item, by the criterion χ^2:

$$\chi^2_{\vartheta} = (N - m) \cdot \frac{S^2_{\vartheta}}{\sigma^2_{\vartheta}},$$

where m is the number of parameters to be estimated, the remaining notation is evident. In experiments with a sharp cone with the cone half-angle equal to 15° at Mach number of ~2 and angles of attack up to 28°, the number of parameters to be estimated is 6 in one launch: two initial conditions; the constant component of the static moment caused by the uncontrollable asymmetry of the object; the coefficients of linear and cubic terms of the static moment and the constant coefficient of the damping moment. The number m increases by 3 when the number of launches to be processed simultaneously is increased by one. Naturally, when introducing the nonlinear term into the expression for the static moment, we should verify its importance, see Equation (40).

The acceptability of the cone lift approximation which is linear in the angle of attack can be judged by criterion F (Fisher's distribution) with the use of experiments with the spheres:

$$F_y = \frac{S^2_{ycone}}{S^2_{ysphere}}, \tag{43}$$

where the estimates of the variances with the corresponding numbers of degrees of freedom are in both the numerator and the denominator.

In the case of the consistent estimates S^2_y for the spheres and the hypothetical variance σ^2_y verification of the adequacy by criterion (43) must yield the same result as application of criterion χ^2 (41).

The cone drag coefficient within the chosen range of the angle of attack varies more than twice in an essentially nonlinear way. In connection with this point the cone dependence $t(x)$ is not appropriate for measurement system verification but is used to estimate the quality of the identification of the nonlinear drag coefficient by means of comparison with known data. The form of the dependence approximating C_x is chosen by parallel checking of the model adequacy in accord with the criterium $F_t = S^2_{t,cone}/S^2_{t,sphere}$ and of the importance of the newly-introduced parameters in accordance with the criterion t (40) with the use of the covariance matrix of the longitudinal motion and an estimate with the corresponding number of degrees of freedom.

The following situation may be encountered: the model is inadequate, but the parameter introduced in order to attain adequacy turns out to be unimportant (as a result the confidence interval of the precedent parameter which was undoubtedly important may also increase). In this case we should try another way "to improve" the model. When the power approximation (8), (9) is used, success can be achieved by omitting the highest power of the angle of attack and including the term of the next power. It is this approach that has been used in the course of the processing of the fourth cone launched with an oscillation amplitude up to 40°. This method clearly demonstrates the efficiency of the joint exploitation of criteria t and F.

As a result of the measurement system verification by means of trajectory data processing with the known motion model we obtain the estimates of the coordinate measurement variances characterized by a certain number of degrees of freedom.

With this data we can begin the study of new objects estimating the adequacy by the criterion F with simultaneous control of the importance of the estimated parameters by the criterion *t*.

Let us consider once more the principal questions. All the methods described allow us at best to conclude that there are no reasons to reject the chosen model. We must keep in mind that the chosen model is not the only one acceptable for the available data. Furthermore, the confidence intervals for the aerodynamic characteristics determined with the use of different "equally" adequate models may not cover each other entirely. Hence an important conclusion follows: the determined confidence intervals include the true values of the aerodynamic characteristics only when the chosen model is close to the trajectory data and adequately reflects the dependence of aerodynamic coefficients on the motion conditions.

It is impossible to prove adequacy in the latter respect on the basis of an analysis of the residuals, i.e., deviations of the measurements from the approximate trajectory. External (with respect to the trajectory data) information is needed to draw a positive conclusion.

The categorical character of the last statement does not exclude the possibility of successful estimation of the nonlinear aerodynamic characteristics but only warns against ungrounded conclusions. These considerations which are trivial for specialists in mathematical statistics are suggested here by the experimenter for experimenters.

The sequence of verification of the measurement system of the ballistic facility and of the subsequent construction of the mathematical model for motion of the body in question described in the present section, in practice encounters difficulties which sometimes cannot be overcome on the basis of the above-presented considerations alone. Thus, for example, the variance estimates of the transverse center of mass coordinate of the flying object and its flight time determined with the use of the "reference" cone trajectory data turn out to be almost an order of magnitude larger than the corresponding values for the spheres (the mean-square errors, hence, differ by a factor of three). The reasons of such essential deviation, naturally, are to be sought in the distinctions of the real processes from their idealization used in the analysis. Let us list some of them.

First we must take into account the flaws of the object projection photography in the parallel light beams, for example, in the installation in [15]. This drawback manifests itself in the essential refraction effect which leads to object shape distortion. As a consequence it is hard to estimate the errors in the coordinate measurements. In contrast to the sphere characterized by practically the same shape distortions in all photos (the velocity varies insignificantly, in the case of elongated oscillating bodies the distortions depend on the value of the angle of attack, on the density gradients in the vicinities of the characteristic point of the image (the base cross-section, the nose of the body).

Another origin of errors resulting in the growth of the coordinate variance estimates using the residual sums of squares of deviations from the model trajectory for bodies of complex shape (in contrast to spheres) may be the fault of the manufacturing of the object models, i.e., deviations in its shape and mass distribution. These errors lead to differences in aerodynamic characteristics. Hence, simultaneous processing of the data from several experiments allows us to construct a mathematical model which is satisfactory on average but differs from the real one for each individual case.

Thus, it is necessary to estimate the mean-square errors of the coordinate measurements determined with the use of "reference" objects similar in shape to the objects under study (to take into account the effect of refraction in analogous conditions) and also to apply similar technologies in manufacturing both the "reference" and the tested objects. The presence of similar features of the shapes for the "reference" and tested objects is also useful regarding the dynamical smearing effect of the image due to the finite exposure time during the object photography.

The ideal sphere shape and its density uniformity practically exclude the effect of both manufacturing flaws and refraction errors on the accuracy of the determination of its center of mass coordinates. Hence, the variance estimates based on the analysis of the sphere trajectories will always be somewhat lower than the values for more complex objects.

Finally, the error growth in the course of the facility use may result from violation of the tuning caused by the shock loadings during launches; misadjustment of the optical systems may increase in time. This fault may be detected by control experiments using spheres which should, therefore, be repeated periodically.

8. Statistical Simulation

Estimates in the preceding sections are based essentially on the application of linear regression analysis methods. However, as has already been stressed, the problem acquires only formal resemblance to the estimation of the linear regression parameters after the linearization of the expression in the residual sum of squares (14) with the use of the linear segment of the Taylor series.

In connection with this point it is necessary to control the applicability of the above-described methods. This can be accomplished by means of the Monte Carlo method, parameter estimates, and measurement variances obtained with the use of the residual sums of squares.

Assuming the estimates obtained to be the true values of the parameters and variances, we can calculate the "exact" trajectories and use them for multiple simulation of the experiment with the use of a computer.

In each new trial we use a new sampling of the normally distributed pseudorandom numbers with the zero average and variance equal to its estimate obtained during real experiment processing. These pseudorandom numbers imitate the errors of the coordinate measurements.

In each trial the estimation of the parameters is performed anew, but the form of the mathematical model remains unchanged. As a result of multiple estimation using model trajectories with noise we obtain a sampling of the estimated parameters. We can process it statistically, viz, to determine the average parameter values, their variances and covariances, and the confidence intervals using Student's distribution.

There are two ways to calculate the confidence intervals for the aerodynamic coefficient dependences on, for example, the angle of attack. The first consists of calculating the covariance matrix for the obtained sampling of the coefficients and in subsequent use of expressions like (37) and (38).

We have used another method. We construct the values of the aerodynamic coefficient functions with a given increment of the angle of attack for each trial. As a result we obtain a sampling of the aerodynamic coefficient values for each argument value, its size being equal to the number of trials. Each sampling has been processed statistically, i.e., we have determined the average, the variance estimate, and the confidence interval.

The results obtained are, as a rule, in good agreement with the data determined by the use of the elements of matrix (34).

However, since such a verification procedure is not proof of the validity of the estimates (35)-(40), the program includes both procedures: estimates (35)-(40) and estimates using the Monte Carlo method.

9. Some Results of Verification of the Measuring System and Estimation of the Aerodynamic Characteristics

The methods described in the preceding sections have been applied to processing of the trajectory data obtained at a ballistic facility [15].

The technique for adjustment of the photographic equipment coordinate system to that of the laboratory (the author did not participate in this part of the research) consisted of aligning the horizontal benchmark lines with the base axis of the facility (coordinate y) and of measuring the distances between the vertical benchmark lines with the use of a special tape-measure placed in the optimal trajectory plane (coordinate x). The base axis was formed by a laser beam with the use of a collimator and changeable diaphragms. A tape-measure made of invar spanned the whole length of the facility. The tape had holes at accurately measured distances (the accuracy was 0.05 mm for the entire tape length of 15 m). The holes in the tape were photographed at all recording points which allowed determination of the distance between the benchmark lines in photo cassettes. When the accuracy of the optical system tuning forming light beams during the photography of the object are taken into account, errors of the benchmark lines adjustment to the unique coordinate system are estimated to be $\sigma_x = 0.1$ mm and $\sigma_y = 0.2$ mm. The deviations during repeated measurements are also within this range. Six launches of spheres at a velocity of about 450 m/sec gave the following average results: $\sigma_x = 0.1$ mm, $\sigma_y = 0.13$ mm. These values have been accepted as the hypothetical standard deviations of the linear coordinates. The linear coordinate errors allow us to estimate the error of the angle measurement for a given distance between the characteristic points of the image. For an object length equal to 50 mm the hypothetical error of the measurement of its angular position is estimated to be about 0.1°.

With these data on hand we began to process the trajectory data for a sharp cone flying at a velocity of about 680 m/sec (M = 2). The hypothetical error of the time measurement at this velocity and $\sigma_x = 0.1$ mm was expected to be approximately 0.15 μsec.

Now we consider the results of the cone trajectory data processing for the purpose of making the initial data measurements more accurate.

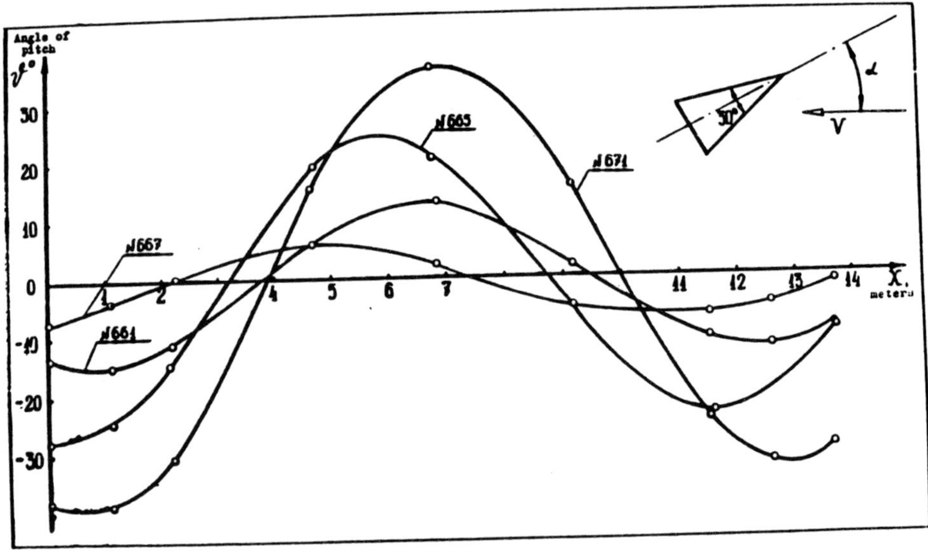

Fig. 2. Angle of pitch variation along the trajectory for four experiments.

The trajectory data for three launches of the sharp cone with a cone half-angle $\theta_c = 15°$ and a base diameter of 30 mm have been processed. The mass of the cone was 45 g, the square of the transverse radius of inertia was approximately 185 mm^2. Experiments were carried out in air under normal conditions. The results of the angle of pitch measurements in these experiments are presented in Figure 2 (661, 665, 667). The maximum value of the angle of pitch was 28° (in experiment 665). In accordance with the data of [14] the aerodynamic moment in this range can be assumed to be linear.

All three experiments were processed together: the mathematical model was chosen to minimize the residuals for three trajectories simultaneously. Analysis of the residuals for launch 667 with the minimum oscillation amplitude detected the presence of systematic errors: the residuals were mainly of the same sign. A similar result was obtained in the course of processing single launch 667 with a linear model of the aerodynamic moment. The mean-square residual is five times the hypothetical value. This fact suggests the presence of asymmetry in cone shape or mass distribution, i.e., the presence of the aerodynamic moment at zero angle of attack. In connection with this point an additional term (a constant, naturally, different in each of the experiments) is introduced into the expression for the aerodynamic moment. An increase of the number of unknown parameters decreased the estimate of the pitch angle mean-square error down to 0.3° with the number of degrees of freedom equal to 14. The mathematical model of the pitch motion in addition to the initial conditions contained three parameters, viz, a constant coefficient of damping moment, the coefficients of the linear, and cubic terms of the static moment. The attempt to "improve" the model by introducing terms with higher powers of the angle of attack was unsuccessful, since the coefficients turned out to be unimportant. In spite of the fact that the observed critical value of the distribution χ^2 essentially exceeded (six times as large) the table value at the 5% level of importance, we were forced to accept the estimate for the pitch angle equal to 0.3° with 14 degrees of freedom in this series of experiments. The reason for the estimate based on the residual sum of squares being essentially larger than the hypothetical value is, apparently, the deviation of the parameters of three cones from the nominal values. As a result we did not succeed in describing three trajectories by a common model.

The situation regarding the model for the center of the mass of motion is similar. The mean-square deviation of the transverse coordinate was 0.26 mm with 15 degrees of freedom while that of the time was 0.6 μsec with 18 degrees of freedom (at a velocity of approximately 700 m/sec this value corresponds to a longitudinal coordinate error of about 0.4 mm).

These estimates being obtained, we added one more experiment (671 in Figure 2) with the oscillation amplitude almost 40° to the series of experiments for simultaneous processing. Adding one term in the expressions for the static moment and the aerodynamic force components, we obtained a satisfactory description of four trajectories using the common mathematical model. Since the observed critical value of the Fisher distribution did not exceed 2 for all three variances (the angle of pitch, coordinate y, and time), there was no reason to reject the hypothesis suggesting the equality of the variances obtained during processing of three and four experiments (the maximum number of degrees of freedom was 23 in the case of four experiments).

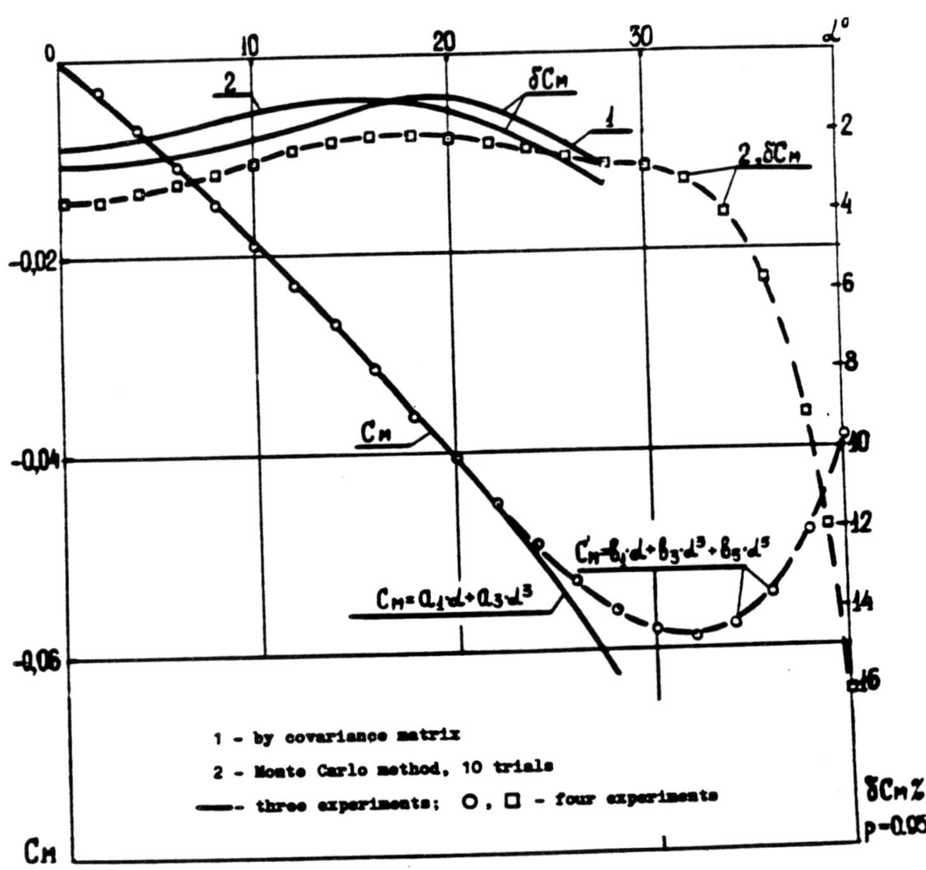

Fig. 3. Static aerodynamic moment for sharp cones.

Let us discuss the results obtained. The dependences of the aerodynamic moment coefficients determined with a series of three and four experiments are shown in Figure 3. A weak nonlinearity is observed in the result for processing of three experiments ($\alpha < 28°$). The relative error for the dependence $C_M(\alpha)$ slowly varies and does not exceed 3%. The "tilt" of the moment dependence manifests itself in the fourth experiment only at $\alpha > 30°$. However, the "tail" of this dependence is determined with far less accuracy (up to 16%). The estimates of variances obtained by means of covariance matrices and the Monte Carlo method are in close agreement.

Processing of three experiments gives a value of the damping moment coefficient equal to -0.2 within the entire range $\alpha < 28°$ (Figure 4). The nonlinear term is unimportant in this range. The 95% confidence interval for the damping moment coefficient equal to -0.2 is shown in Figure 4 by the shaded region (with the slope of dashes to the left).

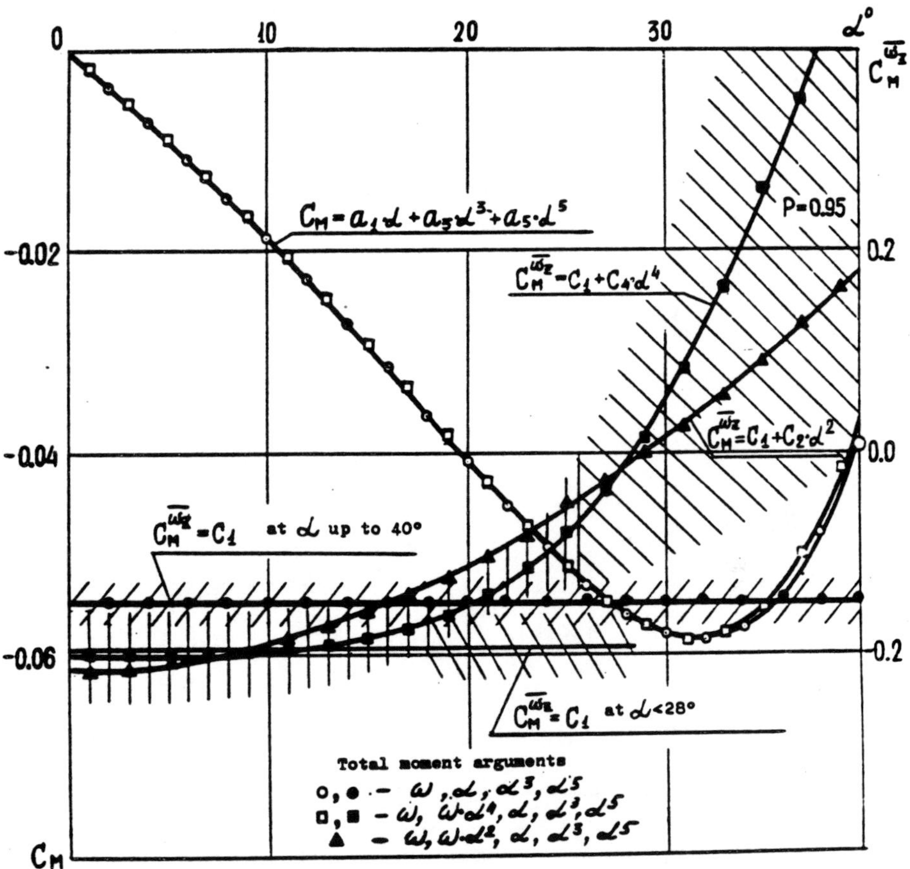

Fig. 4. Damping and static aerodynamic moments: several models.

In the course of processing of four experiments three variants of the damping moment model were estimated. A constant coefficient alone provides the adequacy. However, its value was 10% lower than in three experiments for $\alpha < 28°$. This difference suggests that damping becomes weaker with an increase in the angle of attack and the "effective" value of the constant coefficient decreases. Attempts to successively introduce $\omega\alpha^2$ and $\omega\alpha^4$ as the arguments of the damping moment coefficient result in important coefficients in both cases. However, in the first case the damping coefficient at $\alpha = 0$ was less than 0.2 while the second one practically coincided with the value for the three experiments. The results for processing of three experiments can be considered as *a priori* information when adding the fourth one and, therefore, we prefer the damping dependence at $\omega\alpha^4$. However, it is necessary to admit that the difference between all variants in the vicinity of $\alpha = 0$ is too small for the consideration presented to have other meaning except the methodical one. On the other hand, the following fact is

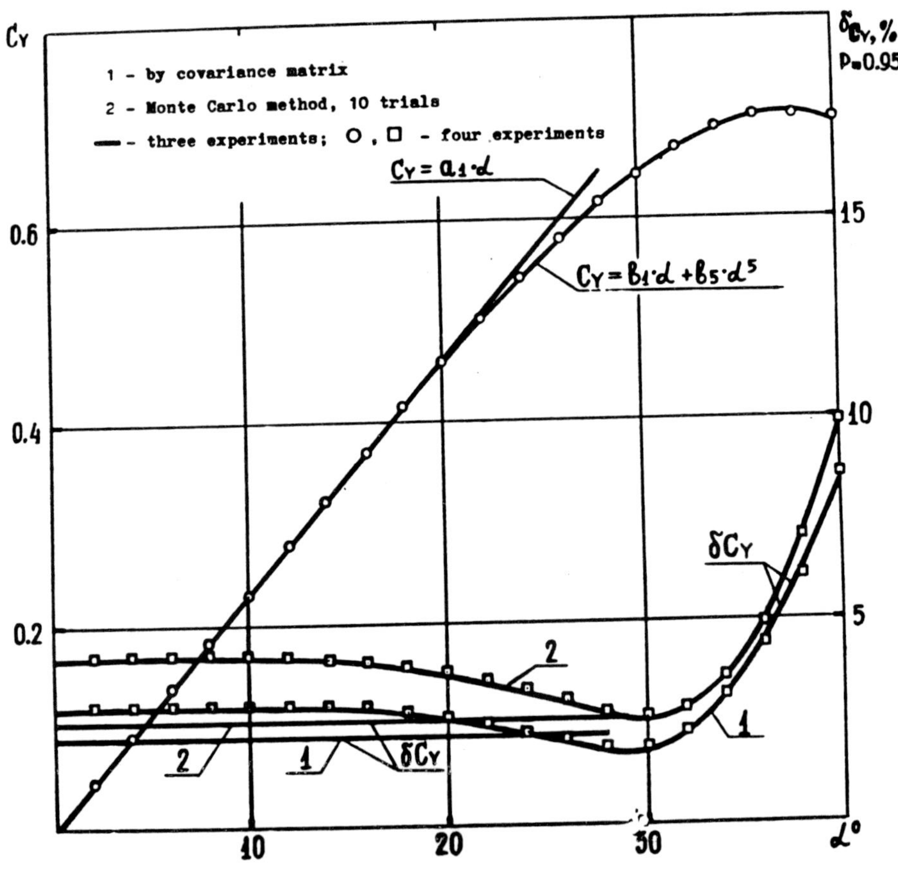

Fig. 5. Lift coefficient dependence (velocity coordinate system).

certain: we succeeded only in detecting the trend of the damping moment decrease with an increase of α but its quantitative estimate for $\alpha > 20°$ is unacceptable. The confidence interval ($P = 0.95$) for the dependence containing $\omega\alpha^2$ for $\alpha > 25°$ is shown in Figure 4 by slanted shading. The interval includes the zero parameter value. The confidence interval for the dependence containing $\omega\alpha^4$ is shown by vertical shading up to $\alpha \sim 25°$; above this value damping becomes unimportant. This is an instructive example for the "piecewise" importance of the parameter.

The lift coefficients in three and four experiments are shown in Figure 5. Since for $\alpha < 20°$ the lift is linear, the fifth power of the angle of attack was found preferable to the third one for the "tilt" description at $\alpha > 30°$ (it provides the lesser residuals). Here, as well as for the moment, the error essentially grows to the end of the α range. It is of interest that for both three and four experiments the constant coefficients for the damping lift turned out to be important (0.03 ±46% and 0.04 ∓30%, respectively). We, however, did not present them in the graph because of large errors.

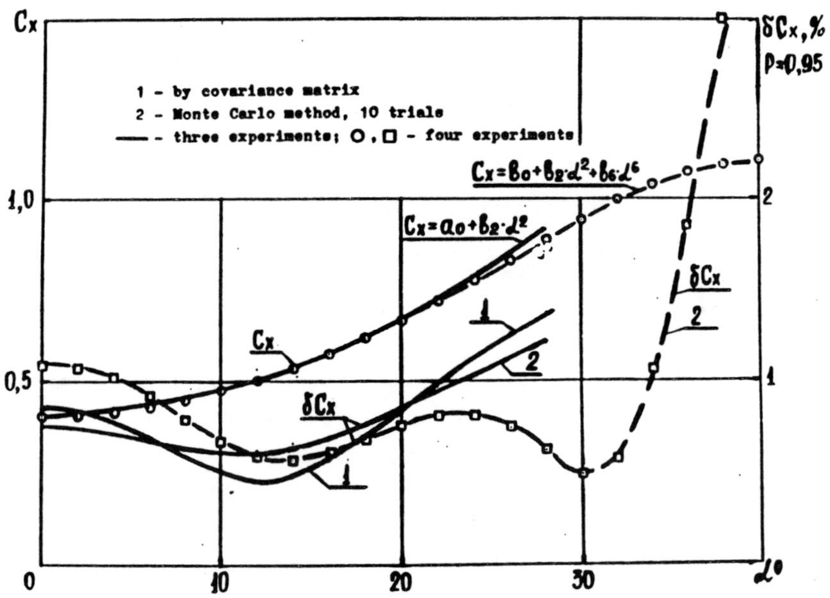

Fig. 6. Drag coefficient dependence (velocity coordinate system).

The drag coefficient dependences for two series of the experiments are shown in Figure 6. As in the lift case, the model with the "omitted" (fourth) power for the angle of attack turned out best. The drag coefficient was determined with better accuracy and the error growth typical for the end of the range is not large (about 3%). The oscillating character of the confidence interval demonstrates clearly the correlation between the determined parameters of the C_x expansion.

10. Conclusion

In conclusion it is necessary to stress the importance of verification of the measurement system for the ballistic facility. Solution of the main problem for the nonlinear estimation, namely, the choice of an adequate mathematical model, depends on the reliability of the verification data. The variance estimates of the initial data being less than the real ones, effects that do not reflect the real physical situation may be introduced into the model. In this case, efficient means of precaution, namely, the methods of mathematical statistics, will provide "the proof" for the delusion.

A second important conclusion is that the approach of Ref. [2] which we have used and attempted to advance, is extremely efficient, especially for the short ballistic range. The possibility of processing simultaneously several trajectories allows us, first, to detect the nonlinear effects which are unobservable in single launch processing (several nonlinear oscillations with different amplitudes can be described separately by sines while a common description requires introduction of the nonlinear effect due to the nonisochronic phenomenon).

Secondly, simultaneous processing of several experiments increases the ratio of the volume for the initial data to the number of parameters to be estimated up to the desired level (the volume of data grows faster than the number of initial conditions for the differential equations does).

Combining the decomposition of the system of the differential equations of motion with the method of successive approximations applied in the present paper diminishes the importance of questions of iteration process convergence. In the course of this work the author has arrived at the conclusion that in the case of a static stable object the only reason for iteration divergence is carelessness in the choice of the initial parameter approximations since the determination of an acceptable linear initial approximation on the basis of data obtained at the ballistic facility does not present significant difficulties.

Acknowledgement

The author is grateful to I. M. Dementjev, A. N. Mikhalev, S. G. Tomson, and V. A. Shirjaev for the trajectory data and information concerning the verification of the ballistic facility; to V. O. Afanasjev and V. I.

Voznjuk for discussion of the principal questions, and to S. N. Mende for the help in coding the auxiliary subroutines.

Nomenclature

C_M, C_x, C_y	coefficients of aerodynamic moment, drag and lift
C_M^ω, C_x^ω, C_y^ω	angular derivatives of aerodynamic coefficients
g	gravity acceleration
l	length of the object
M	Mach number
r	transverse radius of the inertia of the object
S	middle-section area
t	time
V	absolute value of the object velocity
u, v	components for the vector of velocity along axes x and y, respectively
X	drag force
Y	lift force
x, y	axes of the earthbound coordinate system
α	angle of attack
θ	angle of trajectory slope with respect to axis x
ϑ	angle of pitch
ρ	gas density
ω	angular velocity of the object around the transverse axis

The remaining notation is explained in the text.

References

1. N. A. Zlatin, A. P. Krasilshchikov, G. I. Mishin, and N. N. Popov, *Ballistic Facility and its Applications to Experimental Research*, Moscow: Nauka, 1974, 344 pp (in Russian).

2. G. T. Chapman and D. B. Kirk, *AIAA Journal*, no. 4, 1970, pp. 753-758.

3. A. V. Kostrov, *Motion of Asymmetric Ballistic Objects*, Moscow: Mashinostroeniye, 1984, 272 pp (in Russian).

4. A. V. Kostrov and V. V. Gukov, *Kosmicheskie Issledovaniya*, vol. 24, no. 5, 1986, pp. 680-694 (in Russian).

5. V. Kaminskas, *Dynamic Systems Identification by Discrete Observations*, Vilnius: Mosklas, 1982, 244 pp. (in Russian).

6. Y. Bard, *Nonlinear Parameters Estimation*, Moscow: Statistika, 1979,

356   *Mende*

349 pp. (in Russian).

7. A. P. Bedin, G. I. Mishin, and M. V. Chistyakova, In: *Physical-Gasdynamical Ballistic Investigations*, Leningrad: Nauka, 1980, pp. 9-24 (in Russian).

8. I. A. Belov, I. M. Dementeev, C. A. Isaev, et al., *Supersonic Flow Simulation Around Bodies of Revolution with Forward Vortex Zones*, Preprint no. 1033, Physical Technical Institute, Leningrad, 1986, 58 pp. (in Russian).

9. N. P. Mende, In: *Physical-Gasdynamical Ballistic Investigations*, Leningrad: Nauka, 1980, pp. 200-224 (in Russian).

10. G. K. Krug, V. A. Kabanov, G. A. Fomin, and E. S. Fomina, *Planning of the Experiment in Nonlinear Estimation and Image Identification Problems*, Moscow: Nauka, 1981, 172 pp. (in Russian).

11. S. M. Ermakov, A. A. Zhiglayavskii, *Mathematical Theory of Optimal Experiments*, Moscow, Nauka, 1987, 320 pp. (in Russian).

12. K. Hartmann, et al., *Planning of the Experiment in Technological Processes Research*, Moscow: Mir, 1977, 552 pp. (in Russian).

13. V. E. Gmurman, *Probability Theory and Mathematical Statistics*, Moscow: Vysshaya Shkola, 1977, 479 pp. (in Russian).

14. K. G. Artonkin, P. G. Leutin, K. P. Petrov, and E. P. Stolyarov, *Trudy TsAGI*, no. 1413, 1972, 93 pp. (in Russian).

15. I. V. Basargin, I. M. Dementeev, and G. I. Mishin, In: *Aerophysical Investigations of Supersonic Flows.*, Moscow-Leningrad: Nauka, 1967, pp. 168-178 (in Russian).

SUBJECT INDEX